MEDICAL
INTELLIGENCE
UNIT 17

Development of Aneurysms

Richard R. Keen, M.D.

Cook County Hospital
Rush University
Chicago, Illinois, U.S.A.

Philip B. Dobrin, M.D., Ph.D.

Harry S. Truman Memorial Veterans Hospital
University of Missouri
Columbia, Missouri, U.S.A.

LANDES BIOSCIENCE
GEORGETOWN, TEXAS
U.S.A.

EUREKAH.COM
AUSTIN, TEXAS
U.S.A.

DEVELOPMENT OF ANEURYSMS

Medical Intelligence Unit

Eurekah.com
Landes Bioscience
Designed by Judith Kemper
Georgetown, Texas, U.S.A.

Please address all inquiries to the Publishers:
Eurekah.com/Landes Bioscience, 810 South Church Street, Georgetown, Texas, U.S.A. 78626
Phone: 512/ 863 7762; FAX: 512/ 863 0081

ISBN: 1-58706-000-0

While the authors, editors and publisher believe that drug selection and dosage and the specifications and usage of equipment and devices, as set forth in this book, are in accord with current recommendations and practice at the time of publication, they make no warranty, expressed or implied, with respect to material described in this book. In view of the ongoing research, equipment development, changes in governmental regulations and the rapid accumulation of information relating to the biomedical sciences, the reader is urged to carefully review and evaluate the information provided herein.

Library of Congress Cataloging-in-Publication Data

Development of aneurysms / [edited by] Richard R. Keen, Philip B. Dobrin.
     p. cm. -- (Medical intelligence unit)
   Includes bibliographical references and index.
   ISBN 1-58706-000-0 (alk. paper)
1. Aneurysms--Etiology. 2. Aneurysms--Pathophysiology.
I. Keen, Richard R. II. Dobrin, Philip B. III. Series.
   [DNLM: 1. Aneurysm--etiology. 2. Aneurysm--pathology. 3. Arteries--pathology. 4. Vascular Diseases--complications.
WG 580 D489 1999]
RC693.D48 1999
616.1'33--dc21   99-36406
DNLM/DLC   CIP

CONTENTS

EDITORS

Richard R. Keen, M.D.
Department of Surgery
Cook County Hospital
and
Assistant Professor of Surgery
Rush University
Chicago, Illinois, U.S.A.
Chapter 7, 9 and 10

Philip B. Dobrin, M.D., Ph.D.
Chief of Staff
Harry S. Truman Memorial Veterans' Hospital
and
Professor of Surgery and Associate Dean
University of Missouri School of Medicine
Columbia, Missouri, U.S.A.
Chapters 2, 4, 7 and 11

CONTRIBUTORS

B. Timothy Baxter, M.D.
Professor of Surgery
University of Nebraska Medical Center
Omaha, Nebraska, U.S.A.
Chapter 5

Henrik Bengtsson, M.D., Ph.D.
Department of Surgery
Central Hospital
Kristianstad, Sweden
Chapter 1

Michael T. Caps, M.D.
Assistant Professor of Surgery
University of Washington Medical
 Center
Department of Surgery
Division of Vascular Surgery
 and
VA Puget Sound Health Care System
Surgical and Perioperative Care
Seattle, Washington, U.S.A.
Chapter 12

David K. W. Chew, M.D.
Department of Surgery
St. Luke's Roosevelt Hospital
New York, New York, U.S.A.
Chapter 12

Andy C. Chiou, M.D., M.P.H.
Vascular Surgery Fellow
Division of Vascular Surgery
Department of Surgery
Northwestern University Medical
 School
Chicago, Illinois, U.S.A.
Chapter 8

Janet J. Grange, M.D.
Department of Surgery
University of Nebraska Medical Center
Omaha, Nebraska, U.S.A.
Chapter 5

James Knoetgen III, M.D.
Department of Surgery
St. Luke's Roosevelt Hospital
New York, New York, U.S.A.
Chapter 12

Helena Kuivaniemi, M.D., Ph.D.
Associate Professor
Center for Molecular Medicine
Wayne State University
School of Medicine
Detroit, Michigan, U.S.A.
Chapter 13

Toste Länne, M.D., Ph.D.
Department of Vascular
   and Renal Diseases
Lund University, Malmö
University Hospital
Malmö, Sweden
Chapter 3

Peter F. Lawrence, M.D.
Associate Dean for Clinical Programs
   and Professor of Surgery
University of California at Irvine
College of Medicine
Irvine Hall
Irvine, California, U.S.A.
Chapter 2

William H. Pearce, M.D.
Professor of Surgery
Division of Vascular Surgery
Department of Surgery
Northwestern University Medical
   School
Chicago, Illinois, U.S.A.
Chapter 8

Bjorn Sonesson, M.D., Ph.D.
Department of Vascular
   and Renal Diseases
Lund University
Malmö University Hospital
Malmö, Sweden
Chapter 3

William E. Stehbens, M.B. B.S., M.D.,
   D.Phil, F.R.C.P.A., F.R.C.Path
Professor Emeritus
Department of Pathology
Wellington School of Medicine
Wellington South, New Zealand
Chapter 6

M. David Tilson, M.D.
Professor of Surgery
Columbia University
   and St. Luke's Roosevelt Hospital
New York, New York, U.S.A.
Chapter 10

Gerard Tromp, Ph.D.
Assistant Professor
Center for Molecular Medicine
Wayne State University
School of Medicine
Detroit, Michigan, U.S.A.
Chapter 13

PREFACE

Approximately five percent of men over the age of 55 have an abdominal aortic aneurysm, and rupture of abdominal aortic aneurysms has up to 90% mortality. This translates into approximately 15,000 deaths per year in the United States. There are many books that describe the diagnosis and treatment of aortic aneurysms. This book is not one of them. Rather, this text is a compilation of papers devoted to the understanding of the formation and development of aneurysms.

This text is divided into six sections. The first is concerned with the epidemiology and familial aspects of aneurysms. In the first chapter, Henrik Bengtsson examines the epidemiology of abdominal aortic aneurysms in Sweden. In his chapter, Dr. Bengtsson considers the strengths and weaknesses of population screening studies and postmortem studies. In the second chapter, Peter Lawrence and Philip Dobrin examine the epidemiological and familial aspects of peripheral aneurysms of the lower extremity.

The second section of this book is devoted to the physiology and pathophysiology of aortic aneurysms. In Chapter 3, Björn Sonesson and Toste Länne describe the mechanical properties of the normal and aneurysmal abdominal aorta as studied by ultrasonic echo-tracking methods. In Chapter 4, Philip Dobrin examines the contributions of elastin and collagen to the physiology and pathophysiology of the arterial wall. In Chapter 5, Janet Grange and Timothy Baxter describe connective tissue protein synthesis associated with aortic aneurysms.

The third section of the book considers the pathologic aspects of aneurysms. In Chapter 6, William Stehbens describes the pathology of aneurysms and proposes mechanisms of pathogenesis of degenerative lesions. This chapter provides the unique perspective of the experimental pathologist.

The fourth section of the book describes models of aneurysms. In Chapter 7, Philip Dobrin and Richard Keen assemble the wide variety of animal models used to simulate arterial aneurysms. Surgical, enzymatic and metabolic models are described in some detail.

The fifth section of the book examines the role of inflammation. Andy Chiou and William Pearce discuss the role of cytokine-mediated inflammation on the formation of aneurysms. Their work describes molecular studies supporting an inflammatory model. In Chapter 9, Richard Keen examines the possible role for pharmacologic agents and their potential limitations in the prevention and treatment of aneurysms.

The sixth and last section of the book considers the etiology of aneurysms. Four related chapters are included. In Chapter 10, Richard Keen discusses the possible roles of pulse pressure, shear stress and vibrations in the formation of aneurysms. In Chapter 11, Michael Caps and Philip Dobrin discuss the possible roles of infectious agents in the etiology of aneurysms. This chapter focuses particularly on Chlamydia pneumonia and cytomegalovirus. In Chapter 12,

David Chew, James Knoetgen III and David Tilson discuss the possible genetic mechanisms and molecular mimicry in the formation of aneurysms. In the thirteenth and final chapter of the book, Helena Kuivaniemi and Gerard Tromp describe studies to identify the genetic mechanisms in the formation of aneurysms. They focus particularly on a candidate gene approach.

The chapters in this book have been written by investigators who have expert knowledge in their fields. The text is compiled for vascular surgeons, cardiologists, cardiac surgeons, those in training for these professions, and basic scientists in a wide range of disciplines who are interested in the problem of aneurysm formation. It is hoped that the information provided, the references cited, and the many ideas expressed will be useful and will stimulate new and penetrating investigations of aneurysm formation, an under-appreciated vascular disease.

Richard Keen, M.D.
Philip B. Dobrin, M.D., Ph.D.

CHAPTER 1

Epidemiology of Abdominal Aortic Aneurysm (AAA)

Henrik Bengtsson

Epidemiology deals with the frequency and distribution of diseases within the population, and epidemiological methodology also may be helpful in identifying possible causes of a disease. The two basic measurements of the presence of a disease in the population are prevalence and incidence. Prevalence describes the number of patients that have a disease at a certain time while incidence describes how many new individuals will develop the disease during a certain period of time. In order to establish the occurrence of a disease, diagnostic criteria must first be formulated. Narrow diagnostic criteria decrease the risk of including persons without the disease but at the same time increases the risk of classifying those with the disease as being without it. Broad criteria will result in the opposite.

During the last decades interest in AAA epidemiology has increased rapidly. The reason for this is probably the reduced mortality observed following elective surgery and the introduction of modern imaging techniques which makes large scale screening possible. In this chapter the prevalence of AAA and incidence of ruptured AAA will be presented.

Definition of AAA

The word aneurysm is originally Greek and means a widening. Aortic aneurysm thus means a widening of the aorta, and this imprecise definition is often sufficient in the clinical situation where one deals with clinically significant aneurysms. But for epidemiological and scientific purposes a more precise definition must be used in order to differentiate a small aneurysm from the normal aorta. Initially, in vivo studies of the normal abdominal aorta were performed with angiography[1] and later by noninvasive methods such as ultrasonography and computed axial tomography (CAT) scan.[2-4] Large studies of normal populations have now given us a fairly good picture of the normal aorta with respect to age, sex and body habitus.[5-7] On the basis of this increased knowledge of aortic anatomy, several definitions of aortic aneurysms have been suggested, but still no general agreement has been accomplished.[6,8-10] Some of the proposed definitions are based on the fact that the normal aortic diameter varies in different individuals[6] while other definitions relate the aneurysm diameter to more proximal parts of the aorta that are not afflicted with the disease.[9,10] A less sophisticated but more practically useful definition of AAA, especially in screening situations, is to accept a fixed infrarenal aortic diameter, often 30 mm, the dimensions of the normal aorta; larger vessels are defined as aneurysmal.[8] The four most commonly used definitions of small AAA (less than 4 cm in diameter) have been summarized by Moher and are presented in Table 1.1.[11] Moher also has convincingly shown that, in screening investigations, the definition of small AAA has a profound influence on the reported frequency. He demonstrated that the

Development of Aneurysms, edited by Richard R. Keen and Philip B. Dobrin.
©2000 Eurekah.com.

Table 1.1. Definitions of small AAA

- Diameter > 1.5 x normal diameter
- Diameter > 1.5 x suprarenal diameter
- Diameter > 39 mm or > 1.5 x suprarenal diameter
- Diameter > 29 mm

highest frequency could be up to 9 times that of the lowest, depending on the definition being used.[12]

Identification of AAA might be accomplished with different methods. Simple palpation can reveal very large aneurysms but the sensitivity and specificity of this method has been shown to be very low and therefore it is of very little value in clinical situations, and definitely of no value in epidemiological screening.[13] Arterial angiography is an invasive method with certain risks for the patient and furthermore it does not show the wall of the aneurysm if there is a intraluminal thrombus. This method is not useful in epidemiological studies. The two methods of value for epidemiological studies are ultrasonography (US) and CAT scan. Both methods have been evaluated for AAA screening and are considered to measure aortic diameter with a fairly good accuracy but sometimes with a certain underestimation of the diameter with US, as compared with CAT scan. Both methods are observer-dependent with certain inter- and intraobserver variability. This is most obvious in ultrasound examinations.[14-16] US is by far the most widely used method in population screening; this is so because the method is quick, noninvasive, and relatively inexpensive. Furthermore the apparatus is mobile and can be brought tothe patients.[17-19]

Prevalence of AAA

There are four major sources of data on AAA prevalence.[20]

Necropsy Data

These may be used to study prevalence since a careful postmortem examination will reveal an AAA even though it is not ruptured. Unfortunately however, there are many disadvantages when necropsy data are being used. The diagnostic criteria for an aneurysm in the postmortem situation is different from those set up in vivo because the aorta is collapsed. Furthermore most, if not all, published necropsy studies on AAA are retrospective analyses of the postmortem protocols without a clear-cut definition of AAA set up beforehand. But from an epidemiological point of view, the main shortcoming with most necropsy studies is an often low autopsy rate. Another problem concerning necropsy data is that patients with a sudden and unexpected death may be overrepresented since they are more likely to be subjected to postmortem examination.[21] The advantages of necropsy studies are that they usually cover a long period of time and they often report on a great number of cases. Table 1.2 includes necropsy data for AAA frequency in postmortem studies examining individuals dying from 1950 onward.[22-35] The studies are in many ways heterogeneous and therefore difficult to compare. Despite the obvious fact that AAA frequency increases with age it is only in the studies by McFarlane et al and Bengtsson et al that the age- and sex-specific frequency of AAA are calculated.[33,35] In both these studies the AAA frequency shows a similar age and sex pattern but the frequency values are somewhat lower in the study by McFarlane et al. This lower frequency probably is due to the fact that the methods employed excluded symptomatic or previously known AAA in order to minimize the influence of selection bias. By contrast, in the study by Bengtsson et al, a very high necropsy rate reduced the

Table 1.2. Necropsy-based studies of abdominal aortic aneurysm

Author	Period	Number of autopsies	Frequency of AAA %	M/F Ratio	Increasing trend
Manigila and Gregory[22]	1906-51	6,000	0.5		Yes
Fomon et al[23]	1935-54	7,642	1.0		–
Burch and DePasquale[24]	1947-57	26,554	0.6	3.5	–
Halpert and Williams[25]	1949-60	4,000			–
Carlsson and Sternby[26]	1957-61	5,386	1.8	2.3	–
Turk[27]	1963-64	1,886	2.4	3.6	–
Von Müller et al[28]	1970-74	4,705	1.4	3.0	–
Darling et al[29]	1952-75	24,000	2.0	2.6	–
Kuntz[30]	1954-78	35,380	1.7	2.2	–
Rantakokko et al[31]	1959-79	22,765	0.8		Yes
Young and Ostertag[32]	1977-81	3,375	2.5		–
McFarlane et al[33]	1950-84	7,297	1.5*	2.1	Yes
Sterpetti et al[34]	1956-86	44,144	0.6	4.0	Yes
Bengtsson et al[35]	1958-86	45,838	3.2	2.0	Yes

*Symptomatic or known cases excluded.

influence of selection bias. A graphic representation of the age- and sex- specific frequency of AAA in this study is shown in Figure 1.1. Since this study is based on a very high necropsy rate, 85% of all deaths in the community and more than 90% of all hospital deaths, the age- and sex- specific frequency is comparable to the prevalence rate in the community. The frequency of AAA is very low before the age of 55 in men. It then increases rapidly reaching a peak of 5.9% at 80-85 years, and then decreases. In women aneurysms start to appear at 70 years of age and then the frequency increases to about 4.5% in those 90 years of age and older. The age- and sex- standardized frequency of AAA over the 30 year study period is

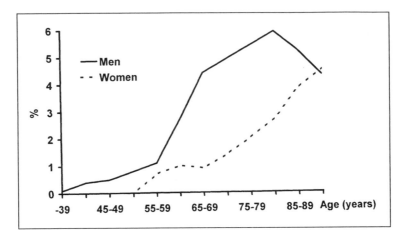

Fig. 1.1. Necropsy-based, sex-specific frequency by age of abdominal aortic aneurysm in Malmö, Sweden. 1958-1986.

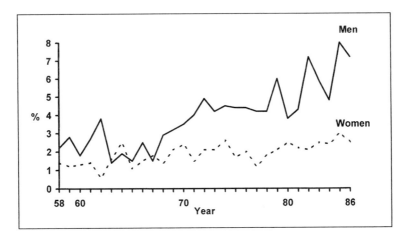

Fig. 1.2. Necropsy-based, age-standardized, sex-specific frequency of abdominal aortic aneurysm in Malmö, Sweden. 1958-1986.

shown in Figure 1.2. There is an increase of the age standardized frequency of AAA the last three decades with an annual average of 4.7% in men and 3.0% in women.

Routine Mortality statistics

These may be used to estimate the incidence of ruptured AAA. Unfortunately the validity of these registers is often very poor due to ambiguous coding systems, changing of coding systems during the study period, infrequent postmortem examinations which leaves the cause of death to the opinion of the doctor writing the death certificate, and simple mistakes in the classification.[36-38] These shortcomings of mortality registers make them of doubtful value in epidemiological studies, but unfortunately the registers are often the only information available. In order to get reliable information on the frequency of ruptured AAA, a very high autopsy rate is mandatory. Surgically treated cases of ruptured AAA also must be included to obtain the total number of ruptures. Unfortunately surgical registers also have been shown to be unreliable as they often underestimate the number of surgically treated ruptured aneurysms.[39] We have found population-based studies of ruptured AAA from 11 different places, 10 in Europe[31,40-49] and one in the USA.[50] The rupture incidence varies greatly in the studies, ranging between 3 and 14 ruptures per 100,000 inhabitants per year. The rates are higher in recent studies from the United Kingdom (UK) compared to those found in Scandinavian studies (Fig. 1.3). Whether these differences are truly regional or due to methodological differences is impossible to state. Where trends have been analyzed over time, the incidence of rupture has been found to increase over the past decades. This finding is in accordance with the results from autopsy studies and analyses of hospital inpatient registers.[35,51] In a study from Malmö, Sweden the total autopsy rate was over 85% thus making the estimation of rupture incidence more reliable.[41] The age-specific incidence rate of ruptured AAA (Fig. 1.4) shows that men suffer from rupture 15-20 years before women. The total number of ruptures (Fig. 1.5) is highest in men 70-79 years of age and is highest in women 80-89 years of age due to the fact that there are more people alive in those age groups.

Hospital In-Patient Statistics

This is another possible way to estimate the number of AAA in the population. This method also has its shortcomings. It tends to incorporate only symptomatic cases and is dependent on factors such as the referral pattern, availability of the diagnostic equipment and the interest of the clinician. Reports from hospital records in Europe and Australia have shown a rapid increase in AAA frequency over recent decades,[51-53] but the crucial question is whether these reports represents a true increase of the disease.

Population Screening Studies

These studies using US and using CAT scan, have the potential to best describe the prevalence of AAA. Although both methods have a certain inter- and intraobserver variability, they generally give an excellent anatomic picture of the abdominal aorta. The prevalence figures calculated, especially for small AAA, are highly dependent on the diagnostic criteria being used.[12] Another difficulty with screening studies is that the population sample examined might not be representative of the overall population.[54] To avoid selection bias, efforts should be made to keep the sample size rate high.

Several screening studies of the general population have been reported in the last decade and in all these studies US has been used as the screening instrument. The identification of previously unknown cases of AAA in the screening programs and the subsequent management, surgery or follow up of these patients might lead in the future to a decrease in the number of ruptured aneurysms, but the benefit of screening for AAA is controversial. Although it is difficult to know whether or not AAA screening should be undertaken on a large scale, population-based studies have given us increased epidemiological information. Screening studies of the general population almost exclusively originate from the UK[10,55-61] and Scandinavia[62-64] (Fig. 1.6) except for one study undertaken in Genoa, Italy[65] and one small study from Buffalo, USA.[66] Only two studies include women thus making estimations of female AAA prevalence more difficult.[60,65] It should be noted that we have practically no knowledge of AAA prevalence in the general population outside the Western World.

Data from the population-based studies are used in Figure 1.7 to plot AAA frequency in relation to age in men and women when an AAA was identified, i.e., when the infrarenal aortic diameter was 3 cm or more. The diagram indicates that females develop aneurysms 10-15 years later than men. This result is in accordance with the autopsy findings reported in Figure 1.1.[29] In Figure 1.8, the age specific frequency of AAA in men is estimated using

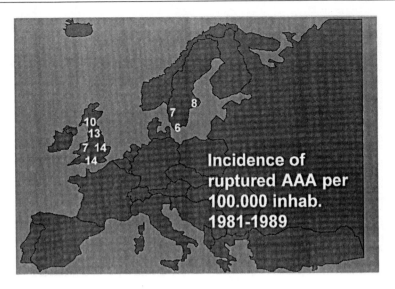

Fig. 1.3. Geographic location and incidence Figs. (per 100.000 inhabitants) of population based studies of ruptured abdominal aortic aneurysms in Europe. 1981-1989.

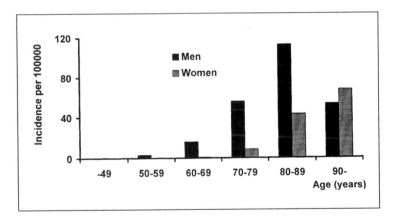

Fig. 1.4. Age and sex-specific incidence of ruptured abdominal aortic aneurysm per 100.000 inhabitants in Malmö, Sweden. 1971-1986.

three different diameters of infrarenal aorta as the definition of aneurysm. The scatter diagram show that among men, aneurysms defined as 3 cm or more in diameter occur earlier and the frequency increases more rapidly than aneurysms defined as 4 cm or more. The reason for this more rapid increase is probably that the expansion rate is highly individual. Some aneurysms increase in diameter rapidly while others may remain unchanged for long periods.[67,68]

Numerous reports on screening for AAA in various risk populations have been published over the last two decades. These studies serve two general purposes. First they may help to identify groups of individuals with a rate of AAA high enough to make screening beneficial. Second, they may help to identify groups at risk for AAA by providing a clue

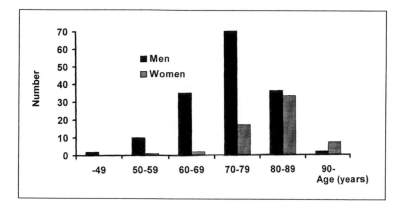

Fig. 1.5. Number of ruptured abdominal aortic aneurysms by age and sex in Malmö, Sweden. 1971-1986.

about their etiology. The risk-group studies are associated with several shortcomings. The number of patients in the study groups is small, often including only about 100 persons; this leads to large confidence limits for the frequency values. Furthermore, the studies frequently incorporate individuals of widely different ages and sometimes the data for men and women are combined. Only a few of the studies include a control group and the definition of AAA varies from study to study. All these factors makes comparisons between different studies difficult, and in many instances, impossible. In order to make all the studies of risk-groups more comprehensible, I have categorized them into four different subgroups.

Close Relatives of AAA Patients

Early reports of familial clustering of AAA[69,70] have raised the possibility of a genetic component in AAA development and have stimulated the performance of several screening studies of close relatives of AAA patients.[11,71-79] In the majority of these studies siblings of patients with AAA have been examined and their offspring have been the focus of investigation in two studies.[71,74] The reason why sons and daughters are less often examined is obvious; they are not old enough to have developed an AAA at the time of diagnosis in their parents. Screening reports of close relatives show a comparatively high frequency of AAA, especially among male probands. The rather coherent sibling studies justifies a compilation of the seven reports where AAA was defined as an aortic diameter of 3 cm or more.[11,71,72,74,76,77,79] The average AAA frequency was 18.6% among men (95% confidence interval, 13-24%) and 3.6% among women (CI 1-6%). The mean age both of male and female siblings was 64 years. The expected AAA frequency in that age group, according to the abovementioned studies of general populations, was 5.5% in men and less than 1% in women.

Patients with Occlusive Arterial Disease

Clinical observation of a coexistence of AAA and atherosclerotic disease have stimulated screening studies of patients selected on the basis of peripheral or coronary artery disease.[80-93] These studies have shown a high frequency of AAA. In the reports where AAA was defined as an aortic diameter of 30 mm or more, the frequency among men ranges between 8 and 17%, while the expected frequency based on general population studies is about 5%. The number of patients in the separate studies is small, and the inclusion criteria in the separate studies are heterogeneous leaving the studies unsuitable for meta-analysis. Despite these

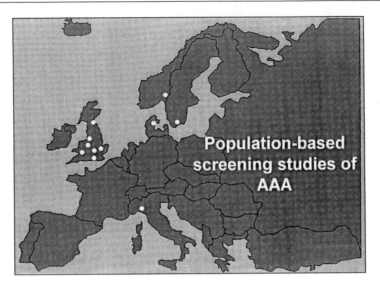

Fig. 1.6. Geographic location of population based screening studies of abdominal aortic aneurysm in Europe.

shortcomings, it is fair to conclude that patients with occlusive peripheral and coronary artery disease constitute a high risk group for the development of AAA.

Patients with Hypertension

Hypertension is a well established risk factor for thoracic aortic aneurysm and also is believed to be a causative factor in the development of AAA. Screening studies of patients selected on the basis of hypertension are few in number.[94-99] The inclusion criteria for the separate studies are rather heterogeneous, the studies are small, and the reported frequencies of AAA reported exhibit great variation. All these factors make it impossible to draw any conclusions concerning the prevalence of AAA in patients with hypertension.

Patients with Miscellaneous Symptoms

Several AAA screening studies have inclusion criteria that do not fit into any of the three previous risk groups. Patients in Japan with familial hypercholesterolemia examined with angiography presented with a very high frequency of AAA.[100] Patients with chronic obstructive pulmonary disease (COPD)[101] and patients with cardiac transplants due to ischemic heart disease[102] also are groups with an increased frequency of AAA. A special risk group studied in Germany is World War II unilateral above-knee amputees.[103] An increased frequency of AAA was found and this finding indicates that altered flow conditions or pressure wave reflections might be of importance for the development of AAA.

Summary

It is known that survivors of ruptured AAA represent "the tip of the iceberg,"[40] but what is hidden beneath the surface has long been a subject for mere speculation. During the last three decades some of these relationships have been brought to the surface and this is so, in part, because the advent of modern imaging and treatment techniques have stimulated epidemiological studies.

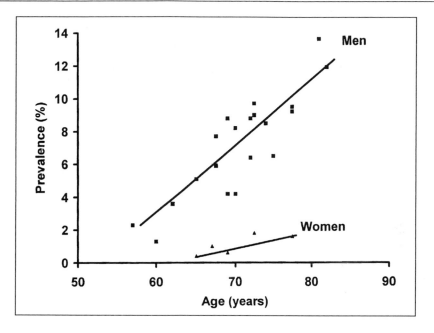

Fig. 1.7. Rate by age and sex of abdominal aortic aneurysm (AAA) defined as abdominal aortic diameter of 30 mm or more. Estimations are based on results from population-based ultrasound screening studies.[10,55-66] Regression lines for male and female rates are included in the diagram.

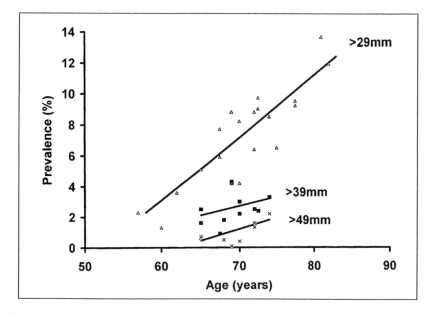

Fig. 1.8. Rate of abdominal aortic aneurysm (AAA) by age in men using three separate definitions. Estimations are based on results from population-based ultrasound screening studies.[10,55-66] Regression lines for the three aneurysm categories are included in the diagram.

Despite increased knowledge of normal aortic anatomy there is still no general agreement as to the definition of AAA. Therefore reports on AAA epidemiology must include a description of the various aortic diameters found. Epidemiological reports on AAA emanate almost exclusively from Western Europe and North America and this leaves us with no knowledge of the situation in the rest of the world. Several sources of information must be used in epidemiological studies of AAA. The most correct information has probably been gained from population-based screening studies, but in order to cover more aspects of AAA epidemiology, other sources must be used.

AAA is predominantly a disease of old age, it starts to appear at about the age of 55 in men and 10-15 years later in women and the prevalence increases with aging; a frequency of over 10% has been found in men older than 80 years. Necropsy studies show that AAA is 2-4 times more common in men than in women. The incidence of ruptured AAA is found to show regional differences and, like the prevalence of AAA, the incidence of rupture increases with age. From necropsy studies it may be concluded that only about 15% of all AAA eventually rupture. Information collected from necropsy and register data indicates that the frequency of AAA has increased 2-4% annually the last three decades. Studies of high risk groups have taught us that there is definitely a genetic component in the development of AAA and that aortic aneurysms have a close relation to atherosclerotic arterial disease.

References

1. Steinberg CR, Archer M, Steinberg I. Measurement of the abdominal aorta after intravenous aortography in health and arteriosclerotic peripheral vascular disease. AJR 1965; 95:703-708.
2. Dixon AK, Lawrence JP, Mitchell RA. Age-related changes in the abdominal aorta shown by computed tomography. Clin Radiol 1984; 35:33-37.
3. Goldberg B, Ostrum B. Ultrasonic Aortography. JAMA 1966; 198:119-124.
4. Horejs D, Gilbert P, Burnsteein S et al. Normal aortoiliac diameters by CT. J Comput Assist Tomogr 1988; 12:602-603.
5. Sonesson B, Länne T, Hansen F et al. Infrarenal aortic diameter in the healthy person. Eur J Vasc Surg 1994; 8:89-95.
6. Johnston W, Rutherford R, Tilson D et al. Suggested standards for reporting on arterial aneurysms. J Vasc Surg 1991; 13:452-458.
7. Kanaoka Y, Ohgi S, Mori T et al. Age-related changes in the abdominal aortic diameter. Vasc Surg 1994; 28:349-358.
8. McGregor JC, Pollock JG, Anton HC. The value of ultrasonography in the diagnosis of abdominal aortic aneurysm. Scott Med J 1975; 20:133-137.
9. Sterpetti AV, Schultz RD, Feldhaus RJ et al. Factors influencing enlargement rate of small abdominal aortic aneurysms. J Surg Res 1987; 43:211-219.
10. Collin J, Walton J, Araujo L et al. Oxford screening program for abdominal aortic aneurysm in men aged 65-74 years. Lancet 1988; Sept 10: 613-615.
11. Moher D, Cole W. Definition and management of abdominal aortic aneurysms: Results from a Canadian survey. Can J Surg 1994; 37:29-32.
12. Moher D, Cole W. Epidemiology of abdominal aortic aneurysm: the effect of differing definitions. Eur J Vasc Surg 1992; 6:647-50.
13. Ellis M, Powell JT, Place J et al. The limitations of ultrasound in surveillance of small abdominal aortic aneurysms. In: Greenhalgh RM, Mannick JA eds. The cause and management of aneurysms. London UK. WB Saunders, 1990:117-121.
14. Gomes MN, Choyke PL. Preoperative evaluation of abdominal aortic aneurysm: Ultrasound or computed tomography? J Cardiovasc Surg 1987; 28:159-166.
15. Thomas PRS, Shaw JC, Ashton HA et al. Accuracy of ultrasound in a screening program for abdominal aortic aneurysms. J Med Screening 1994; 1:3-6.

16. Lederle FA, Wilson SE, Johnson GR et al. Variability in measurement of abdominal aortic aneurysms. J Vasc Surg 1995; 21:945-52.
17. Bengtsson HD, Bergqvist S, Jendteg B et al. Ultrasonographic screening for abdominal aortic aneurysm: Analysis of surgical decisions for cost-effectiveness. World J Surg 1989; 13:266-271.
18. Quill DS, Colgan MP, Sumner DS. Ultrasonic screening for the detection of abdominal aortic aneurysms. Surg Clin North Am. 1989; 69:713-720.
19. O'Kelly TJ, Heather BP. General practice-based population screening for abdominal aortic aneurysms: A pilot study. Br J Surg 1989; 76:479-480.
20. Fowkes FGR. The prevalence of aneurysms. Greenhalgh RM, Mannick JA. eds. The cause and management of aneurysms. London, UK; WB Saunders, 1990; 19-28.
21. McFarlane MJ. The epidemiologic necropsy for abdominal aortic aneurysm. JAMA. 1991; 265:2085-2088.
22. Maniglia R, Gregory JE. Increasing incidence of arteriosclerotic aortic aneurysms. Analysis of six thousand autopsies. Arch Path 1952; 54:298-305.
23. Fomon JJ, Kurzweg FT, Broadway RK. Aneurysms of the aorta: A review. Ann Surg. 1967; 165:557-563.
24. Burch GE, DePasquale N. Study of incidence of abdominal aortic aneurysm in New Orleans. JAMA. 1960; 172:2011-2013.
25. Halpert B, Williams RK. Aneurysms of the aorta. Arch of Path 1962; 74:163-168.
26. Carlsson J, Sternby NH. Aortic aneurysms. Acta Chir Scand 1964; 127:466-473.
27. Turk KAD. The postpartum incidence of abdominal aortic aneurysm. Proc R Soc Med 1965; 58:869-870.
28. Müller T, Buhtz P, Eckstein D et al. Häufigkeit und Symptomatologie des Bauchaortenaneurysmas in einem 5-Jahres-Sektionsgut an der Medizinischen Akademie Magdeburg. Z. Ärtzl. Fortbild. 1979; 73:803-805.
29. Darling RC, Messina R, Brewster DC et al. Autopsy study of unoperated abdominal aortic aneurysms. The case for early resection. Circulation. 1977; 56:161-164.
30. Kunz R. Aneurysmata bei 35 380 Autopsien. Schweiz med WschR. 1980; 110:142-148.
31. Rantakokko V, Havia T, Inberg MV et al. Abdominal aortic aneurysms: A clinical and autopsy study of 408 patients. Acta Chri Scand 1983; 149:151-155.
32. Young R, Ostertag H. Häuftigkeit, Ätiologie und Rupturrisiko des Aortenaneurysmas. Dtsch Med Wochenschr DMW. 1987; 112:1253-1256.
33. McFarlane MJ. The epidemiologic necropsy for abdominal aortic aneurysm. JAMA 1991; 265:2085-2088.
34. Sterpetti AV, Cavallaro A, Cavallari N et al. Factors influencing the rupture of abdominal aortic aneurysms. Surgery 1991; 173:175-178.
35. Bengtsson H, Bergqvist D, Sternby NH. Increasing prevalence of abdominal aortic aneurysms. A necropsy study. Eur J Surg 1992; 158:19-23.
36. Elffors L, Lindahl I. Dör verkligen fler av bukaortaaneurysm? Vilken vägledning ger dödsorsaksstatistiken? Läkartidningen. 1988; 85:106-107.
37. Johansson G, Swedenborg J. Bukaortaaneurysm—många fel i dödsorsaksstatistiken. (In Swedish). Läkartidningen. 1988; 85:928.
38. Drott C, Arfvidsson B, Örtenwall P et al. Förvånande låg precision i dödsorsaksregistret. Läkartidningen. 1991; 88:2137-2139.
39. Berridge DC, Chamberlain J, Guy AJ et al. Prospective audit of abdominal aortic aneurysm surgery in the northern region from 1988 to 1992. Br J Surg 1995; 82:906-910.
40. Armour RH. Survivors of ruptured abdominal aortic aneurysm: The iceberg's tip. Br Med J 1994; 2:1055-1057.
41. Bengtsson H, Bergqvist D. Ruptured abdominal aortic aneurysm: A population-based study. J Vasc Surg 1993; 18:74-80.
42. Budd JS. Management of abdominal aortic aneurysm. Br Med J 1988; 297:484.
43. Dent A, Kent SJS, Young TW. Ruptured abdominal aortic aneurysm. What is the true mortality? Br J Surg 1986; 73:318-323.

44. Drott C, Arfvidsson B, Örtenwall P et al. Age-standardized incidence of ruptured aortic aneurysm in a defined Swedish population between 1952 and 1988: mortality rate and operative results. Br J Surg 1992; 79:175-179.
45. Ingoldby CJH, Wujanto R, Mitchell JE. Impact of vascular surgery on community mortality from ruptured aortic aneurysms. Br J Surg 1986; 73:551-553.
46. Johansson G, Swedenborg J. Ruptured abdominal aortic aneurysms: A study of incidence and mortality. Br J Surg 1986; 73:101-103.
47. Johansson G, Swedenborg J. Little impact of elective surgery on the incidence and mortality of ruptured aortic aneurysms. Eur J Vasc Surg 1994; 8:489-493.
48. Mealy K, Salman A. The true incidence of ruptured abdominal aortic aneurysms. Eur J Vasc Surg 1988; 2:405-408.
49. Thomas RS, Stewart RD. Abdominal aortic aneurysm. Br J Surg 1988; 75:733-736.
50. Bickerstaff LK, Hollier LH, van Peenen HJ et al. Abdominal aortic aneurysms: The changing natural history. J Vasc Surg 1984; 1:6-12.
51. Fowkes FGR, Macintyre CCA, Ruckley CV. Increasing incidence of aortic aneurysms in England and Wales. BMJ 1989; 298:33-35.
52. Norman PE, Castleden WM, Hockey PL. Prevalance of abdominal aortic aneurysm in Western Australia. Br J Surg 1991; 78:1118-1121.
53. Reitsma JB, Pleumeekers HJCM, Hoes AW et al. Increasing incidence of aneurysms of the abdominal aorta in The Nederlands. Eur J Vasc Endovasc Surg 1996; 12:446-451.
54. Janzon L, Hanson BS, Isacsson S-O et al. Factors influencing participation in health surveys. Results from the prospective population study "Men born in 1914", Malmö, Sweden. J Epidemiol Community Health 1986; 40:174-177.
55. Holdsworth JD. Screening for abdominal aortic aneurysm in Northumberland. Br J Surg 1994; 81:710-712.
56. Loh CS, Stevenson IM, Eyes WU et al. Ultrasound scan screening for abdominal aortic aneurysm. Br J Surg 1989; 76:417.
57. Lucarott, M, Shaw E, Poskitt K et al. The Gloucestershire aneurysm screening program: The first 2 years experience. Eur J Vasc Surg 1993; 7:397-401.
58. Morris GE, Hubbard CS, Quick CRG. An abdominal aortic aneurysm screening program for all males over the age of 50 years. Eur J Vasc Surg 1994; 8:156-160.
59. O'Kelly T, Heather BP. General practice-based population screening for abdominal aortic aneurysms: A pilot study. Br J Surg 1989; 76:479-480.
60. Scott RAP, Gudgeon AM, Ashton HA et al. Surgical workload as a consequence of screening for abdominal aortic aneurysm. Br J Surg 1994; 81:1440-1442.
61. Smith FCT, Grimshaw GM, Paterson IS et al. Ultrasonographic screening for abdominal aneurysm in an urban community. Br J Surg 1993; 80:1406-1409.
62. Bengtsson H, Bergqvist D, Ekberg O et al. A population based screening of abdominal aortic aneurysms (AAA). Eur J Vasc Surg 1991; 5:53-57.
63. Kullman G, Wolland T, Krohn-Danckert C et al. Ultrasonografi for tidlig diagnostikk av abdominalt aortaaneurisme. Tidskr Nor Laegeforen. 1992; 112:1825-1826.
64. Lindholt J. Abstract. The International Symposium on Socio-Economic Aspects of Screening for Abdominal Aortic Aneurysms, Viborg, Denmark 1995.
65. Simoni G, Decian F, Baiardi A et al. Screening for abdominal aortic aneurysms and associated risk factors in a general population. Eur J Vasc Endovasc Surg 1995; 10:207-210.
66. Rosenthal TC, Siepel T, Zubler J et al. The use of ultrasonography to scan abdomen of patients presenting for routine physical examinations. J Family Practice 1994; 38:380-385.
67. Bengtsson H, Nilsson P, Bergqvist D. Natural history of abdominal aortic aneurysm detected by screening. Br J Surg 1993; 80:718-720.
68. Stonebridge PA, Draper T, Kelman J et al. Growth Rate of Infrarenal Aortic Aneurysms. Eur J Vasc Endovasc Surg 1996; 11:70-73.
69. Clifton MA. Familial abdominal aortic aneurysms. Br J Surg 1977; 64:765-766.
70. Tilson MD, Seashore MR. Fifty families with abdominal aortic aneurysms in two or more first-order relatives. Am J Surg 1984; 147:551-553.

71. Adams DC, Tulloh BR, Galloway SW et al. Familial abdominal aortic aneurysm: Prevalence and implications for screening. Eur J Vasc Surg 1993; 7:709-712.
72. Adamson J, Powell JT, Greenhalgh RM. Selection for screening for familial aortic aneurysms. Br J Surg 1992; 79:897-898.
73. Bengtsson H, Norrgård Ö, Ängquist KA et al. Ultrasonographic screening of the abdominal aorta among siblings of patients with abdominal aortic aneurysms. Br J Surg 1989; 76:589-591.
74. Bengtsson H, Sonesson B, Länne T et al. Prevalence of abdominal aortic aneurysm in the offspring of patients dying from aneurysm rupture. Br J Surg 1992; 79:1142-1144.
75. Collin J, Walton J. Is abdominal aortic aneurysm familial? BMJ 1989; 299:493.
76. Jaakkola P, Hippeläinen M, Farin P et al. Infrarenal aortic dimensions in siblings of patients with infrarenal abdominal aortic aneurysm or aortoiliac occlusive disease. Eur J Surg 1995; 161:871-875.
77. Fritzgerald P, Ramsbottom D, Burke P et al. Abdominal aortic aneurysm in the Irish population: A familial screening study. Br J Surg 1995; 82:483-486.
78. van-der-Lugt A, Kranendonk SE, Baars AM. Screening for familial occurrence of abdominal aortic aneurysm. Ned Tijdschr Geneeskd. 1992; 136:1910-1913.
79. Webster MW, Ferell RE, St Jean PL et al. Ultrasound screening of first-degree relatives of patients with an abdominal aortic aneurysm. J Vasc Surg 1991; 13:9-14.
80. Allardice JT, Allwright GJ, Wafula JMC et al. High prevalence of abdominal aortic aneurysm in men with peripheral vascular disease: Screening by ultrasonography. Br J Surg 1988; 75:240-242.
81. Andersson AP, Ellitsgaard N, Jorgensen B et al. Screening for abdominal aortic aneurysm in 295 outpatients with intermittent claudication. J Vasc Surg 1991; 25:516-520.
82. Bengtsson H, Ekberg O, Aspelin P et al. Ultrasound screening of the abdominal aorta in patients with intermittent claudication. Eur J Vasc Surg 1989; 3:497-502.
83. Bengtsson H, Ekberg O, Aspelin P et al. Abdominal aortic dilatation operated on for carotid artery stenosis. Acta Chir Scand 1988; 154:441-445.
84. Berridge DC, Griffith CDM, Amar SS et al. Screening for clinically unsuspected abdominal aortic aneurysms in patients with peripheral vascular disease. Eur J Vasc Surg 1989; 3:421-422.
85. Cabellon S, Moncrief CL, Pierre DR et al. Incidence of abdominal aortic aneurysms in patients with atheromatous arterial disease. Am J Surg. 1983; 146:575-576.
86. Carty GA, Nachtigal T, Magyar R et al. Abdominal duplex ultrasound screening for occult aortic aneurysm during carotid arterial evaluation. J Vasc Surg 1993; 17:696-702.
87. Galland RB, Simmons MJ, Torrie EPH. Prevalence of abdominal aortic aneurysm in patients with occlusive peripheral vascular disease. Br J Surg 1991; 78:1259-1260.
88. Karanjia PN, Madden KP, Lobner S. Coexistence of abdominal aortic aneurysm in patients with carotid stenosis. Stroke 1994; 25:627-630.
89. MacSweeney STR, Meara MO, Alexander C et al. High prevalence of unsuspected abdominal aortic aneurysm in patients with confirmed symptomatic peripheral or cerebral arterial disease. Br J Surg 1993; 80:582-584.
90. Nevelsteen A, Kim Y, Meersman . Routine screening for unsuspected aortic aneurysms in patients after myocardial revascularization: A prospective study. Acta Card 1991; XLVI:201-206.
91. Sakalinhasan N. Contribution tho the knowledge if epidemiology and natural history of abdominal aortic aneurysms. Thèse de doctorat en sciences cliniques. Faculti de Médecine. 1994; Université de Liége.
92. Shapira M, Pasik S, Wassermann J-P et al. Ultrasound screening for abdominal aortic aneurysms in patients with atherosclerotic peripheral vascular disease. J Cardiovasc Surg 1990; 31:170-172.
93. Wolf YG, Otis SM, Schwend RB et al. Screening for abdominal aortic aneurysms during lower extremity arterial evaluation in the vascular laboratory. J Vasc Surg 1995; 22:417-423.
94. Allen P I M, Tudway D, Goldman M. Can we improve community mortality from ruptured aortic aneurysm. Br J Surg 1987; 74:332.

95. Gramham M, Chan A. Ultrasound screening for clinically occult abdominal aortic aneurysm. CMAJ 1988; 138:627-629.
96. Lederle FA, Walker JM, Reinke DB. Selective screening for abdominal aortic aneurysms with physical examination and ultrasound. Arch Intern Med 1988; 148:1753-1756.
97. Lindholm L, Ejlertsson G, Forsberg L et al. Low prevalence of abdominal aortic aneurysm in hypertensive patients. Acta Med Scand 1985; 218: 305-310.
98. Sowter MC, Lewis MH. Ultrasonographic screening for abdominal aortic aneurysm in an urban community (letter). Br J Surg 1994; 81:472.
99. Twomey A, Twomey E, Wilkins R A et al. Unrecognized aneurysmal disease in male hypertensive patients. Int Angiol 1986; 5:269-273.
100. Kita Y, Shimizu M, Sugihara N et al. Abdominal aortic aneurysms in familial hypercholesterolemia—case reports. Angiology 1993; 44:491-499.
101. van Laarhoven CHM, Borstlap ACW, van Berge Henegouwen DP et al. Chronic obstructive pulmonary disease and abdominal aortic aneurysms. Eur J Vasc 1993; 7:386-390.
102. Piotrowski JJ, McIntyre KE, Hunter CG et al. Abdominal aortic aneurysm in patient undergoing cardiac transplantation. J Vasc Surg 1991; 14:460-466.
103. Vollmar JF, Paes EHJ, Pauschinger P et al. Aortic aneurysms as late sequelae of above-knee amputation. Lancet 1989; 834-835.

CHAPTER 2

Epidemiological and Familial Aspects of Nonaortic Lower Extremity Aneurysms

Peter F. Lawrence, Philip B. Dobrin

Epidemiology

Although abdominal aortic aneurysms are the most common, and potentially lethal aneurysm, other aneurysms affecting vessels to the lower extremities also can be threatening to life and limb. Patients with a peripheral aneurysm can be asymptomatic, present with mild symptoms of adjacent tissue compression, or may present with exsanguinating hemorrhage caused by rupture. They also can present with thrombosis or embolization. Each aneurysm of vessels to the lower extremities has a different clinical predilection, with iliac artery aneurysms typically behaving much like aortic aneurysms, i.e., with rupture, whereas aneurysms of the femoral and popliteal arteries tend to embolize or undergo thrombosis. Femoral and popliteal aneurysms rarely rupture.

The definition of a peripheral aneurysm is similar as that for such lesions of any other vessel, i.e., a focal dilatation of a vessel which achieves 1.5 times normal vessel diameter.[1] However in spite of this accepted definition, most clinical series that discuss peripheral aneurysms consider diagnosis and treatment when they achieve twice normal dimensions. Iliac artery aneurysms are considered to be clinically relevant when they achieve 3 cm, whether they are of the common, internal, or external iliac arteries. Femoral artery aneurysms must achieve 2 cm and popliteal artery aneurysms 1.5 cm before they are considered to be of clinical importance. In addition, many patients with peripheral lesions are found to have either multiple aneurysms or dilatation of an entire arterial tree, i.e., arteriomegaly.

Prevalence

It is difficult to state the true prevalence of peripheral aneurysms, because they are relatively uncommon and therefore have not been assessed by screening programs. Such screening programs have not been thought to be cost effective for the general population. However it is well known that aneurysms of the femoral and popliteal arteries represent about 14% of all aneurysm repairs, while isolated iliac artery aneurysms represent about 1.2% of all surgical aneurysm repairs. Iliac artery aneurysms which are contiguous with the abdominal aorta occur in 10% of aortic aneurysms.[2] Consequently, iliac aneurysms represent the second most type of these lesions. Aortic aneurysms remain as the most frequent accounting for 67% of all surgical aneurysm repairs. Since epidemiological incidence data for peripheral aneurysms is unavailable, it must be assumed that there is a correlation in patients between the frequency of repairs and the prevalence of the disease.

Clinical Presentation

Once an aneurysm of the iliac, femoral or popliteal arteries is diagnosed, the natural history as for other aneurysms, is one of continued and progressive expansion. Most, but not all patients, develop clinical symptoms by the time the lesion is identified. Dawson et al[3] examined the long-term follow-up of popliteal aneurysms in 50 patients. Twenty-five of these patients were treated nonsurgically. Fifty-seven percent of those patients developed complications, including limb-threatening ischemia. The probability of developing complications increased over time achieving 74% at five years. Consequently, the presence of an aneurysm, even if asymptomatic, is associated with a greater than 50% likelihood of subsequent complications if they are not repaired surgically. Although similar studies have not been performed on iliac artery aneurysms, most studies have indicated that patients rarely have symptoms or rupture from iliac aneurysms that are less than 3 cm. But once these lesions achieve 3 cm diameter, the risk of rupture or complications is high enough to warrant surgical repair.

Age Association

Iliac, femoral and popliteal aneurysms increase in incidence with advancing age. An analysis has been formed of American hospitalized patients with identified aneurysms. Iliac artery aneurysms occurred in 2.82 per 100,000 patients, while femoral and popliteal aneurysms occurred in 3.66 per 100,000 patients.[4,5] Iliac, femoral and popliteal aneurysms were diagnosed very rarely in patients under age 45, but there was a dramatic increase in these aneurysms beginning at approximately age 55. This incidence increased through age 75 indicating that peripheral lesions whether in the iliac, femoral or popliteal regions are diseases of the elderly.

Sex Association

In a pattern similar to abdominal aortic aneurysms, iliac, femoral and popliteal aneurysms are more common in men than in women. The lesions occur approximately seven times more frequently in men than in women. Individual locations have different prevalences. Iliac artery aneurysms are much less prevalent in woman than in men, and femoral and popliteal aneurysms are even less common in women (Table 2.1). Aneurysms of the iliac, femoral and popliteal arteries are related to risk factors for atherosclerosis. Since this occurs with a lower frequency in women than in men, this may be the cause for the greater prevalence of aneurysmal disease in men. If this is true, then one may expect aneurysms to become increasingly prevalent in women as they assume male risk factors regarding diet, smoking, etc. Currently, degenerative aneurysms of the iliac, femoral and popliteal arteries are rarely found in women.[5]

Multiplicity

Dawson's longitudinal study of popliteal aneurysms[3] demonstrated that these lesions were associated with extrapopliteal or contralateral popliteal aneurysms in 50% of patients. In 42% of cases the contralateral popliteal was involved. In addition to peripheral aneurysmal disease, these patients had a 36% incidence of abdominal aortic aneurysms and a 38% incidence of femoral artery aneurysms. Consequently, the presence of a peripheral aneurysm, particularly in the popliteal location, is often associated with other synchronous lesions. This has led to the recommendation that a careful search be made in all patients with peripheral aneurysms for other coexisting lesions.

However, even when coexisting aneurysms are not initially identified, there is a significant risk of developing metachronous aneurysms.[3] In long-term follow-up of patients with popliteal aneurysms, 32% developed another such lesion within five years. This included lesions of the thoracoabdominal aorta, the femoral artery and the contralateral popliteal artery. Up to 40% of these patients required surgery. At 10 years, 49% of patients who initially had no other aneurysmal lesion were found to have developed a metachronous aneurysm, again emphasizing the importance of long-term follow-up and the recognition that peripheral aneurysms are a marker for other aneurysmal disease.

Although most data have been published on popliteal aneurysms, femoral artery aneurysms have a similar natural history.[6] These data are frequently confounded by clinical reports of femoral artery aneurysms that include or combine atherosclerotic aneurysms with pseudoaneurysms related to femoral instrumentation such as cardiac catheterization. Since most pseudoaneurysms have a self-limited natural history with compression or repair, the influence of degenerative or atherosclerotic aneurysms is diluted in these reports. Iliac artery aneurysms frequently are reported with peripheral aneurysms and other nonaortic lower extremity lesions. However, the natural history of iliac aneurysms is very similar to that of abdominal aortic aneurysms. Although the iliac lesions may embolize or thrombose, they more frequently rupture, so that the clinical presentation is frequently life-threatening exsanguination or retroperitoneal hemorrhage. Of interest is the distribution of aneurysms of the iliac arteries. Isolated common and internal iliac artery aneurysms are fairly common, yet external iliac artery aneurysms are quite rare. The explanation for this distribution of vessels is unclear, but has been recognized by many authors and is well established. In addition, since aneurysms of the internal iliac artery are located deep in the pelvis, many patients with the lesion have minimal or vague symptoms until catastrophic rupture occurs.

Familial Aspects

It is well known that abdominal aortic aneurysms exhibit a 15 to 20% familial incidence.[7-10] However, there are no data documenting the familial incidence of peripheral aneurysms or arteriomegaly. Aneurysms of the femoral, popliteal, and isolated iliac arteries are less common than aneurysms of the abdominal aorta,[5] and arteriomegaly is even less common.[5,11] Just as with aortic aneurysms, peripheral aneurysms and the diffuse arterial enlargement that is arteriomegaly carry the risks of embolism, rupture, and thrombosis.[12] However, there are sparse epidemiologic data available to guide screening of relatives of patients with peripheral aneurysms or arteriomegaly. The purpose of this part of the chapter is to examine the prevalence of familial aneurysms in patients with peripheral arterial aneurysms or arteriomegaly and compare this with the incidence of abdominal aortic aneurysms.

Patients diagnosed between 1988 and 1996 with aneurysms of the popliteal, femoral, or iliac arteries or with arteriomegaly were identified at the University of Utah and the Salt Lake City Veterans Affairs Hospital. Patients with instrumentation-induced pseudo-aneurysms, anastomotic false aneurysms, or iliac artery aneurysms that were an extension of an abdominal aortic aneurysm were not included. Each patient's medical record was examined to identify the location of the aneurysm as well as risk factors, including gender, age, hypertension, smoking history, chronic obstructive pulmonary disease, hypercholesterolemia, diabetes and hernia. Each patient (proband) was individually contacted by letter and telephone to confirm the data and to determine if additional aneurysmal disease had developed. In addition, a pedigree was constructed of the patient's first degree relatives, i.e., the patient's parents, siblings, and children. Since aneurysmal disease rarely develops before age 50,[13,14] only first degree relatives that were over the age of 50 were contacted. When aneurysms or arteriomegaly were identified in a patient or a patient's relative, reporting standards of the ISCVS/SVS classification standards[1] were used. Vessels that exhibited "a permanent localized (i.e., focal) dilatation of an artery... an increase in diameter greater than 50% [were] considered evidence of an aneurysm." Arteriomegaly also was defined based on ISCVS/SVS criteria, i.e., as diffuse arterial enlargement involving several separate arterial segments (i.e., nonfocal) with an increase in diameter of greater than 50% of the expected normal diameter. Patients and family members with incomplete medical records who were lost to follow-up, or where insufficient pedigree information was available, were excluded from the study. A concurrent epidemiologic survey was conducted on a group of patients diagnosed with abdominal aortic aneurysms who had been surgically repaired at the University of Utah or Salt Lake City Veterans Hospital between 1988 and 1996 to provide comparison with aortic lesions, and to serve as a control for the methods used in this study. Statistical analysis was performed using Fisher's exact test for computing two tail probabilities and chi-square for comparing multiple groups. Statistical significance was considered at the p <0.05 level.

During the eight year period of study, 300 aortic aneurysms and 124 peripheral arterial aneurysms were treated. This included 19 isolated iliac aneurysms, 18 femoral artery aneurysms and 26 popliteal aneurysms. One hundred forty patients with 86 aortic aneurysms and 40 peripheral aneurysms also were studied. In addition, 14 patients with arteriomegaly had sufficient information to permit construction of a family pedigree. Seven hundred three primary relatives of patients with aneurysms were contacted wherein each patient had an average of five primary relatives who were older than age 50 from whom data could be obtained.

Probands

The distribution of aneurysms in the probands is shown in Table 2.1. As can be seen, 50% of the peripheral aneurysms were in the popliteal location. The mean age of presentation for peripheral aneurysms, arteriomegaly and abdominal aortic aneurysms were similar. Arteriomegaly and peripheral arterial aneurysms occurred predominantly in men with a male-to-female ratio ranging from 9:1 to infinity. Aortic aneurysms had a lower male-to-female ratio of 5.6:1. The male-to-female ratio for peripheral arterial aneurysms was 43:1. This was significantly different (p =0.02) as compared with the 5.6:1 ratio for abdominal aortic aneurysms. However, the male-to-female ratio for arteriomegaly 13:1 was not significantly different (p = .378) than that observed for abdominal aortic aneurysms 5.6:1.

In 12 (28%) of the patients with peripheral arterial aneurysms, there was an anatomically separate abdominal aortic aneurysm identified later. Only four (10%) of the patients with peripheral arterial aneurysms were found to have another synchronous peripheral aneurysm, in spite of routine physical examination or ultrasonic surveillance of the peripheral arteries. Thus, 38% of the patients with peripheral aneurysms were ultimately found to develop a separate aneurysm, the majority being in the abdominal aorta. Examination of the risk factors in the probands (Table 2.2) demonstrated that both the presence of diabetes and hernias were significantly different in the three groups. Hypertension and smoking were present in more than 60% of the probands in all three groups. The 12 patients with synchronous aortic and peripheral aneurysms had no identifiable difference in risk factors (p=ns) as compared with the patients who had isolated aortic or isolated peripheral aneurysms.

First Degree Relatives

A total of 38 (5.4%) of the 703 primary relatives of patients with aneurysms also had an aneurysm with 28 (73%) occurring in the infrarenal abdominal aorta. As shown in Table 2.3, 36% of the probands with arteriomegaly, 22% of the probands with abdominal aortic aneurysm, and 10% of the patients with a peripheral arterial aneurysm had a first

Table 2.1. Distribution of aneurysms in probands

Aneurysms	Total	Mean Age	Men	Women	M/F Ratio	Patients w/ Affected Relatives
Iliac	10	72	9	1	9	0(0%)
Femoral	14	68	14	0	00	2(14%)
Popliteal	20	63	20	0	00	2(10%)
Total Peripheral	40*	67	43	1	43	4(10%)
Arteriomegaly	14	68	13	1	13	5(36%)
AAA	86	70	73	13	5.6	19(22%)

*4 patients had synchronous isolated aneurysms in two locations

Table 2.2. Risk factors in probands

Risk Factors	AAA	Peripheral	AAA w/ Peripheral*	Arteriomegaly	p value
	(n=86)	(n=40)	(n=12)	(n=14)	
Hypertension	43 (50%)	24(60%)	10(83.3%)	9(64%)	N.S.
Smoking History	68(79.1%)	24(60%)	10(83.3%)	10(71.4%)	N.S.
Hypercholesterolemia	26(30.2%)	8(20%)	5(41.7%)	4(28.5%)	N.S.
COPD	16(18.6%)	7(17.5%)	2(16.7%)	0	N.S.
Inguinal Hernia	34(39.5%)	5 (12.5%)	5(41.7%)	6(42.8%)	< 0.01
Diabetes	7(8.1%)	11(27.5%)	1(8.3%)	0	< 0.01

*excluding arteriomegaly patients

degree relative who also had an aneurysm. The risk of any family member developing an aneurysm ranged from 7% for probands with an aortic aneurysm to 2% for probands with a peripheral aneurysm. There was no significant difference in the risk factors such as hypertension, diabetes, or hernia in the first degree relatives who developed aneurysms from those who did not.

Irrespective of the location of aneurysms in the probands, aneurysms in family members occurred predominantly (73%) in the abdominal aorta.[5] The most common site for peripheral aneurysms in family members was the femoral artery. Out of 40 probands with peripheral aneurysms, four had a single relative with an aneurysm; two of these were in the abdominal aortic location, and two were peripheral aneurysms. Of 14 probands with arteriomegaly, five probands had a single relative with an abdominal aortic aneurysm; none of the family members had a peripheral aneurysm. It was not possible to determine how many relatives of probands with arteriomegaly also had arteriomegaly since this is usually diagnosed angiographically. For the 86 probands with abdominal aortic aneurysms, 28 family members had abdominal aortic aneurysms, five had arteriomegaly, and four had peripheral aneurysms.

Table 2.3. Aneurysm risk in first degree relatives

Probands	Any 1° Relative*	Risk/1° Relative
Arteriomegaly (AM)	5/14 36%	5.5%
Abdominal aortic aneurysm (AAA)	19/86 22%	6.8%
Peripheral aneurysm (PA)	4/40 10%	2%
	* PA vs. AAA (p =. 16)	
	PA vs. AM (p =.08)	
	AM vs. AAA (p =.43)	

Table 2.4. Distribution of aneurysms in first degree relatives

Proband Aneurysm	Relatives over age 50	Location of aneurysms in 1° Relatives					
		Ao	Iliac	Fem	Pop	Renal	Total
Iliac (10)	56	0	0	0	0	0	0
Femoral (14)	58	1	0	2	0	0	3
Popliteal (20)	100	1	0	0	0	1	2
Total Peripheral (40†)	202	2	0	2	0	1	5
Arteriomegaly (14)	91	5	0	0	0	0	5
AAA (86)	410	24	0	3	1	0	28

† 4 patients had synchronous isolated aneurysms in two locations

Sex Association

The prevalence of aneurysms was different in men and women for both the probands as well as the first degree relatives. For the probands, 43 out of 44 (98%) of peripheral aneurysms occurred in men, 73 out of 86 (85%) of aortic aneurysms occurred in men, and 13 out of 14 (93%) of cases of arteriomegaly occurred in men. Since 44 out of 140 (31%) of the probands were identified from the Veterans Affairs Hospital, it is not surprising that a large number of probands were male. The first degree relatives (Table 2.5) had a much larger number of females, but even here the frequency of aneurysms in males still persisted. The male-to-female ratios for the first degree relatives for the aneurysms ranged from 3:2 for arteriomegaly to 6:1 for abdominal aortic aneurysm, and 2:0 for peripheral aneurysms. Thus, the individual risk of developing an aneurysm in any location in a family member was greater in men than in women.

Age Association

Aneurysms in first degree relatives were evaluated only in family members greater than age 50. Nevertheless, the incidence of aneurysms in family members increased steadily from ages 56 through 75. This is shown in Table 2.6. Above age 75 there were fewer first degree relatives alive.

Degenerative aneurysms of the peripheral vessels are thought to be of similar etiology to the more common abdominal aortic aneurysms.[15] However, there are several epidemiologic and clinical differences between these aneurysms that suggests that the peripheral

Table 2.5. First degree relative risk of aneurysm

Proband	Total	Aneurysms		Relatives Prevalence		M/F Ratio
		Men	Women	Men	Women	
Iliac	56	0	0	0	0	0
Femoral	58	2	0	3.4	0	∞
Popliteal	100	2	0	2	0	∞
Total Peripheral	204	4	0	2	0	∞
Arteriomegaly	91	3	2	3	2	1.5
AAA	410	24	4	6	1	6

Table 2.6. Age at which first degree relatives manifested an aneurysm

Location of Aneurysms in 1° Relatives	# of Cases	50-55	56-60	61-65	66-70	71-75	76-80	81-85	86-90
AAA	33	1	5	6	3	8	3	5	2
Femoral	2					1		1	
Popliteal	2			1				1	
Renal	1							1	
Total Aneurysms in 1° Relatives	38	1	5	7	3	9	3	8	2

aneurysms might be a different disease. Peripheral aneurysms occur almost exclusively in men, they are frequently bilateral, and they thrombose or embolize rather than rupture.[16,17] However, because they are usually of smaller size and are subject to lower wall stress, it is not surprising that they exhibit a lower frequency of rupture than do abdominal aortic aneurysms. Nevertheless it is unclear whether peripheral aneurysms have the same familial predilection as abdominal aortic aneurysms. Although several authors have attempted to demonstrate genetic evidence for aortic aneurysms,[18-22] candidate genes have not been confirmed. Moreover, aneurysmal disease presents late in life so that both familial as well as genetic factors may play a role. Often patients afflicted with aneurysmal disease are hypertensive, smoke cigarettes and are hypercholesterolemic.

Patients with peripheral aneurysms often have a synchronous peripheral aneurysm or a synchronous aortic aneurysm.[20,22] In our study there was a 10% prevalence of synchronous peripheral aneurysms. This low number was found in spite of routine ultrasonic surveillance and physical examination of both femoral and popliteal arteries in all patients with peripheral aneurysms. However, we did identify aortic aneurysms in 25% of the patients with peripheral aneurysms, suggesting that aortic ultrasonography should routinely be performed in these patients.

This study raises several additional issues. For example, 36% of patients with arteriomegaly had an affected first degree relative. It therefore appears that arteriomegaly is a strong predictor of familial aneurysms. Abdominal aortic aneurysms had a 22% incidence of family members with aneurysms, and patients with peripheral aneurysms had a 10% incidence of familial aneurysms. Since women comprise approximately one-half of family members, and since peripheral aneurysms and arteriomegaly rarely occur in women, this makes first degree male relatives of patients with arteriomegaly at particularly high risk. We compute that a male relative of a patient with arteriomegaly has a 6.7% risk of developing a clinically significant aneurysm. In patients with peripheral aneurysms, 5% of first degree male relatives and 0% of female relatives also developed an aneurysm. The likelihood of any first degree relative of a proband having a clinically significant aneurysm was 5.4%. This number is similar to the 5% prevalence of aneurysms in men over 50 in the European population who developed aneurysms. However, it should be added that many of the aneurysms in the European population were of small size and might not have been treated. All of our patients had an aneurysm large enough to require repair.

Screening of First Degree Relatives

This study shows that there is a gradation of risk in the relatives of probands with aneurysmal disease. Thirty-six percent of probands with arteriomegaly had a first degree relative with an aneurysm, 22% of probands with an aortic aneurysm had a first degree relative with an aneurysm, and 10% of probands with a peripheral aneurysm had a first degree relative with an aneurysm. This raises the clinical question as to if and how the first degree relatives should be screened. As shown in Table 2.4, the great majority of aneurysms in the first degree relatives was located in the abdominal aorta and the femoral arteries. This was true irrespective of the location of the lesion in the probands. Based on this observation, we recommend that, upon reaching age 50, all first degree male relatives of patients with arteriomegaly or an abdominal aortic aneurysm be screened using ultrasonography of the abdominal aorta and iliac arteries, and physical examination of the abdomen and peripheral arteries. A similar strategy, applied selectively, to elderly first degree relatives of probands with peripheral aneurysms also may be useful. When aneurysms occur, they develop most frequently in the abdominal aorta and femoral arteries.

References

1. Johnston KW, Rutherford RB, Tilson MD et al. Suggested standards for reporting on arterial aneurysms. J Vasc Surg 1991; 13:444-450.
2. Goldstone J. Aneurysms of the aorta and iliac arteries. In: Moore WS, ed: Vascular Surgery. A Comprehensive Review. Philadelphia: WB Saunders, 1993:401-422.
3. Dawson I, vanBockel JH, Brand R et al. Popliteal artery aneurysms. Long-term follow-up of aneurysmal disease and results of surgical treatment. J Vasc Surg 1991; 13:398-407.
4. Lawrence PF, LorenzoRivero S, Lyon JL. The incidence of iliac, femoral and popliteal aneurysms in Utah and the United States. J Vasc Surg 1995; 22:409-415.
5. Lawrence PF, Wallis C, Dobrin PB et al. Peripheral aneurysms and arteriomegaly: Is there a familial pattern? J Vasc Surg 1998; 28:599-605.
6. Graham LM, Zelenock GB, Whitehouse WM Jr et al. Femoral artery aneurysms. Arch Surg 1980; 115:502-507.
7. Johansen K, Koepsell T. Familial tendency for abdominal aortic aneurysms. JAMA 1986; 256:1934-1936.
8. Norrgard 0, Rais 0, Angquist KA. Familial occurrence of abdominal aortic aneurysms. Surgery 1984; 95:650-656.
9. Tilson DM, Seashore MR. Fifty families with abdominal aortic aneurysms in two or more first order relatives. Am J Surg 1984; 147:551-553.
10. Majumder PP, St. Jean PL, Ferrell RE et al. On the inheritance of abdominal aortic aneurysm. Am J Hum Genet 1991; 48:164-170.
11. Hollier LH, Stanson AW, Gloviczki P et al. Arteriomegaly: Classification and morbid implications of diffuse aneurysmal disease. Surgery 1983; 93:700-708.

12. Hands LJ, Collin J. Infrainguinal aneurysms: Outcome for patient and limb. Br J Surg 1991; 78:996-998.
13. Wolf YG, Otis SM, Schwend RB et al. Screening for abdominal aortic aneurysms during lower extremity arterial evaluation in the vascular laboratory. J Vasc Surg 1995; 22:417-423.
14. Quill DS, Colgan MP, Summer DS. Ultrasonic screening for the detection of abdominal aortic aneurysms. Surg Clin N Am 1989; 69:713-720.
15. Flanigan DP, Aneurysms of the peripheral arteries. In: Moore WS, ed: Vascular Surgery. A Comprehensive Review. Philadelphia: WB Saunders, 1993:424-434.
16. Graham AR, Lord RSA, Bellemore M et al. Popliteal aneurysms. Aust N Z J Surg 1983; 53:99-103.
17. Barroy JP, Barthel J, Locufier JL et al. Atherosclerotic popliteal aneurysms. Report of one ruptured popliteal aneurysm. Survey and analysis of the literature. J Cardiovasc Surg 1986; 27:42-45.
18. Tilson MD, Seashore MR. Human genetics of the abdominal aortic aneurysm. Surg Gyn & Ob 1984; 158:129-132.
19. Powell JT, Bashir A, Dawson S et al. Genetic variation on chromosome 16 is associated with abdominal aortic aneurysm. Clinical Science 1990; 78:13-16.
20. Cole CW, Thijssen AM, Barber GG et al. Popliteal aneurysms: An index of generalized vascular disease. Can J Surg 1989; 32:65-68.
21. Kuivanicmi H, Watton SJ, Price SJ et al. Candidate genes for abdominal aortic aneurysms. Ann N Y Acad Sci 1996; 800:186-197.
22. Takolander RJ, Bergqvist D, Bergentz SE et al. Aneurysms of the popliteal artery. Acta Chir Scand 1984; 150:135-140.

CHAPTER 3

The Mechanical Properties of the Normal and Aneurysmal Abdominal Aorta In Vivo

Björn Sonesson, Toste Länne

Background

The infrarenal aorta is a site predisposed to aneurysmal widening. The cyclic stress caused by the pulse wave in conjunction with factors which decrease the strength of the wall may lead to dilatation, and ultimately, to rupture. The prognosis of these dilatations is uncertain, and both the diameter and mechanical properties may play a role. By studying the normal aortic diameter and the wall mechanics of the normal and aneurysmal aorta, information basic to the understanding of the pathophysiology may be obtained. Furthermore, wall mechanics play an important role in the regulation of cardiovascular hemodynamics, such as cardiac performance and blood pressure.

This chapter will give some insight into the biological basis for wall mechanics and possible methods to use for in vivo measurements. It will also 1) define normal values of the infrarenal aortic diameter and correlate them with age, sex and body size; 2) define the mechanical properties in the infrarenal aorta and their changes with age and sex, and during sympathetic stimulation; 3) sum up what is known about wall mechanics in the aneurysmal aorta.

The Biology of the Aortic Wall and the Mechanical Properties

The mechanical properties of the aorta are mainly determined by the matrix components of the wall. These are predominately elastin, collagen and smooth muscle cells. Elastin and collagen determine the passive mechanical properties,[1] whereas smooth muscle cells have the potential to contract and relax modulating wall mechanics.[2] The latter is of little practical importance in the aorta.[2-4] The distensible elastin is loadbearing at low pressures and the 100 to 1000 times stiffer collagen is loadbearing at high pressures[1,5] giving rise to the nonlinear pressure diameter curve.[6] Thus, it is clear that the collagen-to-elastin ratio is the principal determinant of wall mechanics.[2,7,8] This ratio varies with localization within the arterial tree, increasing towards the periphery[9] with a concomitant decrease in distensibility.[10] Further, the thickness as well as the architecture of the vessel wall are additional determinants of its mechanical properties.[2,7,8]

Development of Aneurysms, edited by Richard R. Keen and Philip B. Dobrin.

Methods of Studying the Mechanical Properties of Arterial Walls In Vivo

Various methods of examining the mechanical properties of large arteries exist. Most earlier investigations of the mechanical properties of the aorta have been based on in vitro investigations of isolated rings and strips of the aorta.[2] Examination in vivo, however, is preferable in order to avoid the alterations in mechanical properties post mortem as well as the methodological difficulties which arise in respect to smooth muscle relaxation and retraction of excised arteries.[11]

At present the methods for estimating wall mechanics in vivo involve five approaches: 1) direct optical measurement or measurement by means of electromechanical gauges applied directly to the exposed vessel; 2) measurement of pulse wave velocity (PWV); 3) measurement of the decline in arterial pressure during diastole (assuming a Windkessel model); 4) characteristic impedance determined from Fourier analysis of pressure and flow waves recorded simultaneously; 5) determination of fractional diameter change with respect to pressure change using imaging techniques (e.g., angiography, ultrasound, magnetic resonance imaging).

The first approach requires surgically exposed vessels.[12,13] Exposure as well as the mechanical restraint, imposed by the application of a measuring device, decreases distensibility and affects the validity of the method.[2,14]

The second, third, and fourth approaches are indirect methods of estimating wall mechanics. With PWV it is also impossible to assess local vascular properties,[15] and the use of diastolic pressure decline (Windkessel model) is limited, since reflected pressure waves induce significant error in the values obtained.[16] Measurement of characteristic impedance requires arterial catheterization and the use of microflow sensors and microtip manometers makes it complicated to use and difficult to apply in larger investigations.

The fifth approach involves imaging techniques to determine fractional diameter change in respect to pressure change. At present several imaging techniques exist, such as angiography,[17] magnetic resonance imaging,[18] ultrasonography using M-mode,[19] intravascular ultrasound[20] or echo-tracking systems.[21,22] They differ in respect to resolution capacity and ease of use. The values in parenthesis refer to the resolution capacity: angiography (0.5 mm), magnetic resonance imaging (0.2 mm), ultrasonography using M-mode (0.5 mm), intravascular ultrasound (0.1 mm) and echo-tracking systems (< 10 mm). Thus echo-tracking technique offers the best resolution capacity at the moment. Furthermore, it is simple to use and gives reproducible results.[23] The ultrasonic echo-tracking technique will be described in detail below.

Ultrasonic Echo-Tracking

The electronic echo-tracking instrument (Diamove, Teletec AB, Lund, Sweden) used in our investigation was interfaced with a B-mode real-time ultrasound scanner (EUB-240, Hitachi, Tokyo, Japan) and fitted with a 3.5 MHz linear array transducer. An echo-tracking phase-locked loop circuit restores the position of an electronic gate relative to the moving echo. The discrete compensatory steps of the gate yield the echo movement per unit time. The instrument is equipped with dual echo-tracking loops, which makes it possible to simultaneously track two separate echoes from opposite vessel walls. The difference between the signals instantaneously indicates any change in vessel diameter. In the system used the smallest detectable movement is 7.8 mm.[22] The repetition frequency of the echo-tracking loops is 870 Hz, and the consequent time resolution is approximately 1.2 ms.

The distal abdominal aorta is visualized in a longitudinal section between the renal arteries and the bifurcation on the real time image of the ultrasound scanner. Two electronic markers, each representing one tracking gate, are aligned with and locked to the luminal interfaces of the echo image of the anterior and posterior vessel wall at a site 3-4 cm proximal to the aortic bifurcation (Fig. 3.1). The echo-tracker measures the distance between the vessel walls perpendicular to the longitudinal axis of the vessel. Although tilting the transducer away from the longitudinal axis will falsely increase the measured diameter and its change, and this error is minimized, since the vessel is visualized continuously and the transducer can be held parallel to the longitudinal axis of the aorta throughout the recording.

The methodological error of the present echo-tracking system has been tested both in vitro and in vivo. A high degree of agreement is found in vitro in both static and dynamic tests between the measurement obtained via the echo-tracker and an independent optical system. By using a pulsating rubber tube the two independent measuring systems were shown to have a correlation of r = 0.992.[24] Intra-observer variability in the healthy aorta for static aortic diameter is 5%, for pulsatile diameter change 16%,[23] and in the aneurysmal aorta for static diameter 3%,[25] for pulsatile diameter change 22% and for β 18% (unpublished data).

The blood pressure was measured in the left brachial artery (upper arm) by the auscultatory method using a sphygmomanometer immediately after measurement of the pulsatile aortic diameter. The blood pressure thus obtained was assumed equal to systemic blood pressure in the abdominal aorta. Mean arterial blood pressure (MAP) was defined as the diastolic pressure plus one-third of the pulse pressure.

A data acquisition system containing a personal computer type 386 (Express, Tokyo, Japan) and a 12-bit analogue-to-digital converter (Analogue Devices, Norwood, USA) was included for the simultaneous monitoring of the arterial blood pressure and vessel diameter. Pressure was sampled at the same rate as the diameter (Fig. 3.2).

Arterial wall stiffness (β) was expressed as:

Pressure strain elastic modulus (Ep) was defined according to Peterson et al 1960[10] as:

$$Ep = K \times \frac{P \text{ systolic-P diastolic}}{(D \text{ systolic-D diastolic})/D \text{ diastolic}} \qquad \text{\{Eqn. 3.1\}}$$

P systolic and P diastolic are the maximum systolic and end-diastolic blood pressure levels, respectively, in mmHg. D systolic and D diastolic are the corresponding vessel diameters in mm. Ep is measured in N/m^2, K=133.3 is the factor for converting mmHg to N/m^2. Because of the nonlinear pressure-diameter relationship of the arterial wall, Ep is pressure-dependent (see equation 2). Hayashi et al[26] established a relation for calculating stiffness in vitro based on the observed linear relation between the logarithm of relative pressure and distention ratio. This index is called stiffness (β) and characterizes the entire deformation behavior of the arterial wall without pressure dependence in the physiological range. It was later modified and used in vivo by Kawasaki et al.[27]

$$\text{Stiffness } (\beta) = \frac{\ln (P \text{ systolic/P diastolic})}{(D \text{ systolic-D diastolic})/D \text{ diastolic}} \qquad \text{\{Eqn. 3.2\}}$$

ln (P systolic/ P diastolic) is the natural logarithm for the systolic-diastolic pressure ratio. Abbreviations as in Equation 3.1.

The pressure dependence of these indices was tested during an exercise-induced rise in blood pressure.[28,29] An increase in mean arterial pressure of about 40-50% caused a significant increase in Ep. However, the slight increase in β failed to reach statistical significance. Stiffness (β) may therefore be a more useful index of aortic distensibility. Furthermore, the influence of age, sex, aortic diameter and blood pressure on aortic β was analyzed using a multiplicative multiple regression linear regression model on 165 healthy individuals, 84 males and 81

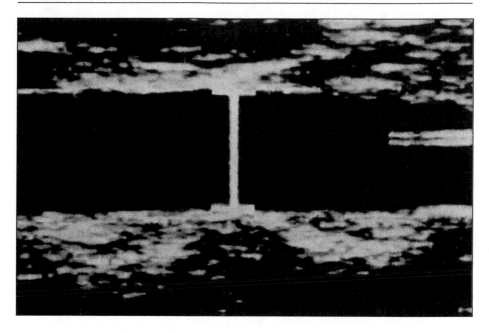

Fig. 3.1. Ultrasonic longitudinal image of the abdominal aorta with the cursors locked to the luminal interface of the echo image of the anterior and posterior wall of the abdominal aorta. The tip of the pressure catheter can be seen in the aortic lumen. Reproduced with the permission from Sonesson B, Länne T, Vernersson E et al. Sex difference in the mechanical properties of the abdominal aorta in human beings. J Vasc Surg 1994; 20:959-69.

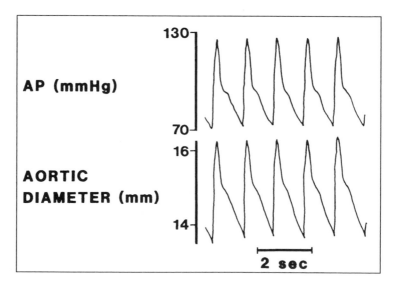

Fig. 3.2. Original tracing of simultaneously obtained aortic blood pressure (AP) and pulsatile diameter change of the distal abdominal aorta in a healthy 24-year old male.

females, 4-79 years of age. β was influenced most by age in both males (91% p < 0.0001) and females (75% p < 0.0001) and only to a very small extent by diameter (males 1% p < 0.01, females 3% p < 0.01), whereas blood pressure analysed with respect to systolic pressure, MAP, pulse pressure, natural logarithm (systolic/diastolic) had no influence at all.

The calculated stiffness (β) of the abdominal aorta is based on the assumption that abdominal aortic pressure is equal to the pressure obtained in the brachial artery (upper arm) by the auscultatory method. This assumption can be questioned, since, to begin with, pressure is determined by auscultation rather than intraarterially. Furthermore it does not take into account the difference in pressure between central and peripheral arteries attributed to wave travel and reflection within the arterial tree.[30] Mean pressure falls gradually as one moves towards the periphery, but the pulse pressure and the peak systolic pressure increase although these differences seem to become less marked later in life.[31] These potential differences in pressure, at the site of pulsatile diameter measurement (aorta) and in the brachial artery obtained by the auscultatory method, could induce an error in the obtained values. However, an evaluation showed that good agreement was found between peak systolic pressure at the two measuring points. Diastolic pressure proved to be consistently overestimated when determined by auscultation. No significant age or sex-related differences were observed. The fact that pulse pressure as measured by the auscultatory method in the brachial artery is less than the actual pulse pressure in the abdominal aorta leads to a systematic underestimation of Ep and β by 15-20%[6] when using the auscultatory method.

The Normal Infrarenal Aortic Diameter

The infrarenal aorta was found to increase steadily in diameter throughout life (Fig. 3.3). This was most pronounced during growth but there was a substantial increase of about 25% in adult males and females thereafter. From about 25 years the diameter was larger in males than in females (p < 0.01), although this difference vanishes if corrected for differences in body surface area between the sexes.

Significant correlations were found between aortic diameter and weight (r = 0.84, p < 0.001), height (r = 0.77, p < 0.001) and body surface area (r = 0.83, p < 0.001). Age followed by body surface area were the factors most influencing aortic diameter in both sexes. It seems obvious that vessel diameter and body size should be closely related, since a larger body has greater metabolic needs and therefore requires larger conduits. However, previous investigations[32-37] have shown inconsistent or weak correlations in contrast to the strong correlation demonstrated by our investigation.[38] The discrepancy in results may be the result of their use of individuals with vascular disease, metastatic tumors, or other patient groups rather than healthy nonsmoking individuals, whereas we studied healthy subjects.

On the basis of the present study it is now possible to predict an individual infrarenal aortic diameter with 95% confidence limits by means of a multiple stepwise regression model if age, sex and body size are known. As an example the predicted aortic diameter for males is shown in Figure 3.4. These nomograms of the infrarenal aortic diameter may be of assistance in determining whether a measured aortic diameter is of pathologic significance or not. For further details and complete tables of predicted infrarenal aortic diameter with 95% confidence limits, see Sonesson et al 1994.[38]

The Mechanical Properties of the Normal Aorta

The simultaneously recorded aortic pressure and diameter curve shows great similarities (Figs. 3.5A and 3.5B). In the young person a dicrotic wave is superimposed on the declining (diastolic) part of both the pressure and diameter curves (Fig. 3.5A), and this is most likely the effect of reflections from bifurcations, vessel branch points and resistance vessels.[39] In

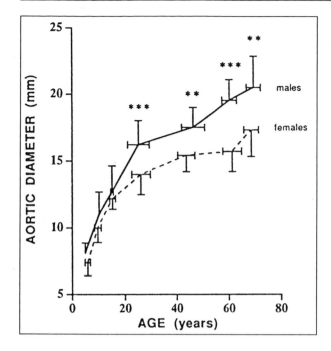

Fig. 3.3. Diameter of the infrarenal aorta in healthy, nonsmoking females, n = 69 (---) and males, n = 76 (—) * $p < 0.05$ **$p < 0.01$, *$p < 0.001$ Reproduced with the permission from Sonesson B, Länne T, Hansen F et al. Infrarenal aortic diameter in the healthy person. Eur J Vasc Surg 1994; 8:89-95.

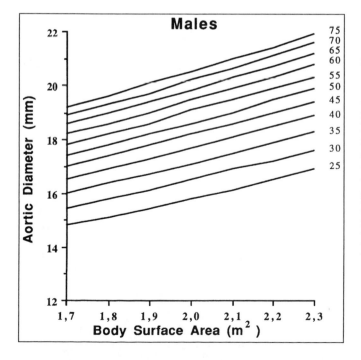

Fig. 3.4. Predicted infrarenal aortic diameter in males. Select the appropriate curve for age marked on the right. Follow the curve to the appropriate body surface area (m^2) on the horizontal axis. The corresponding aortic diameter is shown on the vertical axis. Reproduced with the permission of Sonesson B, Länne T, Hansen F et al. Infrarenal aortic diameter in the healthy person. Eur J Vasc Surg 1994; 8:89-95.

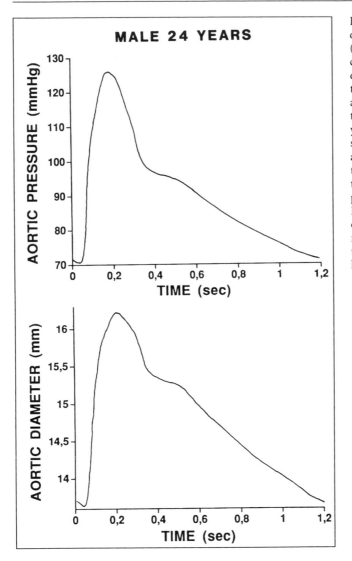

MALE 24 YEARS

Fig. 3.5A. Original tracing of a mean pressure curve (upper tracing) and the corresponding mean diameter curve (lower tracing) of the abdominal aorta based on 9 consecutive cardiac cycles in a 24-year old male. Notice the similarity in the pressure and diameter curves and the dicrotic wave in diastole. Reproduced with the permission of Sonesson B, Länne T, Vernersson E et al. Sex difference in the mechanical properties of the abdominal aorta in human beings. J Vasc Surg 1994; 20:959-69.

the elderly, however, the dicrotic wave is no longer observed, and instead, late systole is accentuated in both the pressure and diameter curves (Fig. 3.5b). The most likely explanation is that the reflected waves move faster in the stiffer older vessels with an earlier interaction in the pulse pressure curve resulting in an accentuation of late systole rather than of diastole.[40]

Although the similarities in the pressure and diameter curves of the young and elderly predominate, small differences do exist here. In the diastolic phase some differences can be seen between the pressure and diameter curves. The amplitude of diameter is seen to exceed the amplitude of pressure at a time when the rate of change in pressure is smaller, and this implies that the arterial wall is more distensible when it is stressed less rapidly.[10] This variation in distensibility, dependent on the rate of change of pressure (or stress), is due to the viscoelastic properties of the arterial wall.[10] The hysteresis of the corresponding pressure-diameter curve (P-D curve) with a smaller diameter during expansion, systole, than retraction, diastole, confirms this deduction (Fig. 3.6).

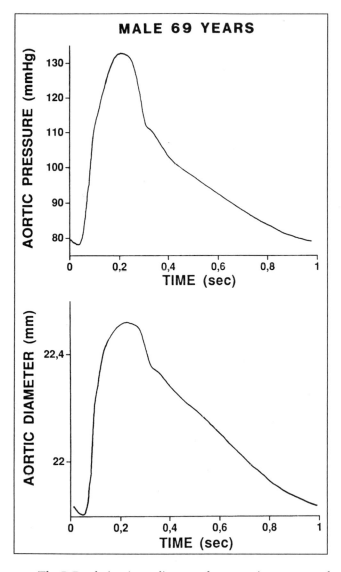

Fig. 3.5B. Original tracing of a mean pressure curve (upper tracing) and the corresponding mean diameter curve (lower tracing) of the abdominal aorta based on 11 consecutive cardiac cycles in a 69-year old male. Compared with the 24-year old's curves (Fig. 3.5A), the dicrotic wave in diastole is absent, and instead an augmentation of late systole is seen. Reproduced with the permission from Sonesson B, Länne T, Vernersson E et al. Sex difference in the mechanical properties of the abdominal aorta in human beings. J Vasc Surg 1994; 20:959-69.

The P-D relation is nonlinear and concave in respect to the pressure axis (Fig. 3.6). It also shows that the aortic wall has a more distensible part at lower pressures and a stiffer part at higher pressures. Although there is no sharp inflection point in the curve, the transition from distensible to stiff behavior occurs above 80 mmHg. These mechanical characteristics have been attributed to the properties of the matrix of the wall, i.e., mainly elastin, collagen and smooth muscle cells. The more distensible elastin is the principle loadbearer at low pressures and small distensions while at high pressures and large distensions both elastin and the stiffer collagen are loadbearing.[1]

The aortic media has the potential to modulate distensibility of the wall by active constriction and relaxation by smooth muscle cells (SMC). This might be of considerable importance because a change of stiffness and buffering capacity of the aorta affects cardiac load and systolic blood pressure. The role of the SMC in the human aorta is unclear. Bader[3]

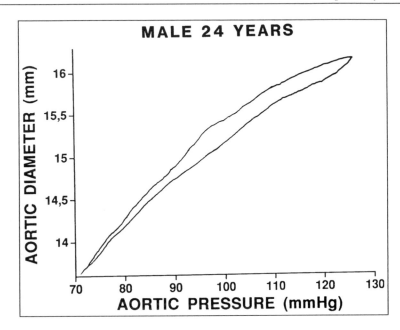

Fig. 3.6. Original tracing of the pressure-diameter (P-D) curve of the abdominal aorta in a 24-year-old male. The fact that the diameter is smaller during expansion than during retraction can be ascribed to the viscoelastic properties of the vessel wall and accounts for the hysteresis of the curves. Furthermore, the P-D relation is nonlinear, revealing that the aortic wall is more distensible at lower than at higher pressures. Reproduced with the permission of: Sonesson B, Länne T, Vernersson E et al. Sex difference in the mechanical properties of the abdominal aorta in human beings. J Vasc Surg 1994; 20:959-69.

as well as Dobrin[2] considered it in a large elastic artery, such as the aorta, to be insignificant judged from in vitro investigations. This is in accordance with in vivo studies from our own laboratory where sympathetic stimulation of the vascular smooth muscle cells was produced by applying lower body negative pressure (LBNP) to healthy individuals increasing peripheral resistance by 85%. However, neither the P-D relation nor Ep or β were changed (Fig. 3.7). This indicates that the SMC in the abdominal aorta do not contribute to wall mechanics.

With increasing age the P-D curves become less steep, i.e., the aorta stiffens and the aortic diameter increases (Fig. 3.8). There are several possible explanations for this. The first is an increase in the collagen-to-elastin ratio. It has been shown that elastin content decreases[41] and collagen content increases with age.[42] Another factor is that the architecture of the wall gradually becomes disarranged. Histological studies have shown a progressive loss of the orderly arrangement of elastic fibers and laminae, which display thinning, splitting, fraying and fragmentation.[43,44] A third factor is that there is an increase in wall thickness with age that is mainly due to increased deposition of collagen and ground substance,[45,46] and calcification of the elastic fibers. The hypothesis has been put forward that the fatiguing effect of cyclic stress induces the progressive degenerative process and stiffening of the wall. The repetitive pulsatile pressure over long periods leads to fractures of the loadbearing elastic fibers. The fractures in turn permit progressive dilatation of the vessel and remodelling of the wall. This transfers stress from the distensible elastin to the stiffer collagen. At the same

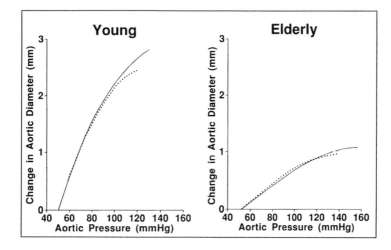

Fig. 3.7. P-D curves in the abdominal aorta during rest (solid line) and during sympathetic stimulation (dotted line) compiled from the young (n = 10) and elderly (n = 9) individuals. With increasing age the P-D curve becomes less steep, i.e., the aortic wall becomes less distensible. No changes of the P-D curves during sympathetic stimulation are indicating that it does not alter wall mechanics. Reproduced with the permission of: Sonesson B, Vernersson E, Hansen F et al. Influence of sympathetic stimulation on the mechanical properties of the aorta in humans. Acta Physiol Scand 1996. Submitted.

time a reparative process is initiated by the cellular elements with deposition of collagen.[31] All these factors lead to a larger and stiffer artery.

Atherosclerosis may be another factor involved in the stiffening of the aortic wall. Atherosclerotic changes are very common and closely related to aging, thus making it difficult to separate changes caused by atherosclerosis from those caused by aging of the wall. A large number of studies, however, favor the opinion that the increase in wall stiffness is independent of atherosclerosis.[31,47-51] One possible explanation is that the distribution of athereosclerosis is usually patchy and, as long as no more than one-quarter of the vessel area is engaged, no change in mechanical properties can be observed.[47] It is possible that the vessel compensates by means of increased distensibility in disease-free areas. However, when medial calcification is more extensive the vessel becomes stiffer.[47] Thus, there is abundant evidence indicating that the mechanical properties of the aorta reflect degenerative changes in the wall and may therefore reflect its biological age.

Wall mechanics can also be estimated from the distensibility index pressure strain elastic modulus (Ep) and stiffness (β) calculated from pulsatile diameter change and blood pressure. This has the advantage that it can be done completely noninvasively and therefore on a larger epidemiologic scale. When comparing the age-related differences in aortic stiffness (β) between males and females in a large cohort of healthy nonsmoking individuals large differences are seen (Fig. 3.9). There is an age-related increase in stiffness, which is exponential in nature and much more pronounced in males than in females (p < 0.0001). Kawasaki[27] and Imura,[52] who used similar echo-tracking techniques, reported a linear increase in aortic Ep and stiffness (β) with age. Their values lie between those of our males and females and are probably the result of the inclusion of both females and males in their study. Our data show the importance of differentiating between sexes. The decreased distensibility appears

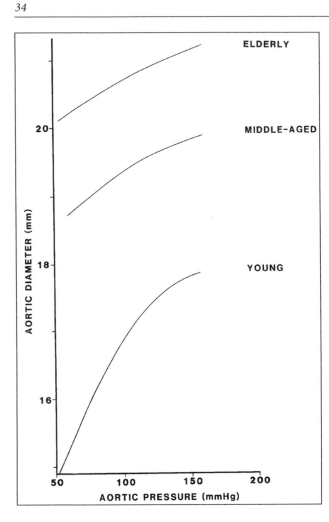

Fig. 3.8. P-D curves of the abdominal aorta compiled from males in the three different age groups, young (n = 5), middle-aged (n = 4) and elderly (n = 4). With increasing age the P-D curve becomes less steep, the aortic diameter increases and the nonlinear behavior becomes less obvious. Thus, the aorta becomes less distensible with age and dilates. Reproduced with the permission from Sonesson B, Länne T, Vernersson E et al. Sex difference in the mechanical properties of the abdominal aorta in human beings. J Vasc Surg 1994; 20:959-69.

at an earlier age in males, implying that the age-related changes in aortic wall structure are delayed in females. Several factors may be involved: structural differences between the sexes in the composition of the aortic wall, hormonal influence and atherosclerosis. Wall mechanics are mainly determined by the collagen-to-elastin ratio, the relative thickness and architecture of the wall. Information as to sex-related differences in these or other biochemical or histological factors is not available for the abdominal aorta at the moment. However, it has been shown experimentally in animals that the collagen-to-elastin ratio in the aortic wall changes when animals are treated with different sex hormones. Estrogen and progesterone decrease the collagen-to-elastin ratio whereas testosterone increases it.[53,54] If this is true for the human aorta as well, female sex hormones may increase aortic compliance and account for some of the sex-related differences in distensibility. Interestingly, a higher elastin content in the thoracic aorta has been found in females than in males.[49] Estrogen substitution in postmenopausal women results in a slightly increased aortic compliance in comparison with untreated individuals. (Gennser, personal communication, 1994).

Differences in atherosclerotic involvement between the sexes may exist although the investigated healthy individuals were clinically free of symptoms. However, atherosclerosis

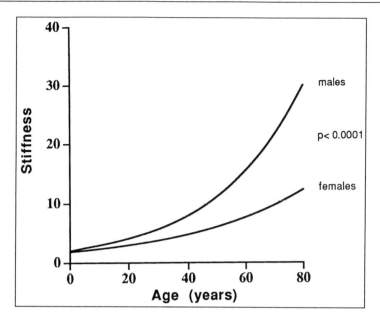

Fig. 3.9. Stiffness (β) in a nonsmoking healthy reference population of 165 males (n = 84) and females (n = 81) in the infrarenal aorta. There is an age-related increase in (β), which is exponential in nature and much more pronounced in males than in females (p < 0.0001). Reproduced with the permission from: Sonesson B, Hansen F, Länne T. Abnormal mechanical properties of the aorta in Marfan's syndrome. Eur J Vasc Surg 1994; 8:595-601.

does not play a major role in the mechanical properties except when it is severe[31,47-51] and is probably not responsible for the sex-related differences in wall mechanics. Further research into the histological and biochemical differences in the male and female abdominal aorta and the influence of hormones may provide an explanation for differences in wall mechanics.

The Diameter and the Mechanical Properties of the Aneurysmal Abdominal Aorta

The nomograms of the predicted infrarenal aortic diameter discussed previously in this chapter may be of assistance in determining whether or not a measured aortic diameter is of pathological significance (Table 3.1). Furthermore, these nomograms may also be of interest when defining AAA. No consensus has been reached yet regarding diagnostic criteria for AAA. It has, however, been proposed by the American Society for Vascular Surgery that an aneurysm is to be defined as a permanent localized dilatation of an artery that exceeds 1.5 times of the expected normal diameter.[55] To use this definition it is necessary to know the normal diameter of the aorta. Defining an aneurysm in relation to the individual predicted aortic diameter may also give a more dynamic view of the aneurysmal expansion.

Even if the size of the aneurysm still remains as the most important predictor of rupture[56] it has been known for a long time that even small AAA may rupture.[57] There is a consensus among vascular surgeons that AAA < 40 mm should be treated conservatively with close follow-up with ultrasonography and that AAA > 55-60 mm should be operated on. However, there is no consensus for those between 40 and 55 mm and now ongoing prospective randomized trials will hopefully in the future give an answer to what is the optimal treatment.

Table 3.1.

	Number	Age (year)	Body surface area (m^2)	AAA (mm)	Predicted AAA size
Males	35	68 + 5	1.9 + 0.2	48 + 5	2.5 + 0.2
Females	16	72 + 8	1.6 + 0.1	48 + 5	3.0 + 0.4
Significance		NS	p < 0.001	NS	p < 0.001

In our own series of electively operated AAA 40-55 mm, no difference in absolute size between the sexes could be seen. However, when relating the size to the predicted normal aortic diameter the females had significantly larger AAA than males.

Aneurysms considered small in terms of absolute diameter may in fact be large in relation to the normal diameter of the aorta in that particular patient. This indicates that the aneurysm size should be considered more individually and dynamically.

Abdominal aortic aneurysm (AAA) in men show an increased wall stiffness (Fig. 3.10).[28] Similar results have been reported in vitro[58,59] and in vivo by other investigators.[60] The ranges of Ep and β in our investigations indicate a large heterogenicity in wall mechanics of the aneurysm. Aneurysms of similar size may have a 10-fold difference in stiffness. There is no obvious reason for this. Only a small part of this could be attributed to the variability of the method used as the coefficient of variation for β in aneurysms was 18% (unpublished data).

Many individuals with AAA are smokers and some are diabetic, and these factors may influence the measured properties. However, we have recently shown that neither smoking (unpublished data) nor diabetes[61] affects the stiffness values in the male aorta of normal dimensions. In females, however, both smoking and diabetes increases aortic stiffness.

The alterations found in the mechanical properties of the aneurysm may be related to the marked changes in structure and composition. Histological examinations of AAA reveal disrupted thin media with lack of elastic fibers, fibrosis and thickening of the adventitia and neointima.[62] Elastin content is decreased[58-60,63] whereas collagen content has been reported increased[59,62-64] or unchanged.[65] These changes may be the result of reported increased proteolytic activity of collagenase[64,66] and elastase[67,68] whereas at the same time lower antiprotease activity has been detected.[69] Together, these factors will raise collagen-to-elastin ratio and alter wall structure resulting in changed wall mechanics.

The question has been raised whether aneurysm formation is the result of a general defect in the vasculature with a preferential localization in the abdominal aorta. Measurement of wall mechanics of vessels of normal dimensions in individuals with AAA may give indication of this. Preliminary results from our laboratory show that wall mechanics are changed in the carotid artery of individuals with AAA.

Another important issue is whether wall mechanics can be related to growth and risk of rupture. The hypothesis has been put forward that the development of AAA can be regarded as a classical case of material failure.[70] Either the applied load is excessive or the tensile strength of the material inadequate, or a combination of both. The pulse pressure increases and the rise of pressure (dp/dt) becomes steeper when the pulse wave is transmitted from the aortic root to the femoral region. Thus, the largest pulsatile load in the aorta is localized to the infrarenal region. This is due to three factors: the tapering of the aorta distally, the progressive stiffening of the aortic wall, especially as it enters the abdomen, and the addition of the wave reflection from the iliac bifurcation to the incoming pressure wave.[70-72]

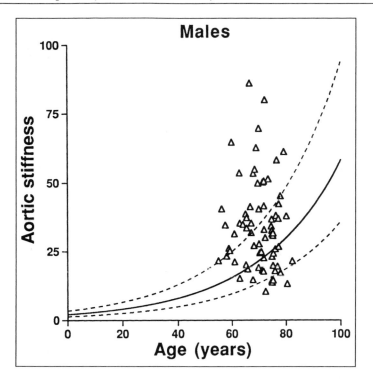

Fig. 3.10. This figure shows that stiffness (β) of the aneurysm was increased compared with the reference population (p < 0.0001). The solid line represents mean and the dotted lines upper and lower 95% confidence interval for the reference population. Reproduced with permission from: Sonesson B, Hansen F, Länne T. Abnormal mechanical properties of the aorta in Marfan's syndrome. Eur J Vasc Surg 1994; 3:595-601.

All these result in a local high-pressure zone. At the same time the distal part of the aorta has a lower capacity to withstand the pulsatile load as compared to other mammalian species. It has fewer elastic lamellae and fewer or no vasa vasorum relative to its wall thickness.[73,74] Elastin is the principle loadbearing protein in these lamellae and failure of some of these might shift the load to the remaining intact lamellae and overloading them and the collagen. This may be the initial step in the dilatation. Dobrin[75] treated arteries with elastase and/or collagenase and showed that elastin failure leads to dilatation while collagen failure leads to additional dilatation and rupture. This shows that collagen provides the tensile strength and prevents excessive distension and rupture of the vessel. Our investigations show that defects in fibrillin/elastin, such as occurs in the Marfan's syndrome, alter aortic wall mechanics with increasing stiffness.[76] This is in accordance with the demonstrated relation between elastin content and arterial stiffness.[60,77] However, in individuals with Ehlers-Danlos syndrome, with defects in collagen giving rise to spontaneous ruptures of large arteries, no change in wall mechanics could be detected in the physiological pressure range (unpublished data). This might indicate that the structural defect in the arterial wall collagen and thus the tendency to vessel fragility cannot be predicted and revealed during normal physiological pressure conditions. Powell et al[78] demonstrated an increased stiffness of AAA with a defect in the collagen III gene compared with AAA where no gene mutations could be demonstrated.

However, they did not report whether or not there were any differences in elastin content between the two groups of AAA. A difference in elastin content might be an alternative explanation to their findings.

When addressing the issue of risk of aneurysm rupture, Laplace's law is commonly discussed. This law describes the relation between intraluminal pressure (p), vessel radius (r), wall thickness (h) and wall tension (T) and is expressed as $T = p \ (r/h)$. The tension increases with increasing intraluminal pressure and aneurysm diameter or with decreasing wall thickness. One would expect that the tension increases and the aneurysm wall becomes stiffer with increasing size of the aneurysm. However, we found no correlation between stiffness (β) and aneurysm size ($r = 0.18$, ns).[28] This indicates that the intrinsic properties of the wall are of great importance in maintaining structural integrity, preventing dilatation and rupture. This is corroborated by the fact that aneurysms of equal size may or may not rupture. Thus, not only size/tension of an AAA, but also factors in the aneurysm wall, at present unknown, are of importance as well in determining the risk of rupture.

The knowledge of the mechanical properties and the diameter of the normal aorta seem to be of importance in further studies of AAA. To follow the mechanical properties of aneurysms and the diameter increase from the predicted normal value in patients not selected unfit for surgery may be one way of gaining information on the natural history of this condition and forecast the risk of rupture.

References

1. Roach MR, Burton AC. The reason for the shape of the distensibility curves of arteries. Can J Biochem Physiol 1957; 35:681-90.
2. Dobrin P. Vascular mechanics. In: Handbook of Physiology. Part I. Peripheral circulation and organ blood flow, Shepard JT, Abbound FM, eds. Baltimore, Williams & Wilkins, 1983; 65-102.
3. Bader H. Anatomy and physiology of the vascular wall. In: Handbook of Physiology. Part II. Circulation, Hamilton WF, Dow P, eds. Washington D.C. American Physiological Society, 1963; 875-6.
4. Sonesson B, Vernersson E, Hansen F et al. Influence of sympathetic stimulation on the mechanical properties of the aorta in humans. Acta Physiol Scand 1996. Submitted.
5. Burton AC. Relationship of structure to function of the tissues of the wall of blood vessels. Physiol Rev 1954; 34:619-42.
6. Sonesson B, Länne T, Vernersson E et al. Sex difference in the mechanical properties of the abdominal aorta in human beings. J Vasc Surg 1994; 20:959-69.
7. Strandness DE, Jr., Sumner DS. Hemodynamics for Surgeons. New York: Grune and Stratton, 1975.
8. Glagov S, Vito R, Giddens DP et al. Micro-architecture and composition of artery walls: Relationship to location, diameter and the distribution of mechanical stress. J Hypertens 1992; 10:101-104.
9. Harkness ML, Harkness RD, McDonald DA. The collagen and elastin content of the arterial wall in the dog. Proc R Soc Lond 1957; 146B:541-51.
10. Peterson LH, Jensen RE, Parnell J. Mechanical properties of arteries in vivo. Circ Res 1960; 8:622-33.
11. O'Rourke MF. Arterial function in health and disease. Edinburgh: Churchill and Livingstone, 1982.
12. Busse R, Bauer RD, Schabert A et al. The mechanical properties of the exposed human carotid arteries in vivo. Basic Res Cardiol 1979; 74:545-54.
13. Greenfield JC, Jr., Patel DJ. Relation between pressure and diameter in the ascending aorta in man. Circ Res 1962; X:778-81.
14. Megerman J, Hasson JE, Warnock DF et al. Noninvasive measurements of nonlinear arterial elasticity. Am J Physiol 1986; 250:181-8.

15. McDonald DA. Regional pulse-wave velocity in the arterial tree. J Appl Physiol 1968; 24:73-8.
16. Hickler RB. Aortic and large artery stiffness. Current methodology and clinical correlations. Clin Cardiol 1990; 13:317-22.
17. Luchsinger PC, Sachs M, Patel DJ. Pressure-radius relationship in large blood vessels of man. Circ Res 1962; XI:882-8.
18. Bogren HG, Mohiaddin RH, Klipstein RK et al. The function of the aorta in ischemic heart disease: A magnetic resonance and angiographic study of aortic compliance and blood flow patterns. Am Heart J 1989; 118:234-47.
19. Arndt JO, Klauske J, Mersch F. The diameter of the intact carotid artery in man and its change with pulse pressure. Pflugers Arch 1968; 301:230-40.
20. Slördahl SA, Piene H, Linker DT et al. Segmental aortic wall stiffness from intravascular ultrasound at normal and subnormal aortic pressure in pigs. Acta Physiol Scand 1991; 143:223-32.
21. Lindström K, Gennser G, Sindberg P et al. An improved echo-tracker for studies on pulse waves in the fetal aorta. In: Fetal Physiological Measurements, Rolfe P, ed. London, Butterworths, 1987; 217-26.
22. Benthin M, Dahl P, Ruzicka R et al. Calculations of pulse-wave velocity using cross-correlations—effects of reflexes in the arterial tree. Ultrasound Med Biol 1991; 17:461-69.
23. Hansen F, Bergqvist D, Mangell P et al. Noninvasive measurement of pulsatile vessel diameter change and elastic properties in human arteries: a methodological study. Clin Physiol 1993; 13:631-43.
24. Sindberg PS. Fetal dynamics and maternal smoking. Application of new ultrasonic methods. Malmö, Sweden: Lund University,1984. Thesis.
25. Länne T, Sandgren T, Mangell P et al. Improved reliability of ultrasonic surveillance of abdominal aortic aneurysms. Eur J Vasc Endovasc Surg 1996. Accepted.
26. Hayashi K, Handa H, Nagasawa S et al. Stiffness and elastic behavior of human intracranial and extracranial arteries. J Biomech 1980; 13:175-84.
27. Kawasaki T, Sasayama S, Yagi SI et al. Noninvasive assessment of the age related changes in stiffness of major branches of the human arteries. Cardiovasc Res 1987; 21:678-87.
28. Länne T, Sonesson B, Bergqvist D et al. Diameter and compliance in the male human abdominal aorta: Influence of age and aortic aneurysm. Eur J Vasc Surg 1992; 6:178-84.
29. Sonesson B, Hansen F, Stale H et al. Compliance and diameter in the human abdominal aorta—the influence of age and sex. Eur J Vasc Surg 1993; 7:690-7.
30. Karamanoglu M, O'Rourke MF, Avolio AP et al. An analysis of the relationship between central aorta and peripheral upper limb pressure waves in man. Eur Heart J 1993; 14:160-7.
31. Nichols WW, O'Rourke MF. Aging, high blood pressure and disease in humans. In: McDonald's blood flow in arteries, Nichols W, O'Rourke MF, eds. London, Edward Arnold, 1990; 398-420.
32. Steinberg CR, Archer M, Steinberg I. Measurement of the abdominal aorta after intravenous aortography in healthy and arteriosclerotic peripheral vascular disease. A J R 1965; 95:703-8.
33. Tilson MD, Dand C. Generalized arteriomegaly. A possible predisposition to the formation of abdominal aortic aneurysms. Arch Surg 1981; 116:1030-2.
34. Cronenwett KL, Garett E. Arteriographic measurement of the abdominal aorta, iliac, and femoral arteries in women with atherosclerotic occlusive disease. Radiology 1983; 148:389-92.
35. Liddington IJ, Heather BP. The relationship between aortic diameter and body habitus. Eur J Vasc Surg 1992; 6:89-92.
36. Pedersen OM, Alasken A, Vik-Mo H. Ultrasound measurement of luminal diameter of the abdominal aorta and iliac arteries in patients without vascular disease. J Vasc Surg 1993; 17:596-601.
37. Pearce WH, Slaughter MS, LeMaire S et al. Aortic diameter as function of age, gender and body surface area. Surgery 1993; 114:691-7.
38. Sonesson B, Länne T, Hansen F et al. Infrarenal aortic diameter in the healthy person. Eur J Vasc Surg 1994; 8:89-95.

39. Remington JW, O'Brien LJ. Construction of aortic flow pulse from pressure pulse. Am J Physiol 1970; 218:437-47.
40. O'Rourke MF, Kelly RP, Avolio AP. The arterial pulse. Philadelphia: Lea and Febiger, 1992.
41. Hornbeck W, Adnett JJ, Robert L. Age dependent variations of elastin and elastase in aorta and human breast cancers. Exp Gerontol 1978; 13:293-8.
42. Faber M, Moeller-Hou G. The human aorta. Acta Pathol Microbiol Scand 1952; 31:377-82.
43. Schlatmann TJ, Becker AE. Histological changes in the normal aging aorta: Implication for dissecting aneurysms. Am J Cardiol 1977; 39:13-20.
44. Schlatmann TJM, Becker AE. Pathogenesis of dissecting aneurysm of the aorta. Comparative histopathologic study of significance of medial changes. Am J Cardiol 1977; 39:21-6.
45. Toda T, Tsuda N, Nishimori I et al. Morphometrical analysis of the aging process in human arteries and aorta. Acta Anat 1980; 106:35-44.
46. Learoyd BM, Taylor MG. Alterations with age in the visco-elastic properties of human arterial walls. Circ Res 1966; 18:278-92.
47. Butcher HR, Newton WT. The influence of age, arteriosclerosis and homotransplantation upon the elastic properties of major human arteries. Ann Surg 1958; 148:1-20.
48. Nakashima T, Tanikava J. A study of human aortic distensibility with relation to atherosclerosis and aging. Angiology 1971; 22:477.
49. Langewouters GJ. Visco-elasticity of the human aorta in vitro in relation to the pressure and age. Netherlands: University of Amsterdam, 1982: 143-160.Thesis.
50. Aviolo AP, Chen S-G, Wang R-P et al. Effects of aging on changing arterial compliance and left ventricular load in a northern Chinese urban community. Circulation 1983; 68:50-8.
51. O'Rourke MF, Aviolo AP, Clyde K et al. High serum cholesterol and atherosclerosis do not contribute to increased arterial stiffening with age. J Am Coll Cardiol 1986; 7:247A.
52. Imura T, Yamamoto K, Kanamori K et al. Noninvasive ultrasonic measurement of the elastic properties of the human abdominal aorta. Cardiovasc Res 1986; 20:208-14.
53. Fisher GM, Swain ML. Influence of contraceptive and other sex steroids on aortic collagen and elastin. Exp Mol Pathol 1980; 33:15-24.
54. Fisher GM, Bashey RI, Rosenbaum H et al. A possible mechanism in arterial wall for mediation of sex differences in atherosclerosis. Exp Mol Pathol 1985; 43:288-96.
55. Johnston KW, Rutherford RB, Tilson MD et al. Suggested standards for reporting on arterial aneurysms. J Vasc Surg 1991; 13:444-50.
56. Szilagyi DE, Smith RF, DeRusso FJ et al. Contribution of abdominal aortic aneurysmectomy to prolongation of life. Am Surg 1966; 164:678-99.
57. Darling RC. Ruptured arteriosclerotic abdominal aortic aneurysms. Am J Surg 1970; 119:397.
58. Sumner DS, Hokanson DE, Strandness DE, Jr. Stress-strain characteristics and collagen-elastin content of abdominal aortic aneurysms. Surg Gynecol Obstet 1970; 147:211-4.
59. He CM, Roach MR. The composition and mechanical properties of abdominal aortic aneurysms. J Vasc Surg 1994; 20:6.
60. MacSweeney STR, Young G, Greenhalgh RM et al. Mechanical properties of the aneurysmal aorta. Br J Surg 1992; 79:1281-4.
61. Rydén Ahlgren Å, Länne T, Wollmer P et al. Increased arterial stiffness in women, but not in men, with IDDM. Diabetologia 1995; 38:1082-9.
62. Baxter TB, Halloran BG. Matrix metabolism in abdominal aortic aneurysms. In: Aneurysms: New findings and treatments, Yao JST, Pearce WH, eds. East Norwalk, Connecticut, Appleton and Lange, 1994; 25-34.
63. Rizzo RJ, McCarthy WJ, Dixit SN. Collagen types and matrix protein content in human abdominal aortic aneurysms. J Vasc Surg 1989; 10:365-73.
64. Menashi S, Campa JS, Greenhalgh RM et al. Collagen in abdominal aortic aneurysm: Typing, content and degradation. J Vasc Surg 1987; 6:578-82.
65. McGee GS, Baxter BT, Shively VP. Aneurysm or occlusive disease—factors determining the clinical course of atherosclerosis of the infrarenal aorta. Surgery 1991; 110:370-5.
66. Busuttil RW, Abou-Zamzam AM, Machleder HI. Collagenase activity of the human aorta. A comparison of patients with and without abdominal aortic aneurysms. Arch Surg 1980; 115:1373-8.

67. Busuttil RW, Rinderbreicht H, Flesher A et al. Elastase activity: The role of elastase in aortic aneurysm formation. J Surg Res 1982; 32:214-7.
68. Cannon DJ, Read RC. Blood elastolytic activity in patients with aortic aneurysm. Ann Thorac Surg 1982; 34:10-5.
69. Cohen JR, Mandell C, Chang J et al. Elastin metabolism of the infrarenal aorta. J Vasc Surg 1988; 7:210-4.
70. Dobrin PB. Pathophysiology and pathogenesis of aortic aneurysms. Surg Clin North Am 1989; 69:687-703.
71. Newman DL, Gosling RG, Bowden NLR. Pressure amplitude increase and matching the aortic iliac junction of the dog. Cardiovasc Res 1973; 7:6-13.
72. Gosling RG, Newman DL, Bowden L. The area ratio of normal aortic junctions. Br J Radiol 1971; 44:850-3.
73. Wolinsky H, Glagov S. Nature of species differences in the medial distribution of aortic vasa vasorum in mammals. Circ Res 1967; 20:409-21.
74. Wolinsky H, Glagov S. A lamellar unit of aortic medial structure and function in mammals. Circ Res 1967; XX:99-111.
75. Dobrin PB, Baker WH, Gley WC. Elastolytic and collagenolytic studies of arteries. Implications for the mechanical properties of aneurysms. Arch Surg 1984; 119:405-9.
76. Sonesson B, Hansen F, Länne T. Abnormal mechanical properties of the aorta in Marfan's syndrome. Eur J Vasc Surg 1994; 8:595-601.
77. Hayashi K, Sato H, Handa H et al. Biomechanical study of the constitutive laws of vascular walls. Exp Mech 1974; 14:440-4.
78. Powell JT, Adamson J, MacSweeny STR et al. Genetic variants of collagen III and abdominal aortic aneurysm. Eur J Vasc Surg 1991; 5:145-8.

CHAPTER 4

Elastin, Collagen, and the Pathophysiology of Arterial Aneurysms

Philip B. Dobrin

The formation of an arterial aneurysm is a classic example of mechanical failure of a cylindrical structure, in this case the arterial wall. Blood vessel walls are composed of connective tissue, smooth muscle cells and endothelial cells. All of these are suspended in a gelatinous ground substance. Elastin and collagen are the predominant connective tissues in the vessel wall, and these play a central role in determining the mechanical properties of the vessel wall.

Elastin

Characteristics and Arterial Composition

The media of large arteries is composed of concentric layers of elastin condensed into elastic lamellae. In addition, fine fibers of elastin may span the spaces between the concentric lamellae. But these spaces are largely filled with circumferentially-oriented smooth muscle cells. The lamellae are, in fact, sheets of elastin curved to form concentric cylinders. Occasional gaps are found in the lamellae which actually are fenestrations in the elastin. Elastin is composed of two major components: a central region consisting of amorphous elastin, and a surrounding sheet of microfibrillar protein. The enzyme responsible for the formation of stable elastin cross-links is found in the microfibrillar proteins.[1] One of the most important of these proteins is fibrillin. Elastin is synthesized by vascular smooth muscle cells, but the turnover of elastin in the arterial wall is extremely low with a half-life years.[1,2] Fischer and Llaurado[3] reported that the elastin content of arteries varies with location. In the dog, most medium size arteries such as the carotid, femoral and mesenteric arteries contain 20-35% elastin by dry weight. The aorta contains 40% elastin by dry weight, while the renal and coronary arteries contain only 15-18% elastin by dry weight. It is of interest that the elastin content of the aorta drops precipitously after the vessel passes through the diaphragm.[4] In other measurements Fischer and Llaurado found no difference between the elastin content of these vessels in renal hypertensive dogs as compared with the same vessels in normotensive animals[5]. These studies showed, as did those of Harkness and coworkers,[7] that the elastin content of the pulmonary artery resembles that of peripheral systemic arteries (about 25% elastin) rather than the thoracic aorta (about 40% elastin). Biochemical studies have demonstrated a decrease in elastin content early in the formation of arterial aneurysms.[2,6,7] Other investigators have reported a 2.3-fold increase in the elastin content of

Development of Aneurysms, edited by Richard R. Keen and Philip B. Dobrin.
©2000 Eurekah.com.

aneurysmal aortas and an even greater 5.7 fold increase in the collagen content in these vessels.[1,7] This results in a relative increase in the collagen-to-elastin ratio and an apparent decrease in the elastin content in aneurysms.

Thermoelasticity

The most striking property of elastin is its characteristic reversible, long-range deformability. This is especially important in the case of elastic arteries where recoil of the wall plays an important role in absorbing the pulsatile energy generated by the heart and in maintaining effective perfusion pressure during cardiac diastole. The property of rubber-like elasticity is rare and is exhibited by very few substances in nature. These usually are termed "elastomers". The properties of elastomers have been summarized by Baumann.[8]

   a. They exhibit rapid and long-range extensibility, elongating to up to 100% of their original length.
   b. They exhibit high elastic moduli (stiffness) and tensile stress at maximum deformations.
   c. They retract rapidly when released, often exhibiting rebound.
   d. Upon release of an elongating load, they return to their original dimensions.

According to thermoelastic theory, typical elastomers behave as though they were composed of long, largely independent polymer chains. Flexible bonds within the chains and only occasional cross-links between the chains permit rapid and extensive elongation under load. Under resting conditions the polymer chains lie in a shortened, crumpled state. When such a system is subjected to stress it extends with straightening of the chains. Upon removal of the stress the chains readily retract, resulting in a reversible elasticity. Heating elastomers increases their kinetic energy. This increases the entropy and remarkably results in shortening of the chains (Gough-Joule Effect). Rubber, the paragon of elastomeric behavior, actually does shorten when it is heated. However rubber differs from proteins in that rubber is formed by long hydrocarbon chains with few interchain bonds, whereas proteins usually possess many radicals capable of forming weak interchain bonds. In order to exhibit elastomeric behavior, therefore, the radicals in the protein structure must be neutralized by a solvent to reduce interchain electrostatic forces or "friction". Rubber is extensible even when dry, whereas proteins are inextensible and brittle unless a solvent is present. Rubber also crystallizes upon stretching[9] and this undoubtedly underlies much of the extension-retraction hysteresis seen with rubber. As early as 1913, McCartney[10] freed bovine ligamentum nuchae of much of its collagen content by digestion with pepsin, and subjected strips of this elastin-rich material to increased temperature. The ligament was found to contract linearly with increasing temperature, extending along the same temperature-length curve with cooling. It is probable that the preparation actually contained considerable amounts of collagen because the thermoelastic behavior of the strips was found to be perfectly reversible only up to 65°C, the temperature at which native collagen exhibits thermal retraction.[11,12] Meyer and Ferri[13] and Wohlisch and coworkers[14] showed that untreated ligamentum nuchae, when extended to 100% of its original length, conform to elastomeric theory by contracting when heated. Because of an apparent increase in internal energy when extended, Meyer and Ferri suggested that crystallization took place at large extensions.[13] However Hoeve and Florey[15] also studied unpurified ligamentum nuchae, but did so in a 30% glycol-water mixture. This mixture was used because in this solvent the volume of the tissue remained constant over a wide range of temperatures thus avoiding complicating changes in cross-sectional area. Their data also supported the conclusion that elastin is a typical elastomer, but calculation of internal energy indicated that crystallization does not occur when it is stretched. This is in agreement with x-ray diffraction data of stretched elastin published earlier.[16] Elastomeric theory also predicts the behavior of elastin obtained from the arterial wall. Lloyd and coworkers[17] studied strips of ox aortic elastin after removal of collagen with formic acid. Load extension curves were obtained in water at different temperatures. The investigators found that, at a given level of extension, the load was 60% higher at 75°C than at 15°C. This indicates a greater resistance to extension at the higher temperature.

Mechanical Elasticity

Many studies have described the isothermic mechanical properties of various elastin preparations. Most of these studies reported the stiffness of the tissue as elastic modulus. Elastic modulus is computed as the ratio of stress to strain (Fig. 4.1).[18] Stress is given by

$$\sigma = \frac{F}{A} \qquad \qquad \{\text{Eqn. 4.1}\}$$

where σ is stress, F is the force exerted by the tissue when it is at equilibrium, and A is the cross-sectional area over which that force is exerted. Strain is given by

$$\varepsilon = \frac{\Delta L}{L_o} \qquad \qquad \{\text{Eqn. 4.2}\}$$

where ε is strain, ΔL is change in dimension, and L_o is the original dimension. Thus, elastic modulus or Young's modulus is given by

$$E = \frac{\sigma}{\varepsilon} = \frac{F/A}{\Delta L/L_0} \qquad \qquad \{\text{Eqn. 4.3}\}$$

At any given level of deformation or strain, a stiff material will manifest greater stress than will a less stiff material. Therefore a stiff material will also exhibit a higher elastic modulus.

However Young's modulus is useful only for a material having a linear stress-strain relationship. For materials having a curvilinear relationship the Young's modulus will give an incorrect assessment of the stress-strain curves because it will not fit the nonlinear curve. Instead it will estimate a secant cutting across the curve as illustrated by the broken line in Figure 4.1. For materials that have a curvilinear stress-strain relationship, an incremental or tangent elastic modulus may be computed as the ratios of change in stress to change in strain over a small interval

$$E_{INC} = \frac{\Delta \sigma}{\Delta \varepsilon} \qquad \qquad \{\text{Eqn. 4.4}\}$$

This computed modulus only applies for the strain and stress conditions at which it was determined.

Returning to the computation of deformation, one may also use extension ratio instead of strain

$$\lambda = L/L_0 \qquad \qquad \{\text{Eqn. 4.5}\}$$

where λ is extension ratio, L is observed length, and L_0 is the original length. The relationship between strain and extension ratio is simply,

$$\lambda = \varepsilon + 1 \qquad \qquad \{\text{Eqn. 4.6}\}$$

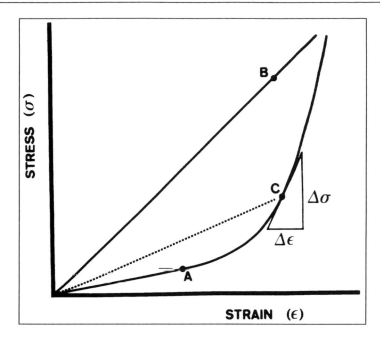

Fig. 4.1. Stress-strain curves for a material having a linear stress-strain curve (origin to point B), and for a material having a nonlinear stress-strain curve (origin to point A to point C). Youngs modulus (σ/ε) is satisfactory for describing the linear curve (origin to point B), but inaccurately describes a nonlinear curve (origin to point C). This gives a secant to the curve. An *incremental* elastic modulus ($\Delta\sigma/\Delta\varepsilon$) more accurately describes the nonlinear curve such as seen at point C. Reprinted with permission from: Dobrin PB. Mechanical properties of arteries. Physiol Rev 1978; 58:397-460.

The units of strain and extension ratio are dimensionless, i.e., centimeters divided by centimeters. The units of stress and of elastic modulus are force (grams) accelerated by gravity 981 cm/sec/sec to give dynes divided by area in centimeters. In the Centimeter-Grams-Seconds (CGS) system, stress and elastic modulus are given in dynes/cm². In the SI system, stress and elastic modulus are given in newtons/m². One may readily convert stress and elastic modulus from the CGS system to the SI system. For mathematical reasons stresses and elastic moduli in CGS happen to be one order of magnitude greater than they are in SI. For example, the elastic modulus for elastin is about 4.0×10^6 dynes/cm² or 4.0×10^5 N/m². For a more detailed explanation of the fundamental aspects of vascular mechanics see ref. 19.

Ligamentum nuchae is one of the most frequently studied elastin-containing tissues. Reuterwall[20] studied the elasticity of bovine ligamentum nuchae. He found that this tissue was extensible up to about 20% of resting length. His data were converted by Burton in his classic article[21] to conventional stress-strain terms. According to these measurements the elastic modulus of ligamentum nuchae is about 4.0×10^7 dynes/cm². This value is higher than most other reports of elastic tissue and may reflect the 9% collagen present in unpurified ligamentum nuchae. Krafka[22] also studied the length-force relationships of strips of ox ligamentum nuchae. He found the elastic modulus to be about 3.0×10^6 dynes/cm² and virtually linear throughout the entire range of extension. Remington[23] did not find the elasticity of ligamentum nuchae to be linear, but rather that it progressed from 1.5×10^6

dynes/cm^2 at moderate extensions to 4.1 x 10^6 dynes/cm^2 at large extensions. Stretching of elastin and the presence of small amounts of residual collagen may have influenced the elasticity at large extensions. Hass[24] extracted elastin from human aortas with 88% formic acid. The thermal retraction of this elastic tissue was found to be greater than that of the intact blood vessel. Burton[21] recomputed Hass's data and determined that the elastic modulus was about 3.0 x 10^6 dynes/cm^2. The tensile stress, i.e., the stress at which aortic elastin broke was found to be between 1.5 and 6.7 x 10^6 dynes/cm^2. The fact that the tensile strength bore no relationship to the amount of the elastic tissue in multiple preparations suggests that a stiff parallel element is present which sustains the load at large extensions and determines the tensile stress of the tissue. As will be shown below, this stiff parallel element is undoubtedly collagen. As also will be seen, collagen plays a critical role in the mechanics and tensile stress of aneurysms.

In the normal artery, the elastic lamellae bear about 2,500 dynes/cm per lamella.[25] This value applies for a large number of mammalian species ranging from the mouse to the elephant. In an interesting experiment in pigs in vivo, the aorta was subjected to graded crush injuries. Aneurysms formed when fewer than 40 elastic lamellae remained intact; this may be compared with 75 intact lamellae in the normal intact artery. The tension borne by the elastic lamellae in the aneurysmal artery was 4,087 to 4,543 dynes/cm per lamella. This may be contrasted with 1,316 dynes/cm/lamella in the intact control arteries[26] illustrating quantitatively the over-loading of the elastic lamellae in the aneurysmal artery. Some authors suggest that failure of elastin is a critical initial step causing the formation of aneurysms.[27] However, failure of collagen appears to play a critical role in the enlargement and rupture of aneurysms. Nevertheless, as will be shown below, failure of elastin rather than of collagen, does play an important role in the formation of aneurysmal tortuosity.

Degeneration of the arterial media occurs with age. This is associated with thinning, fraying and fragmentation of elastin. This is accompanied by moderate arterial dilation and stiffening of the vessel wall.[28-31] Dilatation with stiffening also is seen in vitro following enzymatic degradation of arterial elastin.[32] Elastin is an elastomeric material that can be stretched up to 70% of its original length.[33] But elastin wears with age, especially in the central conduit arteries, and particularly in the aorta. These are precisely the vessels that tend to become aneurysmal. Evidently, fatigue and fracture of elastic fibers can occur over time due to cyclic stress. A comparable phenomenon occurs with rubber. This is a naturally-occurring elastomer which also can be stretched far beyond its original length. Rubber fractures after approximately 1 x10^9 oscillations.[34] Elastin is extremely inert and remains chemically unchanged in the body for many years, but it is prone to structural damage when subjected to oscillating loads for long periods of time.[35,36] With a heart rate of 60-70 beats per minute, 1 x 10^9 oscillations are accumulated after 25-30 years. Histologic examination of the arterial wall reveals that, with age, the media of the aorta and the large conduit arteries exhibit areas of injury to elastin with mucoid degeneration or medial necrosis, and localized defects that are thought to be the cause of aortic dissection and rupture.[35,36] Following the degeneration of elastin that proceeds with age, remodeling occurs with deposition of stiff collagen fibers. The smaller muscular arteries do not exhibit the same degree of medial degeneration with age,[37] as the elastin fibers seem to be protected by vascular smooth muscle cells and collagen fibers. Small muscular arteries are less distensible than the conduit vessels, oscillating only about 5% with cyclic arterial pressures. Thus, aneurysms may form in the large arteries because of failure of elastin shifting the oscillating stress to the stiff collagen fibers. This is critical because it has been demonstrated that failure of collagen is an essential step in the development of aneurysms.[38] Moreover, certain fragments of elastin appear to serve to attract inflammatory cells.[39] This has been seen in the lung[40] and also in the rat aorta with the infusion of elastase,[41] as well as with infusion of a hexopeptide which is a breakdown product of elastin.

Collagen

Characteristics and Arterial Composition

The second major structural component of the arterial wall is collagen. At present, 12 subtypes of collagen have been identified. At least five of these subtypes are found in the vessel wall.[43] Smooth muscle cells are known to synthesize types I, III and V. Of these, types I and III are the major ones found in the vessel wall and are referred to as interstitial collagens. They appear as striated fibers and provide the great stiffness and tensile stress exhibited by blood vessels at high pressures and large deformations. Types I and III collagen are found in the media. In addition, small amounts of type V may be found there as well. Type I collagen is the predominant one found in the adventitia. Type IV is found mainly in basement membranes[43] such as in the subendothelial space. A number of studies have demonstrated that collagen represents between 20 and 50% of the dry weight of the vessel wall.[3-5] Analysis of aneurysmal vessels[7] shows that collagen increases in aneurysms and that the collagen-to-elastin ratio increases with aneurysmal size. The collagen-to-elastin ratio is 1.85:1 in normal, human nonaneurysmal aortas. It increases to 3.75:1 in human aortic aneurysms less than 5 cm, and rises to 7.91:1 in aneurysmal human aortas greater than 5 cm in diameter. Thus the collagen-to-elastin ratio increases with size in aneurysmal vessels.

Thermoelasticity

A characteristic property of collagen is that, upon heating to 60-65° the fiber contracts to one third its original length.[11,12] This thermoelastic property differs from the thermal contraction of elastin in that with collagen the retraction is irreversible. The typical x-ray diffraction patterns of collagen disappear following thermal shrinkage and are not restored by cooling.[44] Following heating, the banded appearance is lost and the shrunken fibers exhibit increased susceptibility to chemical and enzymatic proteolysis.[45] Thermal contraction can be prevented by applying sufficient weight or load to the tissue. An interesting observation is that the load necessary to prevent thermal contraction increases with age. This suggests that an increased number of cross-links may form with age.

Mechanical Elasticity

The elastic modulus of collagenous tendons has been determined. Collagen is structurally different than elastin because it is composed of tightly wound helical chains. These chains are tightly cross-linked thereby limiting their extensibility. Studies of collagen-rich tendon or isolated collagen fibers demonstrate that this protein can only be stretched 2-4%.[42] This may be compared with the 70-100% extensibility of elastin. When elastin is stretched it exhibits an elastic modulus between 1.0 and 4.5 x 10^6, dynes/cm.[2,19,21,23] By contrast, when collagen fibers are stretched their elastic modulus is approximately 1.0 x 10^{10} dynes/cm.[2,42] Thus, stretched collagen is several hundred to 1,000 times as stiff as stretched elastin. Reuterwall[20] reported load-extension data for various human muscle tendons. Burton[21] used these data to determine that the elastic modulus of tendon is about 1 x 10^9 dynes/cm². Verzar[46] reported load-extension data for collagenous rat tail tendon with a cross-sectional area of 0.15 mm². Because the extensions were very small one may assume a constant cross-sectional area and then convert these load-extension data into elastic moduli. At 2% extension the incremental elastic modulus would be computed to be 2.9 x 10^9 dynes/cm²; at 4% extension it would be 5.8 x 10^9 dynes/cm². Skin is composed of a network of collagenous fibers. Conabere and Hall[47] determined the Young's modulus for dry, tanned leather fibers values to be 1×10^{10} dynes/cm^2 over the linear portions of the stress-strain curves. The fibers also exhibited some irreversible extension following the first loading.

Elastin and collagen both contribute to the elastic properties of arteries. Krafka[22] had observed that the stiffness of aortic tissue was less than that of elastin-like ligamentum nuchae. He therefore suggested that elastin and collagen in arteries are arranged as a loose network which bears little load until the collagen fibers are oriented by tissue extension. Reuterwall[20] had observed microscopically that arterial collagen fibers appeared slack at small vessel diameters, but were extended at large diameters. He suggested that elastin may determine vessel properties at small diameters and that collagen may contribute to the stiff properties in the distended vessel. Remington and coworkers[48] studied the putrefaction of rings of aorta over several days to observe the effects of destruction of the muscle and connective tissues. There was a loss of elastic fibers over time, an increase in diameter and also an increase in stiffness. The investigators attributed the increase in stiffness of the dilated vessels to the remaining collagen. In what has proven to be a classic experiment, Roach and Burton[49] studied the contributions of elastin and collagen in human iliac artery. These investigators used crude trypsin, the only available agent at the time of the experiment, to degrade elastin, and they used formic acid to degrade collagen. They concluded that elastin is load-bearing at small dimensions and that medial collagen is load-bearing at large dimensions. Apter[50] noted that the circumference of isolated adventitia of arteries tends to be considerably larger than that of the isolated elastic lamellae. This observation is consistent with the view that medial collagen is load-bearing at high physiological pressures but that adventitial collagen is not load-bearing until extremely large dimensions are achieved. Wolinsky and Glagov[51] and others have provided histologic evidence to support this view. Systematic studies of normal intact arteries show that elastin bears loads in the circumferential, longitudinal and radial directions,[52,53] whereas collagen bears loads almost solely in the circumferential direction.[52]

In Vitro Simulation of Aneurysms

In order to elucidate the connective tissue changes associated with aneurysms, a series of in vitro experiments were performed.[32] Common carotid arteries were obtained from dogs following acute experiments, and external and internal iliac arteries were obtained from human cadavers. The arteries were catheterized at both ends, elongated to in situ length and mounted in a tissue bath with one end of the vessel connected to a force transducer (Fig. 4.2). The vessels were filled with, and immersed in, balanced physiological salt solution, and then pressurized in 25 mm steps up to 300 mm Hg. A ± 20 mm Hg oscillating pressure was superimposed on the mean pressure. Vessel diameter was measured with a displacement transducer which had a movable core that was in contact with the vessels. The core weighed 550 mg reducing transmural pressure by approximately 4 mm Hg. This value was neglected in the mechanical studies described below. Longitudinal force was measured with a Grass FT-10 force transducer. The arteries were relaxed by imposing repeated stepwise pressure cycles between 0 and 300 mm Hg until reproducible pressure-diameter curves were obtained.

The vessels then were assigned randomly for treatment with proteolytic enzymes. Some of the vessels were treated with purified elastase (Worthington ESFF) using doses of 8 U/ml for the dog vessels and 40 U/ml for the larger human vessels. Other vessels were treated with purified collagenase (Worthington CLSPA) at doses of 64 U/ml for the dog arteries and 320 U/ml for the human vessels. Some of the vessels were treated sequentially with both enzymes. These doses of the enzymes were used to disrupt elastin and collagen and demonstrate the mechanical contribution of these connective tissues. They were not intended to simulate circulating levels of human endogenous enzymes.

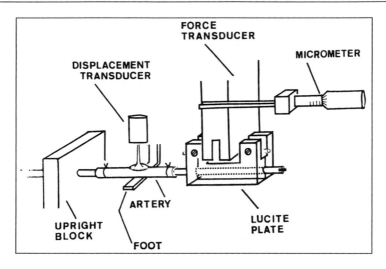

Fig. 4.2. Method used to study arteries in vitro. Modified with permission from: Dobrin PB, Doyle JM. Vascular smooth muscle and the anisotropy of dog carotid artery. Circ Res 1970; 27:105-119.

Histologic examination of the vessels after treatment demonstrated different responses with each of the enzymes. After treatment with elastase, the vessels exhibited fragmented elastic lamellae. After treatment with collagenase they exhibited decreased uptake of Masson's trichrome stain for collagen. These histologic changes were quite specific for each enzyme. Although the enzymes were not exclusively specific for each of the connective tissues, biochemical studies of the collagenase indicate a remarkable degree of specificity of this enzyme.[54]

Circumferential Behavior of Blood Vessels

Canine Arteries

Pressure-diameter curves for 56 dog carotid arteries were obtained (Fig. 4.3). The bottom curve in each panel was that obtained in the relaxed state before treatment. The steep portion of the curve at low pressure reveals marked vessel compliance. This is attributed to the gradual stretching of elastin. The flatter portion of the curve at high pressures is a manifestation of decreased compliance, i.e., increased stiffness. As will be seen, this behavior is the result of loading of stretched elastin and the recruitment of medial collagen. A similar biphasic distensibility pattern also has been observed for human vessels in vivo.[55-57]

The left panel of Figure 4.3 depicts data obtained when the arteries were treated with elastase (upper curve). It may be noted that degradation of elastin caused an increase in vessel diameter and an increase in vessel stiffness (flatter slope). None of these vessels ruptured. This demonstrates that intact elastin is important for providing vessels with normal compliance and normal dimensions, but it does not appear to be critical for maintaining vessel integrity.

The middle panel of Figure 4.3 depicts data for 18 other dog carotid arteries that were relaxed, then treated with collagenase. These vessels dilated less than that seen with elastase. However unlike the elastase-treated vessels, these arteries became more compliant as manifested by their steeper slope. In spite of manifesting only a slight increase in diameter,

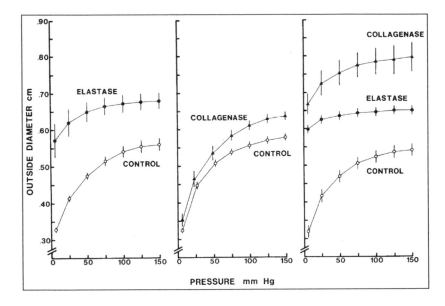

Fig. 4.3. Pressure-diameter curves for 56 dog carotid arteries in relaxed pretreatment condition (control), and treatment with degradative enzymes (elastase or collagenase). Treatment with elastase caused moderate dilatation, but did not cause rupture. Treatment with collagenase caused dilatation and vessel rupture. Reprinted with permission from: Dobrin PB, Baker WH, Gley WC. Elastinolytic and collagenolytic studies of arteries. Implications for the mechanical properties of aneurysms. Arch Surg 1984; 119:405-407.

each one of these vessels ruptured. In fact, the curve labeled collagenase was the last complete curve obtained in the intact vessel. Degradation of collagen may very well have increased the diameter even further but this could not be ascertained because the vessels ruptured.

The right panel of Figure 4.3 presents data for 18 other dog carotid arteries first treated with elastase and then treated with collagenase. Once again it is evident that the elastase-treated vessels dilated and became stiffer as the load was shifted from extensible elastin to stiff collagen. When these elastase-treated vessels were treated with collagenase, they all dilated markedly and ruptured. The data shown in Figure 4.3 demonstrate that collagen is critical for maintaining vessel integrity.

Human Iliac Arteries

Data obtained for human external iliac arteries looked remarkably like that shown for the dog arteries.[54] Figure 4.4 presents data for three human *internal* iliac arteries. Clinically this vessel often becomes aneurysmal, most frequently when there is an associated aneurysm of the abdominal aorta. The left panel shows that treatment with elastase elicited a response comparable to that observed with the dog carotid arteries. The internal iliac artery dilated, became markedly stiffer, but remained intact. The middle panel shows the effect of treatment with collagenase. This vessel dilated, markedly decreasing in stiffness. It also promptly ruptured. The responses of the sequentially treated vessel are shown in the right panel. When treated with elastase this vessel dilated and became stiffer. When treated with collagenase, it dilated dramatically and ruptured.

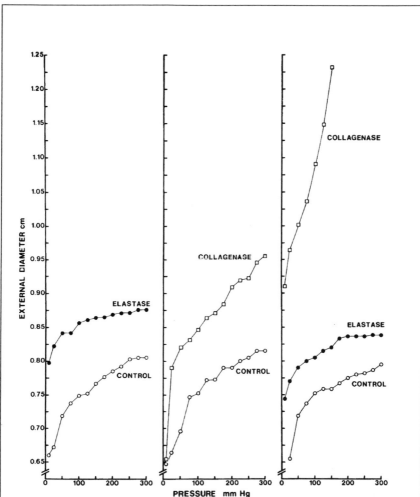

Fig. 4.4. Pressure-diameter curves for three human internal iliac arteries in the relaxed pretreatment condition (control), and after treatment with degradative enzymes. Vessels dilated but did not rupture after treatment with elastase. Vessels dilated profoundly and ruptured after treatment with collagenase. Reprinted with permission from: Dobrin PB, Baker WH, Gley WC. Elastinolytic and collagenolytic studies of arteries. Implications for the mechanical properties of aneurysms. Arch Surg 1984; 119:405-407.

Figure 4.5 shows data for a human aneurysmal internal iliac artery.[36] When treated with elastase for up to 15 hours, this vessel dilated but then reached stable dimensions. It did not rupture. When subsequentially treated with collagenase, it dilated further and promptly ruptured at only 10 mm Hg. Figure 4.6 shows data for another aneurysmal vessel. This vessel was relaxed and then treated only with collagenase. It ruptured at 50 mm Hg precluding the possibility of observing its dimensions at higher pressures. However it dilated to very large dimensions even at 50 mm Hg. All of these data indicate that elastin provides blood vessels with their characteristic distensibility and maintains them at normal dimensions. By

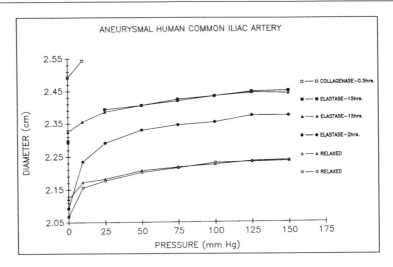

Fig. 4.5. Pressure-diameter curves for an aneurysmal human common iliac artery treated with elastase for up to 15 hours. Vessel dilated to a stable diameter after treatment with elastase. Then dilated further and ruptured following treatment with collagenase. Reprinted with permission from: Dobrin PB, Mrkvicka R. Failure of elastin or collagen as possible critical connective tissue alterations underlying aneurysmal dilation. Cardiovasc Surg 1994; 2:484-488.

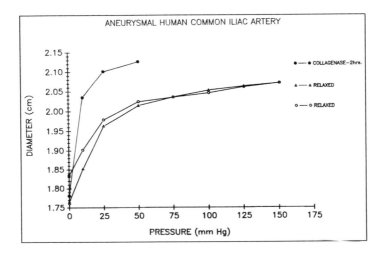

Fig. 4.6. Pressure-diameter curves for an aneurysmal human common iliac artery. This vessel was relaxed, then treated only with collagenase. This vessel ruptured at 50 mm Hg at a very large size for that pressure. Reprinted with permission from: Dobrin PB, Mrkvicka R. Failure of elastin or collagen as possible critical connective tissue alterations underlying aneurysmal dilation. Cardiovasc Surg 1994; 2:484-488.

contrast, collagen provides stiffness at large dimensions and assures the integrity of vessels. Thus, arteries behave mechanically as though they were constructed of a distensible elastic tube surrounded by a loose, protective steel jacket; the inner distensible tube is composed predominantly of elastin while the loose surrounding steel jacket is composed of collagen. This is consistent with the observation that the arterial wall is stiffer in dilated, aneurysmal vessels.[58] It also is stiffer in Marfan's syndrome[54-61] a disease associated with genetic defects in fibrillin[62] and possibly also with defects in collagen.[63] In this regard it is of interest that, a large number of patients with aortic aneurysms also have body wall hernias.[64,65] These data also suggest that the critical element that fails and permits the development and rupture of aneurysms is not elastin, but rather is collagen.[36,66] Studies of aortic tissue following excision demonstrate that aneurysms exhibit tortuous and disorganized elastic fibers,[27] and an increased collagen-to-elastin ratio, and increased ground substance[67] as compared with nonaneurysmal arteries.[1,6,7,67-70] In one study,[7] there was a linear increase in elastin, collagen and total protein content versus vessel circumference. This study found a two-fold increase in elastin content in aneurysmal human aortae, but a three-fold increase in collagen content in the same vessels. In the above study the age of the patients with aneurysms was similar to that of patients with normal aortas. This is important because in some other studies age differences have been a confounding factor. In a recent study of aneurysmal tissue, elastin content was found to be increased 2.3-fold, whereas collagen content was increased 5.7 fold[1] and increased with aortic size.[7] However, the ratio of type I:type III collagen was not altered.[7] Recently it has also been found that the collagen content is increased even in the nonaneurysmal portions of the aorta in patients who have infrarenal aneurysms.[70] This is consistent with the view that there is a constitutional abnormality in the connective tissue in patients with aneurysms. Propranolol has been used to increase the cross-linking of connective tissues,[71] and in fact administration of this agent has been reported to decrease the rate of enlargement of aneurysms in turkeys,[72] in the male blotchy mouse,[73] and in rats where experimental aneurysms were produced by infusion of elastase.[74] Propranolol also has been shown to reduce the rate of aneurysmal enlargement in humans.[75,76] Of course, in all of these cases it is unclear whether the efficacy of propranolol is due to its effect on connective tissue cross-linking, or is due to a decrease in arterial pressure or to decreased rate of rise of arterial pressure, i.e., dP/dt.

Location of Aneurysms

Aneurysms occur most frequently in the infrarenal aorta. This vessel is unique in several respects. It is subject to unusually large oscillating distending forces because it is one of the largest diameter arteries in the body and is exposed to large pulsatile pressures with especially high systolic pressures. Increased pressure occurs in the infrarenal aorta for several reasons: First, the aorta decreases in cross-sectional area because as it gives off branches it decreases in diameter;[67] Second, the aorta becomes stiffer distally. The abdominal aorta possesses a higher collagen-to-elastin ratio than does the thoracic aorta.[3-5] Third, pressure waves passing down the aorta reflect off distal vessels.; these reflections summate with incoming pressure waves to amplify the systolic and pulse pressure in the abdominal aorta while decreasing the diastolic pressure (Fig. 4.7). Wave reflections are least when the cross-sectional area of the iliac arteries is 1.1-1.2 times that of the cross-sectional area of the aorta.[79,80] In man this ratio is 1.1 only in early infancy. This decreases to 0.75 by age 50.[81] This ratio is decreased even further by atherosclerosis which narrows the ostia of the iliac arteries at the bifurcation.[82] Because of wave reflection, the pressure in the infrarenal aorta exhibits an increased rate of pressure rise, i.e., dP/dt. This is known to accelerate thoracic aortic dissections in patients. Increased systolic pressure and increased dP/dt require that the abdominal aortic wall provide increased circumferential retractive force in order to maintain equilibrium. Recently,

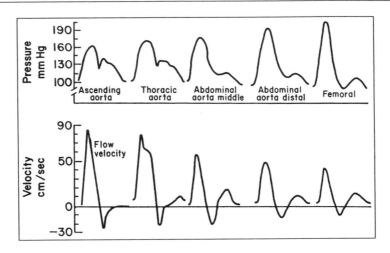

Fig. 4.7. Transformation of pressure pulse as it passes from aortic valve to bifurcation. While flow velocity is progressively damped, pressure wave increases in pulse pressure, systolic pressure and rate of pressure rise. Mean and diastolic pressure decrease slightly. Reprinted with permission from: McDonald DA. Blood Flow in Arteries. Baltimore, Williams and Wilkins 1975:90.

three dimensional computer models were used to simulate aortic flow during resting conditions and exercise.[83] These models demonstrate that, under resting conditions, there is a recirculation flow zone in the region of the infrarenal aorta. This was associated with oscillatory shear stresses, a factor correlated with atherosclerosis.[84] Remarkably, the computer model demonstrated that during vigorous exercise, flow reversal and regions of low and oscillatrory shear stress were eliminated. In spite of the requirement for increased retractive force, the abdominal aorta in man has fewer elastic lamellae per unit thickness than does the aorta in other mammals,[25] and fewer vasa vasorum per unit thickness than other mammalian aortae.[85] The number of vasa present is probably not crucial because although ablation of the vasa vasorum in experimental animals causes medial necrosis, it does not cause the formation of aneurysms.[86,87] Similarly, surgical endarterectomy removes virtually all of the media of atherosclerotic arteries, yet this also does not produce aneurysms.

Certain arteries are especially likely to develop aneurysms. The infrarenal aorta is the vessel most frequently involved, presumably due to the mechanical factors discussed above. The popliteal and femoral arteries also are susceptible perhaps because of the flexure of these vessels with bending of the knee and hip joints. In addition it has been demonstrated that a small amount of poststenotic dilatation occurs in the normal femoral artery where this vessel emerges from under the inguinal ligament.[88] Presumably the ligament exerts some degree of compression, especially with flexion of the hip. Aortic aneurysms may involve the iliac arteries, and isolated aneurysms may develop in the iliac arteries as well. When they do, they occur in the internal iliac artery 10 times more frequently than they do in the external iliac artery.[89,90] At present, there is no known explanation for this distribution. The visceral artery that most often becomes aneurysmal is the splenic artery. This occurs most frequently in women and is most likely to rupture during pregnancy, possibly because of the effects of the hormone relaxin.

Factors Determining the Enlargement of Aneurysms

The stability of aneurysms depends open the balance of distending and retractive forces. The distending force (F_D) in a cylindrical vessel is given by

$$F_D = P_T \times D_i \times L \qquad \{Eqn.\ 4.7\}$$

where P_T is transmural pressure, D_i is internal diameter, and L is vessel length. The retractive force (F_R) exerted by the vessel wall is given by

$$F_R = \sigma_\theta \times 2th \times L \qquad \{Eqn.\ 4.8\}$$

where σ_θ is the stress or force per unit area exerted by the wall in the circumferential direction, th is wall thickness, and L is vessel length. At equilibrium these forces are equal and opposite.

$$F_D = F_R \qquad \{Eqn.\ 4.9\}$$

Any condition which increases the distending force (F_D) raises the likelihood that the vessel will enlarge. Hypertension and increased diameter raise the distending force (F_D). A study employing serial computerized tomography scans demonstrated that regions of the circumference of human abdominal aortic aneurysms that underwent flattening exhibited an increased rate of enlargement.[91] Flattening increases the effective radius of the lesion thereby increasing the distending force (Eqn. 4.7). Flattening occurred most frequently in the left posterolateral region. Computer modeling of aneurysms of different shapes suggests that the wall stress is highly dependent on the maximum diameter and shape of the aneurysm.[92] In turn, this geometry-dependent elevation of wall stress may be a prediction of aneurysmal rupture. Similarly, any condition which decreases the retractive force (F_R) also raises the likelihood that the vessel will enlarge. Failure of connective tissues leading to decreased circumferential stress (σ_θ) and reduction of wall thickness (th) both decrease the retractive force (F_R). With these relationships in mind we may review published data regarding aneurysmal enlargement.

Most aneurysms enlarge gradually with a mean enlargement rate of 4 to 5 mm per year.[75,76,93] However the rate of enlargement varies from patient to patient. Actuarial analysis of the rate of enlargement of aneurysms has demonstrated three independent determinants.[94] These are: 1) antero-posterior dimensions at the time of the diagnosis; 2) elevated diastolic pressure; 3) the degree of chronic obstructive pulmonary disease. The importance of diameter and diastolic pressure are readily apparent as both of these factors contribute to increased distending force in the circumferential direction (F_D). However the degree of chronic obstructive pulmonary disease is less clear. Patients with chronic obstructive pulmonary disease are known to possess increased levels of neutrophil-derived elastase and elevated concentrations of the peptide products of elastin breakdown in their plasma and urine.[95-97] This suggests that the aorta in these patients may enlarge more rapidly because it may be subjected to in vivo degradation. Similarly, patients with known stable aneurysms have been reported to rupture at a higher rate than expected after the patients underwent laparotomy for disease unrelated to the aneurysms. Rupture often occurred within the first 30 days after laparotomy.[98] This is a period of rapid wound healing with collagen remodeling. This is associated with an influx of inflammatory cells. This may alter the mechanical properties of the vessel wall, as these inflammatory cells manufacture and release elastase and collagenase. Interestingly, human aneurysms often contain inflammatory cells in the adventitia and the adventitia-medial junction.[99] The walls of aneurysms also possess measurable levels of elastase and collagenase[100-107] with increased elastase activity found in cigarette smokers.[102] The level of collagenase in human aneurysms correlates linearly with the size of the aneurysm.[100] Mechanically-induced aneurysm-like poststenotic dilatation of arteries also is accompanied by elevated collagenase levels indicating that increased collagenase activity also may be caused by the aneurysm rather than be the cause of it.[108] Recently a model using infusion of crude elastase into the isolated aorta in rats has been developed to produce experimental aneurysms in vivo. This model reliably produces aneurysms.[41,109] This is associated with an influx of inflammatory cells, especially macrophages and T cells[41] and the appearance of endogenous proteinases.[110] The influx of inflammatory cells and proteinases correlates temporally with the rapid enlargement of vessels as they become true aneurysms (Fig. 4.8).[41,110] Analysis of the cellular infiltrate discloses this to be neutrophils, ED-2 positive macrophages, and CD4-positive T-helper cells.[41,110] Macrophages are known to produce collagenases.[110-112] Nonspecific activation of the immune system by infusion of thioglycolate plus plasmin into the isolated aorta also produces a gradually enlarging aneurysm with the appearance of macrophages, but not T-cells.[41] This suggests that the inflammatory cells so often seen in the walls of aneurysms in patients may play an active role in causing progressive enlargement of the lesions. An inflammatory rejection reaction with formation of aneurysms also has been seen in spontaneously hypertensive rats transplanted with vascular allografts.[113] Using the elastase infusion model it was found that treating the animals with methylprednisolone or cyclosporine before and after aortic infusion with elastase permits the aneurysms to enlarge to substantially less than that observed in saline-treated elastase-infusion control animals; however, treatment with corticosteroids or cyclosporine does not completely prevent the aneurysms from developing.[114] Treating the animals for four days with Anti-CD18 monoclonal antibody also slows the enlargement of experimental aneurysms.[115] These observations in animal models suggest that inflammatory processes may play an active role in the formation and progression of human aneurysms as well. A crucial factor—perhaps the most crucial one—is identification of the stimulus for the inflammatory reaction. Human aneurysms also often contain a rich inflammatory infiltrate.[116-118]

The current focus on inflammatory processes appears to conflict with the conventional wisdom that aneurysms are caused by atherosclerotic degeneration of the vessel wall.[119] Tilson and Stansel[120] challenged this long-held view. They observed in a clinical study of 100 consecutive patients who had undergone aortic surgery, that those patients with atherosclerotic occlusive disease were markedly different from those with aneurysmal aortic disease. The patients differed with respect to male:female ratio, age of onset of clinical symptoms, and prognosis following surgery. This suggests that the development of aneurysms may be independent of atherosclerosis, a commonly occurring, and possibly an incidental finding in older patients. Epidemiologic studies demonstrate that, in many cases, aneurysms are familial following a genetic pattern.[121-125] In fact, some patients with aneurysms exhibit widespread arteriomegaly which may progress to diffuse or multiple aneurysms.[126,127]

On the other hand atherosclerosis does alter the properties of the vessel wall. In early atherosclerosis there is an accumulation of lipids and lipoproteins in the intima and sub-intimal areas. This may lead to destruction of the normal wall constituents. Several studies have reported that atherosclerotic arteries exhibit decreased or unchanged mechanical stiffness.[128-132] Lipid is bound in the wall by elastin, and this disrupts elastin fiber formation and the cross-linking of elastic fibers.[133] Moreover, elastin-lipid complexes exhibit increased susceptibility to degradation by elastase.[133] Of course, advanced atherosclerotic disease with the accumulation of collagen in the form of a fibrous cap, and especially with calcification, causes vessels to become very stiff. Based on studies in experimental animals and human specimens it has been proposed that the local response of the media to injury may determine the outcome of atherosclerotic injury. It has been proposed that if the media heals

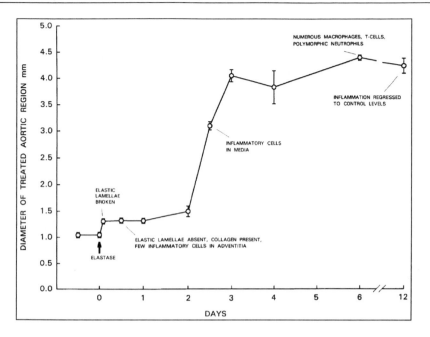

Fig. 4.8. Diameter of abdominal aorta. Two hour infusion of elastase (day 0) produced an immediate 30% increase in diameter. Secondary aneurysmal enlargement of 300% (days 2-3) correlated with influx of inflammatory cells. Reprinted with permission from: Anidjar S, Dobrin PB, Eichorst M et al. Correlation of inflammatory infiltrate with the enlargement of experimental aneurysms. J Vasc Surg 1992; 16:139-147.

completely, then an occluding atherosclerotic lesion may remain; if healing fails to occur and the media becomes atrophic, then an aneurysm may develop.[134] An experiment was performed on cynomolgus and rhesus monkeys fed an atherosclerotic diet to test this hypothesis. After atherosclerosis had developed, the animals were placed on an athero-sclerosis regression diet. Four of 31 cynomolgus monkeys (13%) and 1/107 rhesus monkeys (1%) developed arterial aneurysms. Each had atrophy of the media.[135] These observations are consistent with the view that a small number of patients with atherosclerotic disease may develop aneurysms. Bengtsson and coworkers[136] found that 13.5% of patients who underwent carotid endarterectomy for stenotic atherosclerotic disease also harbored aortic enlargement or an abdominal aortic aneurysm.

Stability of Aneurysms

In light of the numerous factors that accelerate the enlargement and rupture of aneurysms, one may inquire as to why aneurysms, once formed, do not proceed immediately to rupture. Indeed, the distending force (Eqn. 4.7) shows that as diameter increases, even with stable pressures, the force distending the vessel in the circumferential direction increases. In order to maintain equilibrium the retractive force (Eqn. 4.8) also must increase. This becomes ever more difficult as the wall becomes thinner with aneurysmal dilatation. Clearly, in order to provide increased retractive force (Eqn. 4.8), the wall must generate increased retractive stress (s_q). Stress for a cylindrical vessel is given by

$$\sigma_\theta = P_T \times \frac{r_i}{th} \qquad \{Eqn.\ 4.10\}$$

where σ_θ is stress, i.e., force per unit area exerted by the vessel wall at equilibrium, P_T is transmural pressure, r_i is the internal radius of the artery, and th is wall thickness. Why does the unstable wall not proceed instantly to rupture?

There are several answers to this question. First, as aneurysms increase in diameter, they become stiffer,[36,57,60,67] as they recruit collagen fibers that previously had not been loaded. This may be seen in the left panels of Figures 4.3 and 4.4. In each case, treatment with elastase caused dilatation with stiffening of the wall (flattening of the pressure-diameter-curve). Proof that this stiffness is attributable to collagen is given in the right panels of Figures 4.3 and 4.4. These curves show that when the stiff elastase-treated vessels were treated with collagenase they dilated and became more compliant. By comparing the stiffness of collagen with that of the intact wall, and with knowledge of the percentage of the wall occupied by collagen, one may compute that less than 1% of the collagen that is present in the normal intact wall actually is load-bearing. Thus there is a considerable amount of collagen available for recruitment. This computation is illustrated for a distended normal vessel.

Consider a normal artery where it is assumed that all of the wall stiffness is due to collagen, and that due to elastin is neglected.

$$\frac{\text{Stiffness of}}{\text{intact wall}} = \frac{\text{% of wall}}{\text{which is collagen}} \quad \text{x} \quad \frac{\text{Stiffness of}}{\text{collagen}} \quad \text{x} \quad \frac{\text{% of collagen}}{\text{which is load-bearing}}$$

{Eqn. 4.11}

Rearranging one obtains:

$$\begin{array}{l}\text{% of collagen}\\ \text{which is}\\ \text{load-bearing}\end{array} = \frac{\text{stiffness of intact vessel wall}}{\text{stiffness of} \quad \text{x} \quad \text{% of wall}}$$
$$\qquad\qquad\qquad\qquad \text{collagen} \qquad \text{which is collagen}$$

{Eqn. 4.12}

One may solve this equation using an elastic modulus for the distended vessel as 1×10^7 dynes/cm^2 from reference 42, an elastic modulus for collagen as 1×10^{10} dynes/cm^2 from ref. 19. Referring to reference 3 for the collagen content of dog abdominal aorta one obtains a value of 45% of dry defatted tissue weight or 13% of wet tissue weight. Using these values to solve equation 12 one obtains between 0.22% and 0.77% of wall collagen that actually is fully load-bearing in the normal artery. Clearly, there is much collagen in the normal wall that remains to be recruited as an aneurysm develops. Moreover, studies of aneurysmal vessels show that they also have increased quantities of collagen. Therefore, not only does the aneurysmal wall recruit previously unloaded collagen, but it also synthesizes increased quantities of new collagen.[6,7,36,70]

A second reason that aneurysms dilate gradually rather than instantly relates to their shape. As a vessel dilates aneurysmally it takes on a more spherical shape (Fig. 4.9). The distending force for a spherical vessel (F_D) is given by

$$F_D = P_T \times \pi r_i^2$$

{Eqn. 4.13}

where P_T is transmural pressure, and r_i is the internal radius of the sphere. The retractive force (F_R) is given by

$$F_R = \sigma_{sph} \times (\pi r_o^2 - \pi r_i^2)$$

{Eqn. 4.14}

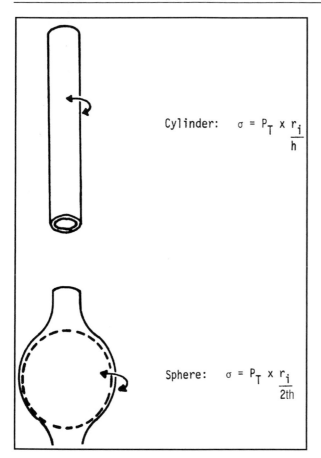

Cylinder: $\sigma = P_T \times \dfrac{r_i}{h}$

Sphere: $\sigma = P_T \times \dfrac{r_i}{2th}$

Fig. 4.9. Wall stress required to maintain equilibrium in a cylinder (top) and a sphere (bottom). Change in shape from a cylinder to a sphere reduces wall stress required by one half. Reprinted with permission from: Dobrin PB, Baker WH, Gley WC. Elastinolytic and collagenolytic studies of arteries. Implications for the mechanical properties of aneurysms. Arch Surg 1984; 119:405-407.

where r_o is the outside radius of the sphere. Equation 4.14 may be factored to become

$$F_R = \sigma_{sph} \times \pi \, (r_o - r_i)\,(r_o + r_i) \qquad \text{\{Eqn. 4.15\}}$$

If one treats the wall as very thin, then r_o approximates r_i. Then equation 15 becomes

$$F_R = \sigma_{sph} \times \pi \, (th \; 2 \; r_i) \qquad \text{\{Eqn. 4.16\}}$$

where th is wall thickness. Since

$$F_D = F_R \qquad \text{\{Eqn. 4.17\}}$$

Equation 4.13 may be set equal to equation 4.16 and solved for the stress (σ_{sph}).

$$\sigma_{sph} = \dfrac{P_t \times r_i}{2 \, th} \qquad \text{\{Eqn. 4.18\}}$$

Thus, changing from a cylinder to a sphere reduces by about 50% the stress (σ) required to maintain equilibrium.

Finally, the lumen of the aneurysm is lined with thrombus. This layer of thrombus usually leaves a luminal stream of normal dimensions through which blood flows. But it is likely that the thrombus itself contributes little directly to static wall mechanics. First, the thrombus acts mechanically like solidified blood; it therefore transmits the distending force exerted by pressure to the vessel wall and does not decrease the value for the diameter in the computation of distending force (Eqns. 4.7 or 4.13). Moreover, the thrombus offers little retractive stress contributing little if any to the retractive force (Eqns. 4.8 or 4.16). In fact, the thrombus may readily be scooped out with little resistance at the time of surgery. All of these observations suggest that the mechanical aspects of the thrombus probably are unimportant in stabilizing the aneurysmal wall. On the other hand, ultrasonographic studies show that the thrombus is about twice as compliant as the aneurysmal wall, and is nearly incompressible.[137] The investigators suggest that the thrombus might act as a cushion protecting the aneurysmal wall.[137] This issue remains to be resolved by direct measurement, not just by mathematical modeling.

Another aspect of the thrombus concerns its possible biochemical role in the stability of aneurysms. The thrombus may be associated with increased plasmin activity.[138] Plasmin acts as an important activator of metalloproteinases in connective tissue.[139,142] Plasminogen activator (t-PA) is found in abdominal aortic aneurysms, both in vivo[141] and after excision,[142] and in macrophages in aneurysms.[107] All of these data are consistent with the observations of Wolf and coworkers[143] who reported a correlation between the proportion of the circumference of the aneurysm lumen covered by laminated thrombus and the rate of enlargement of the aneurysm. Presumably the plasmin in the thrombus leaches out into the vessel wall. In fact, recent light, transmission and scanning electron microscopic studies show that the thrombus is traversed by microscopic interconnected canaliculi which may deliver macromolecules from the thrombus to the aneurysm wall.[144] Therefore, whereas the mechanical aspects of the thrombus are probably unimportant in stabilizing the aneurysmal wall, the biochemical aspects of this layer of clotted blood may actually accelerate enlargement of aneurysms.

In summary, the recruitment of previously unloaded collagen fibers and the synthesis of new collagen fibers helped to stabilize aneurysms. The change from a cylinder to a spherical geometry also tends to stabilize aneurysms. The presence of laminated thrombus in the lumen probably does not provide mechanical stability and may, by biochemical means, facilitate dilatation of aneurysms.

Aneurysmal Tortuosity

Another feature of aneurysms is their tendency to elongate and become bowed. This occurs when the forces applied to a vessel cause it to lengthen but the vessel is prevented from doing so by constraining branches. Under these conditions the segment that is subjected to lengthening between fixed points is forced to buckle. Buckling presents clinically as tortuosity. Tortuosity is seen clinically in aneurysmal and ectatic vessels.[145] It also is seen in aged patients with hypertension.[146,147] In order to understand the mechanics of tortuosity, it is necessary to examine the longitudinal forces acting on a normal, nontortuous vessel. These are shown in Figure 4.10. There are two forces that cause a vessel to lengthen. These are 1) the force attributable to traction (F_T) which pulls on the vessel and holds it at in situ length and 2) the force attributable to pressure (F_P) which pushes on the vessel from within the lumen causing it to lengthen. Almost all vessels are held under longitudinal traction. This develops in the fetus, increases during early development,[148] reaches a maximum in the adult[148] and decreases with age.[150,151] The force produced by traction (F_T) holds vessels under stretch. This is manifested by the retraction that occurs by an artery when it is transected. The longitudinal force due to traction (F_T) depends upon the properties of the vessel and the degree to which it is stretched. It cannot be computed analytically but must be measured at specific conditions. The longitudinal force produced by pressure (F_P) may be determined analytically. It is given by

$$F_P = P_T\, \pi r_i^2 \qquad\qquad \{Eqn.\ 4.19\}$$

where F_P is the longitudinal force due to pressure, P_T is transmural pressure, and r_i is the internal radius of the vessel. As shown in Figure 4.10 the force due to traction (F_T) and the force due to pressure (F_P) add together to give the net force extending the vessel (F_Z). This addition is shown in the middle panel of Figure 4.10. At equilibrium, the force extending the vessel (F_Z) is opposed by a retractive force exerted by the vessel wall (F_R).

The traction force (F_T) and the force due to pressure (F_P) add in a characteristic fashion. This is shown in Figure 4.11. This figure shows data for a vessel pressurized while held by traction at in situ length. At 0 mm Hg, all of the longitudinal force (F_Z) is due to traction (F_T). As pressure is increased the longitudinal force due to pressure (F_P) rises and the vessel extends very slightly. This causes a decrease in the force due to traction (F_T). Thus, the force due to pressure (F_P) rises while that due to traction (F_T) declines. These two changes nearly cancel one another so that the net longitudinal force extending the vessel (F_Z) remains almost constant. Indeed, the net longitudinal force (F_Z) really does not rise steeply until pressures of about 175 mm Hg are encountered. The interaction between the traction force (F_T) and pressure force (F_P) is the reason that arteries exhibit virtually constant length in vivo in spite of variations in arterial pressure. It is evident that if a vessel is to become tortuous, one or more of three scenarios must occur. First, arterial pressure must rise to very high values thereby increasing the pressure force (F_P) and the net longitudinal force (F_Z). Alternatively, there must be a marked decline in the traction force (F_T). The latter does in fact occur with aging.[150,151] Finally, longitudinal extension can occur if the retractive force exerted by the vessel wall (F_R) decreases. Two of these events actually do occur with the development of aneurysms. First, the force due to pressure (F_P) markedly increases because of enlargement of the radius. As shown by Equation 4.13, the force due to pressure is a function of the square of the radius. This is very important because as an aneurysm increases in diameter the longitudinal force due to pressure (F_P) increases geometrically. For example, the longitudinal force due to pressure is 16 times as great in an 8 cm aortic aneurysm as it is in a normal 2 cm diameter aorta.

A second reason that aneurysms tend to become tortuous is because there is a decrease in the longitudinal retractive force (F_R). In order to identify the connective tissue responsible for longitudinal retraction, experiments were performed by mounting arteries in vitro and extending them in random sequence to 20, 40, 60, or 80% longitudinal extension.[152] Normal in situ length for the carotid artery in the dog ranges from 27-60% extension beyond retracted length, depending on the position of the cervical spine.[19] Data were obtained with the vessel extended before treatment and again after treatment with either elastase or collagenase. Figure 4.12 shows data obtained for 24 vessels in the control state (open circles) and again after treatment with elastase. The symbols show means ± standard errors. Treatment with elastase produced a marked and statistically significant decrease in longitudinal retractive force (F_R) at every vessel length. Figure 4.13 shows data for 24 other dog carotid arteries in the control state (open symbols) and following treatment with collagenase (closed symbols). These data show that collagen provides negligible longitudinal retractive force (F_R). Thus from the reduction in retractive force following treatment with elastase (Figure 4.12) and the lack of reduction in retractive force with collagenase, these data demonstrate that it is elastin and not collagen which provides longitudinal retractive force (F_R).

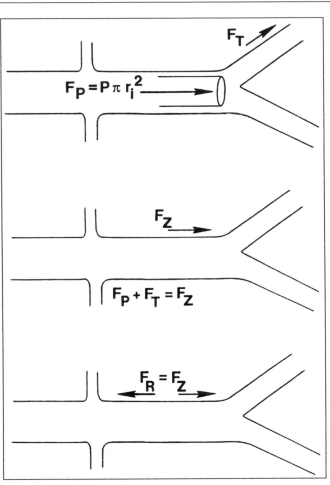

Fig. 4.10. Forces acting longitudinally on an artery. F_P is force due to pressure, F_T is force due to traction, F_Z is the sum of F_P and F_T. F_R is the retractive force exerted by the stretched vessel. At equilibrium F_Z = F_R. Reprinted with permission from: Dobrin PB, Baker WH, Schwarcz TH. Mechanisms of arterial and aneurysmal tortuosity. Surgery 1988; 104:568-571.

The data shown in the preceding two figures illustrate the retractive forces observed in unpressurized dog carotid arteries. In order to examine this in pressurized vessels, human iliac arteries were obtained and these vessels were mounted in vitro.[153] They were extended to in situ length or slightly longer or shorter than in situ length, and then were pressurized. The vessels were treated with elastase or collagenase while longitudinal forces exerted by these isometrically held vessels were recorded. Figure 4.14 shows data for three such vessels in the control state (open circles) and again after treatment with elastase. As the pressure was elevated, the longitudinal retractive force declined for both the control and the elastase-treated vessels. As the data points fell below the broken horizontal line of equilibrium the vessels came under compression and began to buckle. As can be seen in this figure for three vessels extended to three slightly different lengths, all vessels tend to approach compression at very high pressures, and all vessels were more prone to compression and buckling after they had been treated with elastase. Once again this supports the conclusion that elastin provides a major portion of longitudinal retractive force (F_R). Figure 4.15 shows data for three other human iliac arteries subjected to pressure in the control state (open circles) and after treatment with collagenase (closed circles). Treatment with collagenase produced a slight decrease in longitudinal retractive force, much smaller than that observed following treatment with

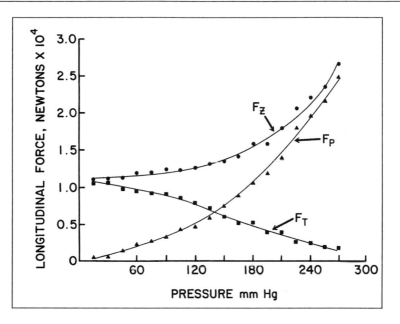

Fig. 4.11. Pressure-longitudinal force values for a vessel held at in situ length and subjected to pressure. At 0 mm Hg all of the longitudinal force (F_Z) is due to traction (F_T). As pressure is increased, the force due to pressure (F_P) rises while the force due to traction (F_T) declines. As a result the net longitudinal force (F_Z) remains nearly constant until very high pressures are imposed. Constant F_Z prevents vessel lengthening or shortening. Reprinted with permission from: Dobrin PB, Baker WH, Schwarcz TH. Mechanisms of arterial and aneurysmal tortuosity. Surgery 1988; 104:568-571.

elastase. Thus, one may conclude that for both the extended but unpressurized dog carotid arteries (Figures 4.12 and 4.13) and for the pressurized human iliac artery held at in situ length (Figures 4.14 and 4.15), elastin provides most of the longitudinal retractive force with only a very small portion contributed by collagen. The conclusion that aneurysmal tortuosity is associated with failure of elastin is consistent with scanning electron microscopic observations of aneurysmal arteries[27] and with histology light studies of congenitally-defective ectatic vessels.[145] Throughout the above discussion no mention has been made of vascular smooth muscle. It is unlikely that vascular smooth muscle plays much of a direct mechanical role in the enlargement or rupture of arterial aneurysms, for the stiffness of vascular smooth muscle is about the same as elastin.[154] Moreover, vascular muscle occupies only a very small proportion of the volume of the wall of large conduit arteries,[19,21] and it is precisely large conduit arteries that are prone to become aneurysmal. However, vascular smooth muscle cells may be thought of as stabilizing the vessel wall by virtue of their ability to synthesize elastin and collagen.[155,156]

Bioengineering Aspects of Aneurysms

In the preceding discussion of mechanics, incremental infinitesimal strain theory was used. This provided the computations of elastic modulus, stress for a cylinder and for a sphere, etc. This computational method assumes that the deformations are extremely small. Clearly the deformations that actually occur to blood vessels are very large. In order to deal with this problem small increments of deformation and stress are taken (Fig. 4.1), and from

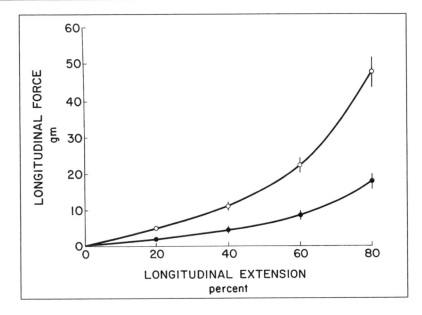

Fig. 4.12. Length-longitudinal force data for 24 dog carotid arteries extended to four lengths without pressurization. Open symbols show data in control state, closed symbols after treatment with elastase. Elastase causes a reduction in the longitudinal retractive force (F_R). Data show that longitudinal retractive force (F_T) is provided by elastin. . Reprinted with permission from: Dobrin PB, Baker WH, Schwarcz TH. Mechanisms of arterial and aneurysmal tortuosity. Surgery 1988; 104:568-571.

this the elastic properties are computed over specific ranges of strain. Hence the term incremental infinitesimal strain theory. Another method for computing of vessel wall mechanics is the use of *finite-deformation* analysis. This method was developed for studying elastomeric materials such as rubber, i.e., those which behave as long chains which may stretch during material extension. Obviously this is quite similar to the behavior of elastin. Applying finite-deformation analysis utilizes the empirical computation of the energy stored in deforming a material. The work performed in producing this deformation is expressed as strain-energy density, i.e., the work stored per unit volume of material. The derivative of the strain-energy density expression then is determined to obtain the stress at any given level of deformation, i.e., the stress-strain relationship. Several investigators have used finite-deformation analysis to evaluate vascular mechanics. Most have found that they could describe strain-energy density for the arterial wall as an exponential[157-159] or polynomial function.[160,161]

A method commonly used in engineering is finite element analysis. This method is a systematic technique for formulating the complex equations describing the elastic behavior of materials. This method identifies a theoretical cube of material of the material under study. If one were to apply forces to each of the eight corners of the cube one would produce deformations, the magnitudes of which depend on the elastic properties of the material. The investigator then determines the work done in producing these deformations. This is comparable to the strain-energy stored by the tissue. The investigator then obtains an approximate relationship between displacement and applied forces. Cubic elements then are combined to reconstruct the whole material. Often a large number of simultaneous equations are required to solve the force-displacement relationships, but often these equations

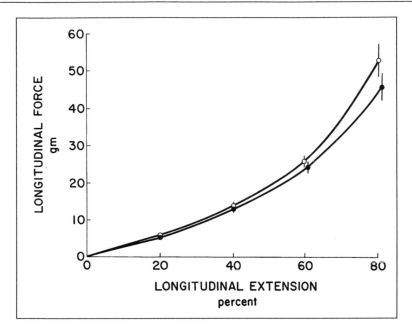

Fig. 4.13. Length-longitudinal force data for 24 dog carotid arteries extended to four lengths without pressurization. Open symbols show data in control state, closed symbols after treatment with collagenase. Collagenase causes no reduction in the longitudinal retractive force (F_R). Data show that collagen provides negligible longitudinal retractive force (F_R), whereas elastin provides most of the longitudinal retractive force (Fig. 4.12). . Reprinted with permission from: Dobrin PB, Baker WH, Schwarcz TH. Mechanisms of arterial and aneurysmal tortuosity. Surgery 1988; 104:568-571.

are algebraic expressions rather than partial differential equations. These algebraic expressions can be solved using computers. Several investigators have used finite element methods to analyze the properties of aneurysms. These analyses determined that the maximum stress in the wall of aneurysms increases with diameter.[161,162] They also found that stresses are generally greater on the inner surface of an aneurysm than on the outer surface[162] and that atherosclerotic plaques increase stress concentrations and may increase maximal wall stress.[161] They also found that circumferential stresses are greater than longitudinal stresses,[162] and that cylindrically shaped constant thickness model aneurysms had higher maximum circumferential stress than did spherical model aneurysms of the same diameter.[163] It is noteworthy that most of these findings are similar to those obtained by incremental infinitesimal strain theory as discussed previously in this chapter.

Finite element methods also have been used to assess the stresses in the aortic wall following patch repair of a coarctation. These studies demonstrated that the major variable affecting stress in the aortic wall after repair was patch geometry. If the patch was permitted to balloon outward, the aortic wall stress increased disproportionately. Maximum wall stress occurred in the aorta opposite the patch.[164] Finally, finite element methods have been used to compute flow characteristics in axisymmetric and lateral saccular aneurysms.[164-167] These investigations demonstrated that a vortex grows rapidly in size and fills the whole area of the lumen of the aneurysm during portions of the cardiac cycle. During this time the center of the vortex moves from the proximal end of the aneurysm to the distal point of the

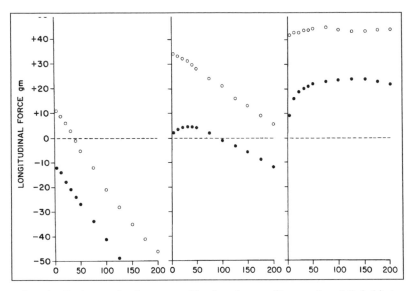

Fig. 4.14. Longitudinal retractive force exerted by three human iliac arteries while held at constant length and subjected to pressure. As pressure is increased, retractive force declines. Open circles show data in control state, closed circles show data after treatment with elastase. Elastase caused a reduction in retractive force (F_R). This shows that elastin provides longitudinal retractive force (F_R). Reprinted with permission from: Dobrin PB, Schwarcz TH, Mrkvicka R. Longitudinal retractive force in pressurized dog and human arteries. J Surg Res 1990; 48:116-120.

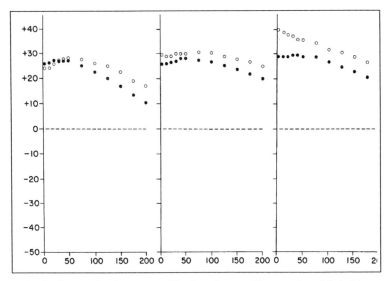

Fig. 4.15. Longitudinal retractive force exerted by three human iliac arteries while held at constant length and subjected to pressure. Open circles show data in control state, closed circles show data after treatment with collagenase. Collagenase caused only a slight reduction in retractive force. These data show that collagen provides only a small portion of longitudinal retractive force (F_R), whereas elastin provides most of the longitudinal retractive force (Fig. 4.14). Reprinted with permission from: Dobrin PB, Schwarcz TH, Mrkvicka R. Longitudinal retractive force in pressurized dog and human arteries. J Surg Res 1990; 48:116-120.

aneurysm. Transient flow reversal occurs at the end of ejection period, passing between the wall of the aneurysm and the centrally located vortex. Finite element analysis demonstrated that the appearance and disappearance of the vortex in the dilated portion of the aneurysm resulted in complex distributions of local pressures and shear forces. In lateral saccular aneurysms a separation vortex occurred which was distorted by the geometric shape of the aneurysm.[166]

References

1. Baxter TB and Halloran BG. Matrix metabolism in abdominal aortic aneurysms. In: Aneurysms; New Findings and Treatments. Yao JST, Pearce WH eds. East Norwalk, CT. Appleton and Lange 1994; 25-34.
2. Powell JT. Dilatation through loss of elastin. In: The Cause and Management of Aneurysms. Greenhalgh RM and Mannick JA, eds. WB Saunders, London. 1990; 89-104.
3. Fischer GM, Llaurado JG. Collagen and elastin content in canine arteries selected from functionally different vascular beds. Circ Res 1966; 19:394-399.
4. Harkness ML, Harkness RD and McDonald DA. The collagen and elastin content of the arterial wall in the dog. Proc Roy Soc London. 1957; 146B:541-551.
5. Fischer GM and Llaurado JG. Connective tissue composition of canine arteries: Effects of renal hypertension. Arch Path 1967; 84:95-98.
6. Powell J and Greenhalgh RM. Cellular, enzymatic, and genetic factors in the pathogenesis of abdominal aortic aneurysms. J Vasc Surg 1989; 9:297-304.
7. Sakalihasan N, Heyeres A, Nusgens BV et al. Modifications of the extracellular matrix of aneurysmal abdominal aorta as a function of their size. Eur J Vasc Surg 1993; 7:633-637.
8. Baumann P. The present state of our knowledge in the field of elastomers. Chem & Indust 1959; III:1498-1504.
9. Guth E. Muscular contractions and rubber-like elasticity. Ann NY Acad Sci 1947; 47:715-766.
10. McCartney JE. Heat contractions of elastic tissue. Quant J Exp Biol 1913; 7:103-114.
11. Gustavson KH. The chemistry and reactivity of collagen. New York Academic Press 1956:343.
12. Piez KA and Gross J. The amino acid composition of some fish collagens: The relation between composition and structure. J Biol Chem 1960; 235:995-998.
13. Meyer KH and Ferri C. Die elastischen eigenschaften der elastischen und der kollagenen fasern and ihre molekulare deutung. Pfleug Arch ges Physiol 1937; 238:78-90.
14. Wohlisch E, Weitnauer H, Grunig W et al. Thermodynamische analyse der dehning des elastischen gewebes von standpunkt der statistisch-kinetischen theorre der kantschukelastizitat. Kolloid Z 1943; 104:14-24.
15. Hoeve CA and Florey PJ. The elastic properties of elastin. J Am Chem Soc 1958; 80:6523-6526.
16. Astbury WT. The molecular structure of the fibers of the collagen group. J Internat Soc Leath Tr Chem 1940; 24:69-92.
17. Lloyd DJ, Jordan LD, Garrod M. A contribution to the theory of the structure of protein fibers with special reference to the so-called thermal shrinkage of the collagen fiber. Tr Faraday Soc 1948; 44:441-451.
18. Dobrin PB. Mechanical properties of arteries. Physiol Rev 1978; 58:397-460.
19. Dobrin PB. Vascular mechanics. In: Handbook of Physiology Part I.Peripheral Circulation and Organ Blood Flow. Shephard JT and Abboud FM eds. Williams and Wilkins, Baltimore MD. 1983; 65-102.
20. Reuterwall OP. Uber dir elastizitat der gefaswand und die methode ihrer naheren prufung. Acta Med Scand Suppl 1921; 2:1-175.
21. Burton AC. Relation of structure to function of the tissues of the wall of blood vessels. Physiol Rev 1954; 34:619-642.
22. Krafka J. Comparative study of the histophysics of the aorta. Am J Physiol 1939; 125:1-14.

23. Remington JW. Extensibility behavior and hysteresis phenomena in smooth muscle tissues. In: Tissue Elasticity. Remington JW ed. Am Physiol Soc. Washington DC. 1957; 138-153.
24. Hass GM. Elastic tissue III.Relations between the structure of the aging aorta and the properties of the isolated aortic tissue. Arch Path 1943; 35:29-45.
25. Wolinsky H and Glagov S. Lamellar unit of aortic medial structure and function in mammals. Circ Res 1967; 20:99-111.
26. Zatina MA, Zarins CK, Gewertz BL et al. Role of medial lamellar architecture in the pathogenesis of aortic aneurysms. J Vasc Surg 1984; 1:422-428.
27. White JV, Haas K, Phillips S et al. Adventitial elastolysis is a primary event in aneurysm formation. J Vasc Surg 1993; 17:371-381.
28. O'Rourke MF, Kelly RP. Wave reflection in the systematic circulation and its implications in ventricular function. J Hypertens 1993; 11:327-337.
29. Virmani R, Avolio AP, Mergner WJ, Robinowitz M, Herderick EE, Cornhill JF, Guo S-Y, Liu TH, Ou DY, O'Rourke MF. Effect of aging on aortic morphology in populations with high and low prevalence of hypertension and atherosclerosis. Am J Pathol 1991; 139: 1119-1129.
30. O'Rourke MF, Avolio AP, Lauren PD, Yong J. Age-related changes of elastic lamellae in the human thoracic aorta. J Am Coll Cardiol. 1987; 9:53A. Abstract.
31. Bader H. Dependence of wall stress in the human thoracic aorta on age and pressure. Circ Res 1967; 20:354-361.
32. Dobrin PB, Baker WH, Gley WC. Elastolytic and collagenolytic studies of arteries. Implications for the mechanical properties of aneurysms. Arch Surg 1984; 119:405-408.
33. Dobrin PB. Elastin, collagen and the pathophysiology of aneurysm. In: Development of Aneurysm. Keen RR, Dobrin PB, eds. Austin, Texas. RG Landes. In press.
34. Caldwell SM, Merrill RA, Sloman CM, Yost FL. Dynamic fatigue life of rubber. Industrial and Engineering Chemistry. 194; 12:19-23.
35. Larson EW, Edwards WP. Risk factors for aortic dissection: a necropsy study of 161 cases. Am J Cardiol 1984; 53:849-855.
36. Glagov S, Vito R, Giddens DP, Zarins CK. Microarchitecture and composition of arterial walls: relationships to location, diameter and distribution of medial stress. J Hypertens 1992; 10:S101-S104.
37. Boutouyrie P, Luarent S, Benetos A, Gireud XJ, Hoeks AP, Safar M. Opposing effects of aging on distal and proximal large arteries in hypertensives. J Hypertens 1992; 10 (suppl 6):S87-S91.
38. Dobrin PB and Mrkvicka R. Failure of elastin or collagen as possible critical connective tissue alterations underlying aneurysmal dilatation. Cardiovasc Surg 1994; 2:484-488.
39. Senior RM, Griffin GL, Mecham RP. Chemotactic activity of elastin derived peptides. J Clin Invest 1980; 66:859-862.
40. Senior RM, Connoly NL, Cury JD et al. Elastin degradation by human alveolar macrophages: A prominent role of metallopeoteinase activity. Am Rev Respir Dis 1989; 139:1251-1256.
41. Anidjar S, Dobrin PB, Eichorst M et al. Correlation of inflammatory infiltrate with the enlargement of experimental aneurysms. J Vasc Surg 1992; 16:139-147.
42. Stromberg DD and Wiederhielm CA. Viscoelastic description of a collagenous tissue in simple elongation. J Appl Physiol 1969; 26:857-862.
43. Gay S and Miller EJ. Disposition of the collagens in the connective tissues. In: Collagen in the Physiology and Pathology of Connective Tissue. Gay S and Miller EJ eds. Gustav Fisher Verlag, New York. 1978; 57-62.
44. Bear RS. Long x-ray spacings of collagen. J Am Chem Soc 1942; 64:727.
45. Banga I. Structure and function of elastin and collagen. Akad Kiado, Budapest 1966; 224.
46. Verzar F. Das Altern des kollagens. Helv Physiol Pharm Acta 1956; 14:207-221.
47. Conabere GO and Hall RH. Physical properties of individual leather fibers I. Evaluation of Young's modulus of elastic extension and percentage permanent set of leather fibers. J Int Soc Leath Tr Chem 1946; 30:214-227.

48. Remington JS, Hamilton WF and Dow P. Some difficulties involved in the prediction of the stroke volume from the pulse wave velocity. Am J Physiol 1945; 144:536-545.
49. Roach MR and Burton AC. The reason for the shape of the distensibility curves of arteries. Can J Biochem Physiol 1957; 35:681-690.
50. Apter JT. Correlation of visco-elastic properties with microscopic structure of large arteries IV. Thermal responses of collagen, elastin, smooth muscle, and intact arteries. Circ Res 1967; 21:901-918.
51. Wolinsky H and Glagov S. Structural basis for the static mechanical properties of the aortic media. Circ Res 1964; 14:400-413.
52. Dobrin PB and Canfield TR. Elastase, collagenase, and the biaxial elastic properties of dog carotid artery. Am J Physiol 1984; H124-H131.
53. Dobrin PB and Gley WC. Elastase, collagenase and the radial elastic properties of arteries. Experientia 1985; 41:1040-1042.
54. Bond MD and Van Wart HE. Characterization of the individual collagenases from clostridium histolyticum. Biochemistry 1984; 23:3085-2091.
55. Länne T, Stale H, Bengtsson H et al. Noninvasive measurement of diameter changes in the distal abdominal aorta in man. Ultrasound in Med Biol 1992; 18:451-457.
56. Länne T, Sonesson B, Bergqvist D et al. Diameter and compliance in the male human abdominal aorta:Influence of age and aortic aneurysm. Eur J Vasc Surg 1992; 6:178-184.
57. Sonesson B, Hansen F, Stale H et al. Compliance and diameter in the human abdomina aorta-the influence of age and sex. Eur J Vasc Surg 1993; 7:690-697.
58. Sumner DS, Hokanson DE and Strandness DE Jr.. Stress-strain characteristics and collagen-elastin content of abdominal aortic aneurysms. Surg Gynec Obstet 1970; 147:211-214.
59. Yin FCP, Brin KP, Ting C-T et al. Arterial hemodynamic indexes in Marfan's syndrome. Circulation 1989; 79:854-862.
60. Hirata K, Triposkiadis F, Sparks E et al. The Marfan syndrome: Abnormal elastic properties. J Am Cell Cardiol 1991; 18:57-63.
61. Sonesson B, Hansen F and Länne T. Abnormal mechanical properties of the aorta in Marfan's syndrome.
62. Kainulainen K, Pulkkinen L, Savolainon A et al. Location on chromosome 15 of the gene defect causing Marfan syndrome. N Eng J Med 1990; 323:935-939.
63. Kontusaari S, Kuivaniemi H, Tromp G et al. A single base mutation in Type III procollagen that converts the codon for glycine 619 to arginine in a family with familial aneurysms and mild bleeding tendencies. Ann NY Acad Sci 1990; 580:556-557.
64. Sterick CA, Long JR, Jamasbi B et al. Ventral hernia following abdominal aortic reconstruction. Amer Surg 1988; 54:287-289.
65. Lehnert B and Wadoub F. High coincidence of inguinal hernias and abdominal aortic aneurysms. Ann Vasc Surg 1992; 6:134-137.
66. Tilson MD, Elefteriades J and Brophy C. Tensile strength and collagen in abdominal aortic aneurysm disease. In: The Cause and Management of Aneurysms. Greenhalgh RM and Mannick JA eds. WB Saunders, London. 1990; 97-104.
67. He CM and Roach MR. The composition and mechanical properties of abdominal aortic aneurysms. J Vasc Surg 1994; 20:6-13.
68. Menashi S, Campa JS, Greenhalgh RM et al. Collagen in abdominal aortic aneurysm: typing, content and degradation. J Vasc Surg 1987; 6:578-582.
69. Rizzo RJ, McCarthy WJ, Dixit SN et al. Collagen types and matrix protein content in human abdominal aortic aneurysms. J Vasc Surg 1989; 10:365-373.
70. Baxter BT, Davis VA, Minion DJ et al. Abdominal aortic aneurysms are associated with altered matrix proteins of the nonaneurysmal aortic segment. J Vasc Surg 1994; 19:797-803.
71. Boucek RJ, Gunia-Smith Z, Nobel NL et al. Modulation by propranolol of the lysyl crosslinks in aortic elastin and collagen of the aneurysm-prone turkey. Biochem Pharmacol 1983; 32:275-280.
72. Simpson CF, Kling JM and Palmer RF. The use of propranolol for the protection of turkeys from the development of (-aminoproprionitrile induced aortic rupture. Angiology 1968; 19:414-420.

73. Brophy CM Tilson JE and Tilson MD. Propranolol delays the formation of aortic aneurysms in the male Blotchy mouse. J Surg Re 1988; 44:687-691.

74. Slaiby JM, Ricci MA, Gadowski GR et al. Expansion of aortic aneurysms is reduced by propranolol in a hypertensive rat model. J Vasc Surg 1994; 20:178-183.

75. Leach SD, Toole AL, Stern H et al. Effect of (-adrenergic blockade on the growth role of abdominal aortic aneurysms. Arch Surg 1988; 123:606-609.

76. Gadowski GR, Pilcher DB and Ricci MA. Abdominal aortic aneurysm expansion rate: effect of size and beta-adrenergic blockade. J Vasc Surg 1994; 19:727-731.

77. Patel DJ, deFreitas FM, Greenfield JC et al. Relationship of radius to pressure along the aorta in living dogs. J Appl Physiol 1963; 18:1111-1117.

78. McDonald DA. Blood Flow in Arteries. Williams and Wilkins, 1974. Baltimore. AUTHOR: Please complete this reference.

79. Womersly JR. Oscillating flow in arteries 2:The reflection of the pulse wave at junctions and rigid inserts in the arterial system. Phys Med Biol 1958; 2:313-323.

80. Newman DL, Gosling RG, Bowden NLR et al. Pressure amplitude increase and matching the aortic iliac junction of the dog. Cardiovasc Res 1973; 7:6-13.

81. Gosling RG,, Newman DL, Bowden NLR et al. The area ratio of normal aortic junctions. Brit J Radiol 1971; 44:850-853.

82. Gosling RG and Bowden NLR. Changes in aortic distensibility and area ratio of normal aortic junctions with the development of atherosclerosis. Atherosclerosis 1971; 14:231-240.

83. Taylor CA, Hughes JR, Zarins CK. Effect of exercise on hemodynamic conditions in abdominal aorta. J Vasc Surg 1999; 29:1077-1089.

84. Moore JE, Xu C, Glagov S, Zarins CK, Ku DN. Fluid wall shear stress measurements in a model of the human abdominal aorta: Oscillatory behavior and relationship to atherosclerosis. Atherosclerosis 1994; 110:225-240.

85. Wolinsky H and Glagov S. Nature of species differences in the medial distribution of aortic vasa vasorum in mammals. Circ Res 1967; 20:409-421.

86. Wilens SL, Malcolm JA and Vasquez JM. Experimental infarction (medial necrosis) of the dogs aorta. Am J Pathol 1965; 47:695-711.

87. Heistad DD, Marcus ML, Larsen GE et al. Role of vasa vasorum in nourishment of the aortic wall. Am J Physiol 1981; 240:H781-H787.

88. Lord JW Jr, Rossi G, and Paula G. The inguinal ligament: Its relation to post-stenotic dilatation of the common femoral artery. Bull NY Acad Sci 1979; 55:451-462.

89. Schuler JJ and Flanigan DP. Iliac artery aneurysms. In: Aneurysms: Diagnosis and Treatment. Bergen J and Yao JST eds. Grune and Stratton, New York. 1982; 469-485.

90. McCready RA, Pairolero PC and Gilmore JC. Isolated iliac artery aneurysms. Surgery 1983; 93:688-699.

91. Veldenz H, Schwarcz TH, Endean ED et al. Morphology which predicts rigid growth of small abdominal aortic aneurysms. Ann Vasc Surg 1994; 8:10-13.

92. Vorp DA, Raghavan ML, Webster MW. Mechanical wall stress in abdominal aortic aneurysm: Influence of diameter and asymmetry. J Vasc Surg 1998; 27:632-639.

93. Bernstein EF and Chan EL. Abdominal aortic aneurysms in high risk patients. Ann Surg 1984; 200:255-263.

94. Cronenwett J, Murphy T, Zelenock G et al. Actuarial analysis of variables associated with rupture of small abdominal aortic aneurysms. Surgery 1985; 98:472-483.

95. Ichikawa Y, Nonomiya H. Kogatt H et al. Erythromycin reduces neutrophils and neutrophil-derived elastolytic-like activity in lower respiratory tract of bronchiolitis patients. Amer Rev Respir Dis 1992; 146:196-203.

96. Kucich U, Christner P, Lippmann M et al. Utilization of peroxidase-antiperoxidase complex an enzyme-linked immunosorbent assay of elastin-derived peptides in human plasma. Amer Rev Respir Dis 1985; 131:709-713.

97. Schriver EE, Davidson JM, Sutcliffe MC et al. Comparison of elastin peptide concentrations in body fluids from healthy volunteers, smokers, and patients with chronic obstructive pulmonary disease. Amer Rev Respir Dis 1992; 145:762-766.

98. Swanson RJ, Littooy FN, Hunt TK et al. Laparotomy as a precipitating factor in the rupture of intra-abdominal aneurysms. Arch Surg 1980; 115:299-304.

99. Koch AE, Haines GK, Rizzo RJ et al. Human abdominal aortic aneurysms. Immunophenotypic analysis suggesting an immune mediated response. Am J Pathol 1990; 137:1197-1213.

100. Busuttil RW , Abou-Zamzam AM and Machleder HL. Collagenase activity of the human aorta: Comparison of patients with and without abdominal aortic aneurysms. Arch Surg 1980; 115:1373-1378.

101. Busuttil RW, Heinrich R and Flesher A. Elastase activity: The role of elastase in aortic aneurysm formation. J Surg Res 1982; 32:214-217.

102. Cannon DJ and Read R: Blood elastolytic activity in patients with aortic aneurysm. Ann Thorac Surg 1982; 34:10-15.

103. Brown SL, Blackstrom B and Busuttil RW. A new serum proteolytic enzyme in aneurysm pathogenesis. J Vasc Surg 1985; 2:393-399.

104. Cohen JR, Mandell C, Margolis I et al. Altered aortic protease and antiprotease activity in patients with ruptured abdominal aortic aneurysms. Surg Gynec Obstet 1987; 164:355-358.

105. Dubick MA, Hunter GC, Perez-Lizano E et al. Assessment of the role of pancreatic proteases in human abdominal aortic aneurysms and occlusive disease. Clin Chim Acta 1988; 177:1-10.

106. Heron GS, Unemori E, Wong M et al. Connective tissue proteinases and inhibitors in abdominal aortic aneurysms. Involvement of the vasa vasorum in the pathogenesis of aortic aneurysms. Arterioscler Thromb 1991; 11:1667-1677.

107. Newman KM, Jean-Claude J, Li H et al. Cellular localization of matrix metalloproteinases in the abdominal aortic aneurysm wall. J Vasc Surg 1994; 20:814-820.

108. Zarins CK, Runyon-Hass A, Lu CT et al. Increased collagenase activity in early aneurysmal dilatation. J Vasc Surg 1986; 3:238-248.

109. Anidjar S, Salzmann JL, Gentric D et al. Elastase-induced experimental aneurysms in rats. Circulation 1990; 82:973-981.

110. Halpern VJ, Nackman GB, Gandhi RH et al. The elastase infusion model of experimental aneurysms: Synchrony of induction of endogenous proteinases with matrix destruction and inflammatory cell response. J Vasc Surg 1994; 20:51-60.

111. Louie JS, Weiss J, Ryhanen L et al. The production of collagenase by adherent mononuclear cells cultured from human peripheral blood. Arthritis Rheum 1984; 27:1397-1404.

112. Wahl LM and Lampel LL. Regulation of human peripheral blood monocyte collagenase by prostaglandins and anti-inflammatory drugs. Cell Immunol 1987; 105:411-422.

113. Petersen MJ, Abbott WM, H'Doubler PB et al. Hemodynamics and aneurysm development in vascular allografts. J Vasc Surg 1993; 18:955-964.

114. Dobrin PB, Baumgartner N, Chejfec G et al. Inflammatory aspects of experimental aneurysms: Effect of methylprednisolone and cyclosporine. Ann NY Acad Sci (In Press)

115. Ricci MA, Strindberg G, Slaiby JM et al. Anti-CD18 monoclonal antibody slows experimental aortic aneurysm expansion. J Vasc Surg 1996; (In Press)

116. Koch AE, Haines GK, Rizzo RJ et al. Human abdominal aortic aneurysms: Immunophenotypic analysis suggesting an immune-mediated response. Am J Path 1990; 137:1199-1213.

117. Brophy CM, Reilly JM, Walker-Smith GJ et al. The role of inflammation in nonspecific abdominal aortic aneurysm disease. Ann Vasc Surg 1991; 5:229-233.

118. Keen RR, Nolan KD, Cipollone M et al. Interleukin-1(induces differential gene expression in aortic smooth muscle cells. J Vasc Surg 1994; 20:774-786.

119. Glagov S. Pathology of large arteries. In: Blood Vessels and Lymphatics in Organ Systems. Abramson DI and Dobrin PB, eds. Academic Press, New York. 1984; 39-53.

120. Tilson MD and Stansel HC. Differences in results for aneurysms vs occlusive disease after bifurcation grafts. Results of 100 elective grafts. Arch Surg 1980; 115:1173-1175.

121. Clifton MA. Familial abdominal aortic aneurysms. Brit J Surg 1977; 64:765-766.

122. Norrgard O, Rais O and Angquist K-A. Familial occurrence of abdominal aortic aneurysms. Surgery 1984; 95:650-656.

123. Johansen K and Koepsell T. Familial tendency for abdominal aortic aneurysms. JAMA 1986; 258:1934-1936.
124. Tilson MD and Seashore MR. Fifty families with abdominal aortic aneurysms in two or more first order relatives. Am J Surg 1984; 147:551-553.
125. Tilson MD and Seashore MR. Human genetics of the abdominal aortic aneurysm. Surg Gynec Obstet 1984; 158:129-132.
126. Tilson MD. Generalized arteriomegaly: A possible predisposition to the formation of abdominal aortic aneurysms. Arch Surg 1981; 116:1030-1032.
127. Hollier LH, Stanson AW, Goviczki P et al. Arteriomegaly: Classification and morbid implications of diffuse aneurysmal disease. Surgery 1983; 93:700-708.
128. Band W, Goedhard WJA and Knoop AA. Comparison of effects of high cholesterol intake on viscoelastic properties of the thoracic aorta in rats and rabbits. Atherosclerosis 1973; 18:163-172.
129. Pynadath TI and Mukherjee DP. Dynamic mechanical properties of atherosclerotic aorta: A correlation between the cholesterol ester content and the viscoelastic properties of atherosclerotic aorta. Atherosclerosis 1977; 26:311-318.
130. Haut RC, Garg BD, Metke M et al. Mechanical properties of the canine aorta following hypercholesteremia. J Biomech Eng 1980; 102:98-102.
131. Hudetz AG, Mark G, Kovach AGB et al. Biomechanical properties of normal and fibrosclerotic human cerebral arteries. Atherosclerosis 1981; 39:353-365.
132. Farrar DJ, Bond MG, Sawyer JK et al. Pulse wave velocity and morphologic changes associated with early atherosclerosis progression in the aorta of cynomolgus monkeys. Cardiovasc Res 1984; 18:107-118.
133. Kagan HM, Milbury PE and Kramsch DM. A possible role for elastin ligands in the proteolytic degradation of arterial elastic lamellae in the rabbit. Circ Res 1974; 44:95-103.
134. Zarins CK and Glagov S. Aneurysms and obstructive plaques. Differing local responses to atherosclerosis. In: Aneurysms, Diagnosis and Treatment. Bergan JJ and Yao JST eds. Grune & Stratton 1982; 61-82.
135. Zarins CK, Xu C and Glagov S. Aneurysmal enlargement of the aorta during regression of experimental atherosclerosis. J Vasc Surg 1992; 15:90-101.
136. Bengsston H, Ekberg O, Aspelin P et al. Abdominal aortic dilatation in patients operated on for carotid stenosis. Acta Chir Scand 1988; 154:441-445.
137. Vorp DA, Mandarino WA, Webster MW, Gorcsan J III. Potential influence of intraluminal thrombus on abdominal aortic aneurysm as assessed by a new noninvasive method. Cardiovasc Surg 1996; 4:732-739.
138. Jean-Claude J, Newman KM, Li H et al. Possible key role for plasmin in the pathogenesis of abdominal aortic aneurysms. Surgery 1994; 116:472-478.
139. Paranjpe M, Engel L, Young N et al. Activation of human breast carcinoma collagenase through plasminogen activator. Life Sci 1980; 26:1223-1231.
140. Murphy G, Ward R, Gavrilovic J et al. Physiological mechanisms for metalloproteinase activation. Matrix 1992 (suppl 1); 224-230.
141. Tromholt N, Jorgensen SSJ, Hesse B et al. In vivo demonstration of focal fibrinolytic activity in abdominal aortic aneurysms. Eur J Vasc Surg 1993; 7:675-679.
142. Reilly JM, Sicard GA and Lucore CL. Abnormal expression of plasminogen activators in aortic aneurysmal and occlusive disease. Surgery 1994; 19:865-872.
143. Wolf VG, Thomas WS, Brennan FJ et al. CT scan findings associated with rapid expansion of abdominal aortic aneurysms. J Vasc Surg 1994; 20:529-538.
144. Adolph R, Vorp DA, Steed DL, Webster MW, Kameneva MV, Watkins SC. Cellular content and permeability of intraluminal thrombus in abdominal aortic aneurysm. J Vasc Surg 1997; 25:916-926.
145. Oschsner JL, Hughes JP, Leonard GL et al. Elastic tissue dysplasia of the internal carotid artery. Ann Surg 1977; 185:684-691.
146. Brown GE and Roundtree LG. Right sided carotid pulsations in cases of severe hypertension. JAMA 1925; 84:1016-1019.

147. Parkinson J, Bedford DE and Almond S. The kinked carotid artery that simulates aneurysm. Br Heart J 1939; 1:345-361..
148. Dobrin PB, Canfield TR and Sinha S. Development of longitudinal retraction of carotid arteries in neonatal dogs. Experientia 1975; 31:1295-1296.
149. Bergel DH. The static elastic properties of the arterial wall. J Physiol 1961; 156:445-457.
150. Band W, Goedhard WJA and Knoop AA. Effects of aging on dynamic viscoelastic properties of the rat's thoracic aorta. Pfleug Arch 1972; 331:357-364.
151. Learoyd BM and Taylor MG. Alterations with age in the viscoelastic properties of human arterial walls. Circ Res 1966; 18:278-292.
152. Dobrin PB, Baker WH and Schwarcz TH. Mechanisms of arterial and aneurysmal tortuosity. Surgery 1988; 104:468-571.
153. Dobrin PB, Schwarcz TH and Mrkvicka R. Longitudinal retractive force in pressurized dog and human arteries. J Surg Res 1990; 48:116-120.
154. Dobrin PB and Doyle JM. Vascular smooth muscle and the anisotropy of dog carotid artery. Circ Res 1970; 27:105-119.
155. Leung DYM, Glagov S and Mathews MB. Cyclic stretching stimulates synthesis of matrix components by arterial smooth muscle cells in vitro. Science 1976; 191:475-477.
156. Gordon D and Rekhter MD. Cell proliferation and collagen gene expression in human intimal hyperplasia. In: Intimal Hyperplasia. Dobrin PB cd. RG Landes, Austin TX. 1994; 211-228.
157. Doyle JM and Dobrin PB. Finite deformation analysis of the relaxed and contracted dog carotid artery. Microvasc Res 1971; 3:400-415.
158. Fung YCB. Elasticity of soft tissues in simple elongation. Am J Physiol 1967; 213:1532-1544.
159. Fung YCB, Fronek K and Patitucci P. Pseudoelasticity of arteries and the choice of its mathematical expression. Am J Physiol 1979; 237:H620-H631.
160. Vaishnav RN, Young JT, Janicki JS et al. Nonlinear anisotropic elastic properties of the canine aorta. Biophys J 1972; 12:1008-1027.
161. Inzdi F, Boschetti F, Zappa M et al. Biomechanical factors in abdominal aortic aneurysm rupture. Eur J Vasc Surg 1993; 7:667-674.
162. Mower WR, Baratt LJ and Sneyd J. Stress distributions in vascular aneurysm: Factors affecting risk of aneurysm rupture. J Surg Res 1993; 55:155-161.
163. Stringfellow MM, Lawrence PF and Stringfellow RG. The influence of aorta-aneurysm geometry upon stress in the aneurysm wall. J Surg Res 1987; 42:425-433.
164. McGiffin DC, McGiffin BE, Galbraith AJ et al. Aortic wall stress profile after repair of coarctation of the aorta. Is it related to subsequent true aneurysm formation? Thor Cardiovasc Surg 1992; 104:924-931.
165. Fukushima T, Matsuzawa T and Homma T. Visualization and finite element analysis of pulsatile flow in models of the abdominal aortic aneurysm. Biorheology 1989; 26:109-130.
166. Matsuzawa T. Finite element analysis in three-dimensional flow through saccular aneurysm. Front Med Biol Eng 1993; 5:89-94.
167. Taylor TW and Yamaguchi T. Three-dimensional simulation of blood flow in an abdominal aortic aneurysm-steady and unsteady flow cases. J Biomech Eng 1994; 116:89-97.

CHAPTER 5

Matrix Protein Synthesis and Genetic Mutations of Collagen, Elastin and Fibrillin

Janet J. Grange, B. Timothy Baxter

Introduction

Aneurysm formation has long been considered a degradative process mediated by proteolytic agents. Recent work suggests that disordered synthesis of matrix proteins may be an important component of the pathogenic process. Collagen and elastin are the major matrix proteins of the vascular wall, and their precise relationship determines its mechanical properties. This relationship is maintained by the balance between protein synthesis and degradation. The diseased aorta maintains its capacity for synthesis and, in fact, has been shown to be far more synthetically active than normal aorta. This synthesis is discoordinate, however, perturbing the critical relationship between the matrix components and compromising the integrity of the vascular wall. Which factors initiate and sustain these metabolic changes is the subject of intense investigation. Collagen and elastin synthesis is complex with many layers of regulation which are only beginning to be defined. The familial tendency to develop aortic aneurysms suggests that genetic factors are somehow important, but a specific genetic defect with a clearly defined role in the pathogenesis of the common abdominal aortic aneurysm (AAA) has not been identified. What is clear is that understanding matrix protein synthesis will be critical not only in defining the pathogenesis of vascular aneurysms, but also in developing strategies to inhibit their growth.

Biochemical and Genetic Features of Matrix Proteins

In order to withstand the pulsatile force from blood within its lumen and to propagate this blood distally, the aorta must be both strong and elastic. These demands are met through the organization and diverse mechanical properties of the major aortic matrix proteins, collagen and elastin.

Collagen

Collagens are synthesized by medial smooth muscle cells (SMC) and adventitial fibroblasts, and they are responsible for the aorta's tensile strength. Collagen types I and III, present in a ratio of approximately three to one respectively, are by far the most abundant. Both types are made up of three polypeptide chains called alpha chains which are coiled around one another in a right handed triple helical conformation. Each alpha chain is unique

---

Development of Aneurysms, edited by Richard R. Keen and Philip B. Dobrin.
©2000 Eurekah.com.

and specific to the particular collagen type; type I collagen is made up of two identical alpha-1 chains and an alpha-2, chain while type III is made of three identical alpha-1 chains.

Alpha chains are characterized by a highly ordered sequence of amino acids which ultimately determines the integrity of the collagen molecule. Every third amino acid in the alpha chain is a glycine so that the molecular formula for any alpha chain can be approximated as (Gly-X-Y)n. The occurrence of glycine, the smallest amino acid, in every third position is essential as this position occupies the restricted space in the center of the triple helix. Accordingly, replacement of any of the glycines by a larger amino acid will produce a local interruption in the helical structure. Likewise, proline occurs in about 20% of the Y positions, and 4-hydroxylation of these proline residues is critical to the thermal stability of the collagen molecule.[1-3]

Once the alpha chains are synthesized and folded into triple helical procollagen molecules, they are secreted into the extracellular space. There, they undergo further modification which converts them into collagens and incorporates them into stable cross-linked fibrils. This extensive posttranslational processing makes collagen biosynthesis extremely complex. At least nine specific enzymes and several nonspecific ones are involved, with opportunities for regulation and dysregulation at each step.[1]

The genes directing collagen synthesis are dispersed over several chromosomes but are believed to have evolved from a single 54 base pair (bp) progenitor unit.[4] The coding sequences of the collagen genes are interrupted by numerous large introns resulting in an elaborate structure much larger than the mature mRNA transcript. Most of the exons code for the triple-helical domain of the alpha chain and they range in size from 45bp to 162bp. Transcription is regulated by an upstream collagen promotor which recognizes a number of transcription factors including NF-1, Sp1, and a CCAAT-binding site.[5] Tissue specific activation is thought to involve the interaction of one or more of these ubiquitous binding factors with other more remote and as yet unidentified sites.[6]

Elastin and the Microfibrils

The elastin lamellae of the aortic media is formed from elastin fibers synthesized by resident smooth muscle cells. The elastin fiber is composed of amorphous tropoelastin deposited on a skeleton of microfibrillar proteins. Its remarkable elasticity is due to the unique arrangement of alternating hydrophobic and hydrophilic domains which form beta spirals that readily stretch.[7] The hydrophobic domains contain many valine, glycine and proline residues while lysine predominates in the hydrophilic domains. These lysine residues may be converted to allysine by oxidation leading to the formation of stable elastin crosslinks, desmosine and isodesmosine.[8] Once crosslinked, elastin becomes insoluble and highly resistant to degradation with a biological half life of 70 years.

The human elastin gene is present as a single copy and contains 34 exons in 45 (kilobases) kb of genomic DNA. As with collagen, there is a high intron/exon ratio (18:1) The hydrophilic and hydrophobic domains are segregated on separate exons. The center of the gene contains seven repeats of a hexapeptide (GVGVAP) which is thought to be responsible for the beta spiral of the tropoelastin molecule.[9]

The microfibrils are an integral part of the elastin fiber, forming the scaffolding upon which tropoelastin is deposited. They consist primarily of two highly homologous forms, fibrillin-1 and fibrillin-2. Both are high molecular weight glycoproteins encoded by different genes, FBN1 and FBN2, on two different chromosomes. Other related microfibrillin proteins have also been found in elastin, though they have been less well characterized. Synthetic regulation of these microfibrillar proteins must be coordinated with elastin synthesis in order to form a functional elastin fiber.

Rapid synthesis and accumulation of elastin fibers is known to occur in the fetus. Formation of new elastin fibers in adults has not been observed leading many to believe that the smooth muscle cell loses its capacity for elastin synthesis postnatally. However, low levels of tropoelastin mRNA are detected in the adult aorta, suggesting that some level of elastin synthesis continues into adulthood.[10] The number of elastin lamellae present in the aortic media is remarkably constant with age, so this new elastin synthesis probably contributes to existing elastin fibers rather than forming new ones. The inability of adult tissue to form new fibers from synthesized elastin may relate to an inability to precisely coordinate the synthesis and assembly of the different components of the elastin fiber.

The regulation of elastin synthesis occurs at different levels depending on the stage of development. Prenatally, regulation occurs at the pretranscriptional level. Postnatally, selective destabilization of elastin mRNA, a posttranscriptional phenomenon, appears to play a critical role in synthetic regulation.[11,12] This transition from transcriptional to posttranscriptional regulation after birth may explain why the adult smooth muscle cell becomes refractory to factors which stimulate elastin synthesis in fetal cells.[13] A highly conserved untranslated region at the 3' end of the elastin gene is implicated in this destabilization process.[14]

Organization of Matrix Proteins in the Vascular Wall

The integrity and function of the vascular wall is critically dependent on the content and organization of its matrix proteins. Collagen and elastin form lamellae surrounding the smooth muscle cells of the vascular media. The content of both of these proteins relative to luminal surface area has been shown to decrease from the proximal to the distal aorta.[15] The relative decrease in collagen and elastin content is similar from the thoracic to the suprarenal aorta so the two proteins maintain a remarkably constant proportion along those segments. This relationship changes at the level of the infrarenal aorta where the elastin content drops dramatically compared to the collagen content (Figs. 5.1, 5.2) Due to their unique biochemical structures, elastin imparts viscoelastic properties to the vascular wall while collagen provides tensile strength. By altering the relationship between those two proteins, the distensibility and strength of the vessel wall is necessarily changed. Interestingly, this change in the relative proportion of collagen and elastin occurs in the infrarenal aorta, the level at which both atherosclerotic occlusive disease (AOD) and abdominal aortic aneurysm (AAA) commonly occur.

Matrix Proteins in Aortic Aneurysms

Aneurysm formation is accompanied by characteristic changes in the normal relationship between matrix proteins. These changes reflect an alteration in the balance between protein synthesis and degradation. The process of aneurysmal dilatation has long been considered a degradative process likened to the bulge on a worn tire. Under this paradigm, one would expect an aneurysm to have an attenuated wall containing a lesser proportion of protein than normal. By clinical observation, however, the wall of the aneurysm is typically thickened rather than attenuated suggesting increased matrix synthesis. This clinical impression has been confirmed by data demonstrating that significant matrix synthesis does, in fact, accompany aortic dilatation (Fig. 5.3). Notably, both total collagen and total elastin are increased. However, the relative elastin increase (2.5x) is smaller than the increase in collagen (4.9x) or total protein (8.4x) In other words, elastin becomes redistributed over a larger area and diluted by the disproportionate increase in other proteins. This phenomenon would explain the well documented decrease in AAA elastin concentration in terms of disordered synthesis rather than simply selective elastolysis.

Changes in protein content noted in AAA are translated into characteristic changes in the thickness of the vascular intima, media and adventitia as illustrated in Figure 5.4.[17] The

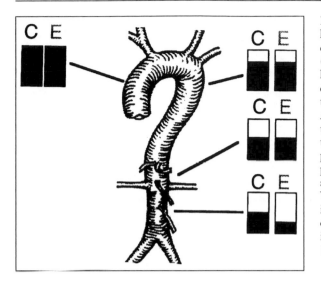

Fig. 5.1. Relative content of collagen and elastin along the course of the aorta in mg/cm². (C=collagen E=elastin) Collagen and elastin content decrease proportionately from the thoracic to the suprarenal aorta. At the infrarenal level, the elastin content decreases more than the collagen content, altering the proportion of these two matrix proteins. Reprinted with permission from: Halloran BG, Davis VA, McManus BM et al. Localization of aortic disease is associated with intrinsic differences in aortic structure. J Surg Res 1995;59:17-22.

Fig. 5.2. Light photomicrographs of Movat's pentachrome stain of the thoracic arch (A), supraceliac abdominal (B), and infrarenal aorta(C). Elastin fibrils stain black. There is a decrease in the number of elastin lamellae and medial thickness from proximal to distal aorta.

lack of attenuation of the adventitia in AAA indicates that collagen synthesis keeps pace with luminal dilatation. The significant thinning of the media suggests that it is stretched as the aorta grows without a compensatory increase in matrix protein synthesis. Plaque thickness in AAA appears equal to AOD, but considering the increased luminal area in AAA, there must be a very large increase in plaque volume which likely accounts for the majority of the increase in matrix proteins in this disease.

The clinical localization of aneurysmal disease to the infrarenal aorta might suggest that aneurysm formation is related to a focal dysregulation of matrix protein synthesis. However, investigation of collagen and elastin content in the proximal, nonaneurysmal aorta from patients with AAA reveals diffuse changes in the relationship of these matrix proteins. Collagen content is increased without a commensurate increase in elastin in both the affected

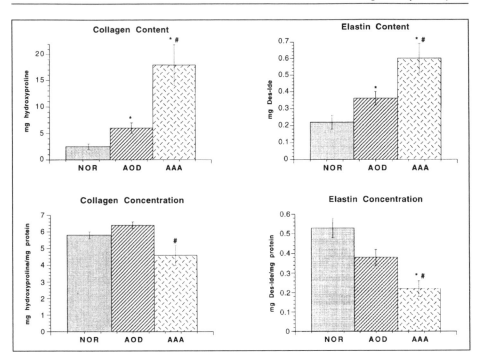

Fig. 5.3. Total collagen and elastin content (mg) and concentration (mg/mg protein) in a complete 1 cm transverse ring of the infrarenal aorta. NOR (normal), n=8; AOD (atherosclerotic occlusive disease), n=9; AAA (abdominal aortic aneurysm), n=9. *Differs from NOR, #differs from AOD; p<.05

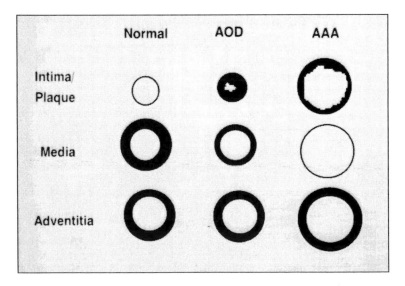

Fig. 5.4. Schematic representation of the changes occurring in the layers of the aortic wall with the development of atherosclerosis (AOD) and aneurysm (AAA). Reprinted with permission from: Halloran BG, Baxter BG. Pathogenesis of aneurysms. Sem Vasc Surg 1995;8:85-92.

and unaffected aorta.[18] This finding, together with the observation that aortic aneurysms are often associated with both diffuse arteriomegaly and peripheral aneurysms, suggests that a systemic abnormality in the regulation of collagen and elastin synthesis may play a role in the pathogenesis of this disease.

While studying changes in matrix protein content provides some insight into the pathogenesis and response to aneurysm formation, it does not provide a dynamic measure of matrix synthesis because of the longevity of these proteins. The rates of synthesis of types I and III collagen as well as elastin are controlled at the posttranscriptional level so that changes in mRNA levels generally reflect changes in the rate of synthesis.[19-21] This is because the messenger mRNA becomes destabilized. In the case of elastin, the steady state levels are decreased. Northern blot analysis of total RNA extracted from normal and aneurysmal aortas showed a marked increase in procollagen mRNA in AAA compared to normal while detecting no difference in tropoelastin mRNA (Fig. 5.5). Northern blot analysis is not the optimal technique with which to study tropoelastin expression because of the low copy number of its transcript. Thus, we have recently studied elastin gene expression in normal, AOD, and AAA aortas using a much more sensitive technique, quantitative competitive-polymerase chain reaction (QC-PCR). This analysis demonstrated that tropoelastin was decreased to similar levels in both AAA and AOD as compared to normal (unpublished data). This decrease in the diseased vessel might be explained by decreased transcript stability, downregulation at the gene level, or a relative depletion of smooth muscle cells expressing elastin.

Factors Modulating Matrix Synthesis

Histologically, AAA is characterized by a prominent inflammatory infiltrate consisting of both macrophages and T lymphocytes. These cells elaborate growth factors and cytokines which are known to regulate a number of vascular SMC functions including matrix protein synthesis.[22,23] One such growth factor is transforming growth factor beta (TGF-β). TGF-β is known to be a particularly potent fibrogenic peptide and it is found in association with atheromatous plaque.[23] We have investigated the effects of TGF-β on human aortic smooth muscle cell 1α(1) procollagen expression and found a moderate increase in procollagen expression (36%).[24] This upregulation was delayed by 24 hours suggesting that synthesis of some intermediary protein was required for the effect. Our group and others have shown that TGF-β induces expression of platelet derived growth factor (PDGF). Using PDGF-AB antibodies, we have demonstrated that PDGF is, indeed, a cofactor in procollagen upregulation by TGF-β (Fig. 5.6).

Other cytokines elaborated by these infiltrating inflammatory cells regulate collagen synthesis through the complex interaction of opposing effects. Interleukins 4 and 10 have been shown to modestly increase the synthesis of collagen types I and III by vascular SMCs.[5,23] Interleukin 1 may also induce local SMC expression of PDGF, thus stimulating collagen synthesis indirectly. On the other hand, activated T lymphocytes secrete IFN-γ which profoundly suppresses both basal and stimulated collagen gene expression by vascular SMC.[23] The proinflammmatory cytokines TNF-α, IL-1β and IL-6 elaborated by activated macrophages have also been shown to downregulate procollagen synthesis. These effects appear to be mediated by tyrosine kinase and protein kinase C. Prostaglandin E_2 is also thought to play an important intermediary role in this downregulation of procollagen.[25]

Defining the complex interactions and net effects of the immune response on matrix protein metabolism will be critically important to our understanding of aneurysm pathogenesis. To that end, we have compared the inflammatory response in AOD to that in AAA and have made some provocative observations. The inflammatory cells present in AAA tissue are essentially the same as those found in AOD; both contain predominantly T lymphocytes and macrophages. And yet, the two disease processes differ remarkably in matrix

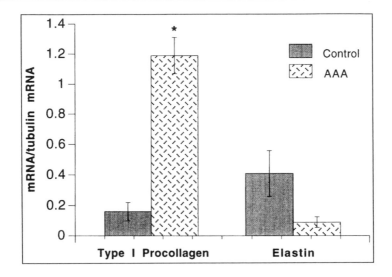

Fig. 5.5. 1α(1) procollagen and tropoelastin mRNA relative to tubulin mRNA levels in normal control aorta and aneurysmal aorta (AAA) (*differs from control; p < 0.05).

protein metabolism. The exuberant collagen synthesis and increased proteolytic activity that typifies AAA is not found in AOD. Instead, AOD exhibits a moderate, early increase in matrix protein content without a sustained increase in procollagen expression or proteolytic activity.[26] This difference in matrix protein metabolism may be related to the specific mix of cytokines elaborated by the infiltrating lymphocytes. T lymphocytes are known to exhibit two distinct patterns of cytokine expression.[27] These different cytokine patterns modulate matrix protein synthesis in specific ways to produce typical disease phenotypes. Likewise, our research suggests that AOD and AAA exhibit characteristic cytokine patterns that may modulate matrix protein metabolism in different ways resulting in distinct disease entities. In other words, the factors that determine which cytokines are elaborated by the infiltrating inflammatory cells may also determine which disease process predominates, AAA or AOD. These factors are likely genetic in origin, predisposing individuals to an aneurysmal or occlusive response to atherosclerotic injury.

Genetic mutations Associated with Aneurysms

A series of studies of the families of patients with aortic aneurysms has revealed a relatively high incidence of affected members. A particularly careful study used ultrasound to examine the aortas of 87 asymptomatic brothers and sisters of 32 patients who had undergone surgery for AAA. An AAA, defined as greater than 2.9 cm at the level of the celiac axis, was diagnosed in 29% of the brothers and 6% of the sisters.[28] This strong familial association suggests that aneurysm formation has a genetic component. Although it is unclear whether the pattern of inheritance is dominant or recessive, there is agreement that it is likely related to a single gene.

Genes regulating the synthesis of collagen and elastin have been the subject of intense investigation in search of the genetic link in AAA.

Vascular aneurysms present commonly as part of a constellation of connective tissue defects in a number of well known clinical syndromes associated with defective matrix protein synthesis. Ehlers-Danlos syndrome type IV results from a specific defect in the type III

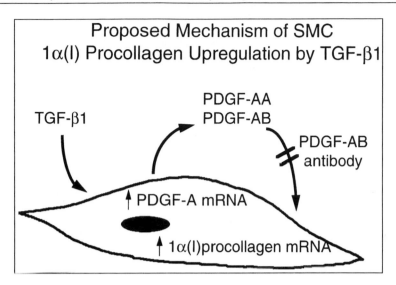

Fig. 5.6. Proposed mechanism for TGF-β1 mediated upregulation of 1α(1) procollagen in human arterial smooth muscle cells. TGF-β1 induces expression of PDGF-A mRNA, consequently increasing PDGF-AA and PDGF-AB protein levels. Neutralizing antibodies to PDGF-AB attenuates the TGF-β1 mediated upregulation of procollagen expression suggesting that PDGF acts as a cofactor in promoting increased procollagen gene expression. Reprinted with permission from: Halloran BG, So BJ, Baxter TB. Platelet-derived growth factor is a cofactor in the induction of 1 alpha(I) procollagen expression by transforming growth factor beta 1 in smooth muscle cells. J Vasc Surg 1996;23:767-773.

collagen gene and is characterized by vascular fragility and aneurysm formation along with other connective tissue abnormalities.[2] Marfan's syndrome has been linked to a defect in the gene for fibrillin-1 and its clinical manifestations include aortic aneurysms in addition to ocular, skeletal, and cardiac abnormalities.[29] Notably, aneurysms associated with Marfan's syndrome and other defects in fibrillin-1 synthesis tend to localize to the thoracic region rather than to the abdominal region as in AAA.[30]

Isolated abdominal aneurysmal disease in the absence of other connective tissue abnormalities has been associated with an identifiable genetic defect in collagen synthesis in only a few sporadic cases. One such case involved a family with a remarkable history of early onset AAA with a high frequency of rupture. Careful genetic analysis identified a mutation that resulted in the substitution of arginine for glycine in type III procollagen synthesis affecting the thermal stability of the synthesized protein. Likewise, two patients with multiple aneurysms but no other generalized connective tissue disorder have been found to have an abnormality in the biosynthesis of type III procollagen.[31]

While these cases are interesting, they are not representative of the largest group of AAA patients: those with late onset, isolated AAA developing in the setting of atherosclerosis and unassociated with other connective tissue abnormalities. In this group, the search for a specific defect in procollagen synthesis has been unrewarding. That is not to say that this presentation of AAA does not have a genetic component. Rather, the defect may occur in a gene which is indirectly related to matrix metabolism, such as a gene which determines how the cells of the vascular wall respond to injury.

Summary

Aortic smooth muscle cells and fibroblasts elaborate a complex and highly ordered interstitial matrix composed primarily of elastin and collagen. The precise relationship of these two proteins determines the functional integrity of the vascular wall. Aneurysmal disease is characterized by a dramatic increase in the synthesis of collagen without a commensurate increase in elastin synthesis, altering this critical relationship. The factors that initiate and sustain this discoordinate synthesis are currently under investigation. The prominent inflammatory infiltrate that characterizes AAA histologically implicates the immune system in the pathogenic process. The cytokines and growth factors elaborated by these infiltrating cells interact to regulate collagen and elastin synthesis in a complex fashion that is incompletely understood. Genetic factors clearly play some role in the development of aneurysms, but a specific genetic defect which accounts for the majority of AAA has not been identified.

Recent developments in molecular biologic and genetic techniques provide powerful tools with which to advance our current understanding of matrix metabolism in the diseased aorta. The particular pattern of cytokines elaborated by the inflammatory infiltrate present in the atherosclerotic aorta may be important in determining whether disease progresses to AOD or AAA. Further efforts to more precisely define the inflammatory response to vascular injury in both AAA and AOD, and then to determine the genetic basis for that response, will be important in elucidating a mechanism for aneurysm formation. Pharmacologic strategies may then be developed to modulate matrix protein synthesis in the developing aneurysm in a way that inhibits its growth.

References

1. Kivirikko KI, Myllyla R. The posttranslational processing of procollagens. Ann NY Acad Sci 1985; 460:187-201.
2. Kivirikko KI. Collagens and their abnormalities in a wide spectrum of disease. Ann Med 1993; 25:113-126.
3. Barnes MJ. Collagens of normal and diseased blood vessel wall. In: Nimni ME, ed. Collagen, Vol. I. Biochemistry. Boca Raton, FL: CRC Press, Inc., 1988. pp. 271-291.
4. Ramirez F, Bernard M, Chu M-L et al. Isolation and Characterization of the Human Fibrillar Collagen Genes. Ann NY Acad Sci 1985; 460:117-129.
5. Rossert JA, Garret LA. Regulation of type I collagen synthesis. Kidney Intern 1995; 47:S-34-S38.
6. Brenner BA, Rippe RA, Rhodes K et al. Fibrogenesis and type I collagen gene regulation. J Lab Clin Med 1994; 124:755-760.
7. Christiano AM, Uitto J. Molecular pathology of the elastic fibers. J Invest Dermatol 1994. 103:53S-57s.
8. Kagan HM. Posttranslational modification of elastin: Lysine oxidation and cross-linking. In: Robert L, Hornebeck W, eds. Elastin and Elastases, Vol. I. Boca Raton, FL.: CRC Press, 1989. pp. 109-127.
9. Urry DW, Long MM. Conformations of the repeat peptides of elastin in solution. CRC Crit Rev Biochem 1976; 4:1-45.
10. Mesh CL, Baxter BT, Pearce WH et al. Collagen and elastin gene expression in aortic aneurysms. Surgery 1992; 112:256-262.
11. Hinek A, Botney MD, Mecham RP et al. Inhibition of tropoelastin expression by 1,25-dihydroxyvitamin D3. Conn Tiss Res 1991; 26:155-166.
12. Johnson D, Robson P, Hew Y et al. Decreased elastin synthesis in normal development and in long-term aortic organ and cell cultures is related to rapid and selective destabilization of mRNA for elastin. Circ Res 1995; 77:1107-1113.

13. Fazio MJ, Olsen DR, Kuivaniemi H et al. Isolation and characterization of human elastin cDNAs, and age-associated variation in elastin gene expression in cultured skin fibroblasts. Lab Invest 1988; 58:270-277.
14. Boedtker H, Fuller F, Tate V. Structure of collagen genes. Int Rev Conn Tiss Res 1983; 10:1-63.
15. Halloran BG, Davis VA, McManus BM et al. Localization of aortic disease is associated with intrinsic differences in aortic structure. J Surg Res1995; 59:17-22.
16. Minion D, Davis V, Nejezchleb P et al. Elastin is increased in abdominal aortic aneurysms. J Surg Res 1994; 57:443-446.
17. Halloran BG, Baxter BT. Pathogenesis of aneurysms. Sem Vasc Surg 1995;8:85-92.
18. Baxter BT, Davis VA, Minion DJ et al. Abdominal aortic aneurysms are associated with altered matrix proteins of the nonaneurysmal aortic segments. J Vasc Surg 1994; 19:797-803.
19. Burnett W, Eichner R, Rosenbloom J. Correlation of functional elastin messenger ribonucleic acid levels and rate of elastin synthesis in the developing chick aorta. Biochemistry 1980; 19:1106-1111.
20. Keeley FW, Hussain RA, Johnson DJ. Pattern of accumulation of elastin and the level of mRNA for elastin in aortic tissue of growing chickens. Arch Biochem Biophys 1980; 282:226-232.
21. Parks W, Secrist WH, Wu LC et al. Developmental regulation of tropoelastin isoforms. J Biol Chem 1988; 282:226-232.
22. Koch A, Haines R, Rizzo R. Human abdominal aortic aneurysm: Immunopheotypic analysis suggesting an immune-mediated response. Am J Pathol 1990; 137:1199-1213.
23. Amento E, P., Ehsani N, Palmer H et al. Cytokines and growth factors positively and negatively regulate interstitial collagen gene expression in human vascular smooth muscle cells. Arterio Thromb 1991; 11:1223-1230.
24. Halloran BG, Byung JS, Baxter BT. Platelet-derived growth factor is a cofactor in the induction of a1(I) procollagen expression by transforming growth factor-ß1 in smooth muscle cells. J Vasc Surg 1996;23:767-773.
25. Sudbeck BD, Parks WC, Welgus HG et al. Collagen-stimulated induction of keratinocyte collagenase is mediated via tyrosine kinase and protein kinase C activities. J Biol Chem 1994; 269:30022-30029..
26. McGee GS, Baxter BT, Shively VP et al. Aneurysm or occlusive disease-factors determining the clinical course of atherosclerosis of the infrarenal aorta. Surgery 1991; 110:370-376.
27.. Mosmann TR, Coffman RL. TH1 and TH2 cells: Different patterns of lymphokine secretion lead to different functional properties. Annu Rev Immunol 1989; 7:145-173.
28. Bengtsson H, O Norrgard K, Angquist A et al. Ultrasonographic screening of the abdominal aorta among siblings of patients with abdominal aortic aneurysms. Br J Surg 1989; 76:589-591.
29. Dietz HC, Pyeritz RE. Mutations in the human gene for fibrillin-1 (FBN-1) in the Marfan syndrome and related disorders. Hum Molec Genet 1995; 4:1799-1809.
30. Francke U, Berg MA, Tynan K et al. A Gly1127Ser mutation in an EGF-like domain of the Fibrillin-1 gene is a risk factor for ascending aortic aneurysm and dissection. Am J Hum Genet 1995; 56:1287-1296.
31. Deak SB, Ricotta JJ, Mariani TJ et al. Abnormalities in the biosynthesis of type III procollagen in cultured skin fibroblasts from two patients with multiple aneurysms. Matrix 1992; 12:92-1000.

CHAPTER 6

Pathology and Pathogenesis of Degenerative Atherosclerotic Aneurysms

William E. Stehbens

Introduction

Aneurysms are persistent, localized pathological dilatations of the lumen of blood vessels and cannot be considered merely as enlargements of the external diameter. Ectasia, on the other hand, may be defined as a diffuse and persistent, frequently nonuniform dilatation of a long segment of blood vessel in the absence of increased functional demand. It is usually manifested after maturation with extreme cases regarded as arteriomegaly being actually a diffuse expression of mural weakness or even a diffuse incipient aneurysm. Aneurysms may arise from ectatic vessels: The essential feature of aneurysmal dilatation is the localized enlargement when compared with the adjacent vessel proximally or distally and not by comparison with the statistically normal vessel diameter at a specified site.

Aneurysmal dilatations of large vessels have attracted most attention because of the devastating hemorrhage following rupture and because small vessel aneurysms are difficult to study, though crucial information obtained from large affected arteries is applicable to small caliber vessels. When perivascular tissue pressure and strength of the vessel wall can resist the pressure within the lumen, there should be no dilatation irrespective of aneurysm size, parent vessel diameter or blood pressure. The physical strength of the wall is of paramount importance for aneurysmal dilatation occurs only when the vessel wall, weakened by some pathological process, can no longer withstand the internal pressure.

Aneurysms are sometimes classified according to shape which depends on the extent of mural weakness, their anatomical location and the presence of nonyielding perivascular anatomical structures fortuitously providing external support or resistance to expansion. Those arising from the side of a vessel are often called lateral saccular aneurysms. Those originating within the fork of arterial bifurcations or branchings are berry aneurysms (mostly of cerebral arteries). Aneurysm size is determined variously by hemodynamic stress intensity, reparative capabilities of the wall, physicochemical characteristics of individual mural constituents and external support, with shape, size and location significant aspects of any aneurysm. Local factors are responsible for localization and play an active role in their natural history.

Development of Aneurysms, edited by Richard R. Keen and Philip B. Dobrin.
©2000 Eurekah.com.

Etiology is the all-important determinant of aneurysmal dilatation, progression and complications, both clinical and pathological. Consequently aneurysms are usually subdivided according to etiology, it being acknowledged that trauma, infections and neoplasms can disrupt the wall sufficiently to produce aneurysms with immunological reactions allegedly responsible in some instances. Sevitt[1] contended that nonseptic thromboemboli induce aneurysmal dilatation at sites of impaction but provided no pathogenetic mechanism and his illustrations, rather than being of aneurysms, were of oblique sections through the origin of small branches!

The most common cause of human aneurysms is atherosclerosis. The two most common varieties are those involving the abdominal aorta and berry aneurysms of cerebral arteries. Considerable controversy has raged over their etiology and even more concerning the etiological role of atherosclerosis, defined here as encompassing the degenerative and reparative changes in response to hemodynamically induced bioengineering fatigue consequent upon the repetitive passage of pulse waves and lesser vibrations generated over a lifetime by the flowing blood at the most susceptible configurations e.g., forks, junctions, curvatures and vessel wall irregularities.[2,3]

Any discussion of aneurysm pathology should differentiate between cause and pathogenesis. Atherosclerosis, being a specific disease entity, has but one cause (the sole prerequisite without which the disease cannot occur) though many factors may aggravate or amelioriate its pathogenesis.[4] Its pathogenesis is thus multifactorial but not the etiology. It follows that atherosclerotic aneurysms also have but one cause and though other diseases may be superimposed, they will coexist with atherosclerosis, one or another possibly predominating.

Currently some authors consider atherosclerosis to be an inflammation[5] because (i) monocytes enter thickened intima in early infancy and childhood to phagocytose cell debris and ultimately become foam cells and (ii) later a few lymphocytes and plasma cells are observed in the overtly atherosclerotic intima with more pronounced perivascular infiltrations about adventitial vasa vasorum.[2] The contention neglects to take into account the extensive degenerative changes preceding leukocyte immigration and whilst these cells participate in inflammation, their presence is not diagnostic of an inflammatory disorder. They are commonly seen about degencrative diseases. Early invasion of cerebral infarcts by polymorphonuclear leukocytes followed by transformation of monocytes and microglia to foam cells and perivascular accumulation of chronic inflammatory cells in no way indicates the lesion is primarily an inflammatory disorder, even if inflammation is part of organization and repair. Neither does their presence in and about tumors make neoplasms inflammatory disorders. For a review of other hypotheses of atherogenesis see references 2 and 3.

This chapter deals with those aneurysms of importance by virtue of their frequency or the evidence they provide regarding etiology and pathogenesis.

Acquired Mural Weakness

Spontaneous fractures of bones incurred during performance of daily tasks are regarded as pathological fractures attributable to undue bone fragility. By analogy spontaneous intimal tears, ulcerations, dissections, ectasia and aneurysmal dilatations are similarly due to pathological loss of tensile strength. These complications developing at physiological levels of blood pressure can be potentiated by hypertension. Intravascular pressures 35-56 times greater than normal are required to rupture healthy arteries.[6] Human aortae from subjects 20-40 years of age do not usually rupture at pressures under 1000 mmHg and seemingly require a pressure of 3000 mmHg,[7] a level never attained even in severe hypertension. Thus structural and functional changes in vessel walls account for the loss of tensile strength, ultimately the causative basis of the pathological complications of atherosclerosis and its clinical manifestations.[2,3] Neither the literature nor the pathology of the disease supports Tilson's claim[8] that atherosclerosis is not usually associated with loss of tensile strength.

Loss of arterial elasticity and tensile strength in atherosclerosis with age is a long recognized phenomenon.[9-11] Thoma, injecting aortas under a pressure of 160 mmHg, found that intimal thickenings rather than encroaching on the lumen usually bulged outwards deforming the external contour.[12] More advanced lesions encroach on the lumen as seen angiographically and particularly about branch ostia. Histologically the underlying medial thinning has been interpreted as deriving from pressure atrophy of the media. However the attenuated media stretched over the external bulging of the wall would be subjected to repetitive expansile and longitudinal stress with passage of each pulse wave deteriorating in the same manner as the aneurysm wall. Qualitative changes in the vessel wall support Thoma's explanation[13] of medial weakness or angiomalacia as the most plausible reason.

Progressive mural weakness of arterial walls at all sites with age[14] has been correlated with increased collagen content and brittleness.[15] A connective tissue disorder has been invoked to explain the fragility of atherosclerotic arteries at surgery[16] particularly when inserting prostheses for aortic aneurysms and intimal fragility is readily demonstrated at autopsy. These changes are unlikely to be due to age per se because the rate of degeneration of pulmonary arteries is dissimilar to that of systemic arteries. On morphological grounds, atherosclerosis progressively affects the vessel wall from within outwards with no layer escaping and with severity greatest in the innermost part of the wall. If the entire vessel wall were simultaneously affected to the same extent, direct transmural rupture would be more common than partial rupture (intimomedial tears) and mural dissection.

Aortic tensile strength is reputedly greatest in the young, least in those aged over 50 years and greater transversely than longitudinally.[7] This observation correlates with the transverse orientation of most tears of the internal elastic lamella (IEL) and dissecting aneurysms and with the longitudinal orientation of smooth muscle cells (SMCs) in compensatory intimal proliferation. Investigation has revealed decreased resistance to physical rupture and dissection from the ascending aorta and arch distally with age[7,17,18] and less intimal tensile strength over torn coronary plaques than neighboring walls.[19,20]

Loss of tensile strength and yielding explains progressive diffuse increase in internal diameter (ectasia) of the aorta, coronary and cerebral arteries after maturation[21] and are usually associated with tortuosity due to increase in length between two relatively fixed points, as instanced by the cervical internal carotid artery. Carotid arteries subjected to increasing internal pressure in vitro initially show a very slight increase in diameter though none at pressures from 250-300 mmHg or even to 2500 mmHg. This indicates greater adventitial strength because arteries rupture at low pressures on incising the adventitia.[22] The finding correlates with enzymatic degradation experiments in which vessels in vitro were exposed to "high" concentrations of elastase and/or collagenase. Dobrin[23] concluded that elastin bears loads in circumferential, longitudinal and radial directions, whilst collagen contributes predominantly in the circumferential direction at pressures above 80 mmHg. All arteries incubated with collagenase ruptured, indicating the overall importance of collagen which is mainly in the adventitia.[24,25] Dobrin deduced that elastin loss leads to some dilatation and elongation due to diminished retractive force and therefore deficient collagen would potentiate dilatation and aneurysmal dilatation.[25] However whilst elastin insidiously undergoes degeneration, other elements including collagen exposed to the same hemodynamic stresses simultaneously undergo similar degeneration even if at different rates.

More pronounced or rapid loss of tensile strength could well explain the emergence of the belief that atherosclerosis is of two types, one more prone to exhibit ectasia and the other with a propensity to proliferation and stenosis.[26] Such responses would have to rely on a balance of physicochemical changes in mural constituents and the nature of local hemodynamic stresses. Negating this dualistic approach is that (i) coronary heart disease (CHD) and cerebrovascular disease are common causes of death in patients surgically treated for abdominal aortic aneurysm,[27] (ii) atherosclerosis is ubiquitous and advanced in the abdominal aorta in all subjects over 60 years and (iii) both atrophic and proliferative lesions of atherosclerosis occur within the same vessel and circulatory bed whilst individual circulatory beds may display atrophic lesions, ectasia and aneurysm depending on flow conditions (e.g., feeding an arteriovenous (AV) shunt or a left to right shunt in coronary arteries).[2,3] The propensity to ectasia and aneurysms is exemplified by younger subjects, in all probability resulting from an unduly rapid loss of tensile strength, and does not preclude atherosclerosis as the cause of aneurysms. Halpern et al[28] reported CHD in 48% of their abdominal aortic aneurysm patients. Most instances of atherosclerotic CHD and cerebral thromboembolism are based on intimal tears or ulceration consequent upon mural fragility, the thromboembolism being a secondary phenomenon and most abdominal aortic aneurysms occur after 60 years of age in the presence of advanced aortic and iliofemoral atherosclerosis.

Macroscopically ectasia, tortuosity, intimal tears, cracks and ulceration correlate with atherosclerosis severity and are most pronounced in the abdominal aorta. Such tears, like pathological stress fractures of bone developing under physiological conditions, are similarly indicative of acquired mural fragility and reflect the culmination of the cumulative loss of tensile strength of atherosclerotic arteries.

Gross loss of arterial tensile strength is a feature of lathyrism and copper deficiency and leads to extensive rents in the vessel wall. In humans transmural arterial rupture is rare though very likely in the Ehlers-Danlos syndrome (type IV). Complete transmural rupture is generally preceded by an aneurysm or dissecting aneurysm with tears and ulceration in reality instances of partial rupture usually associated with a variable degree of dissection. The depth of the tears varies, such mural disruptions supporting the concept that the inner part of the wall is weaker than the outer. Further support derives from the observation that an external impact on the arterial wall tends to lacerate or fracture the intimal surface rather than the outer coats.[29] Moreover internal carotid artery endarterectomy can relieve atherosclerotic obstruction resulting from ulceration and thrombosis. Following repair the residual medial remnants and adventitia usually function satisfactorily for 10 years or more, substantiating the greater intimal fragility. Intimomedial tears, dissection and aneurysms are thus the consequence of loss of mural tensile strength and interrelated, one with another.

Predominating in the middle aged and elderly these phenomena are considered integral features of atherosclerosis and indicative of slow, insidious progressive development of mural fragility (weakness) over 5-6 decades. With such progressive loss of elasticity and tensile strength concomitantly with increasing fibrosis and ectasia, microscopic changes in the aortic wall are to be expected.

Early in utero the aorta possesses a very distinct IEL (Fig. 6.1). Medial elastic laminae are slightly thinner and continuous with occasional branching. At birth the IEL exhibits fragmentation, thinning and often a moth-eaten appearance with variable intimal thickening. Medial elastic laminae also exhibit sporadic fragmentation at any depth (Fig. 6.2). Whilst medial laminae increase in number during maturation, the IEL concomitantly undergoes further disintegration, differentiation between intimal proliferation and media frequently being indistinct such that the innermost prominent elastic lamina is not necessarily the intimal IEL.[30] Newly formed noncontinuous elastic laminae of variable thickness and length develop in the thickening intima (Fig. 6.3). Medial elastic laminae become increasingly fragmented, a feature better visualized in longitudinal than transverse section and lipid accumulation and mineralization are apparent from infancy.

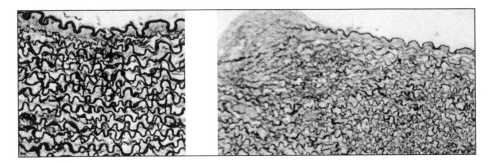

Fig. 6.1. (A) Thoracic aorta of a fetus showing thick well-defined IEL and continuous medial elastic laminae. (B) Same aorta with intimal thickening near a small segmental branch. Note discontinuity of IEL and disorganized elastic laminae beneath intimal thickening. Verhoeff's elastic stain and eosin. A X 280, B X 112.

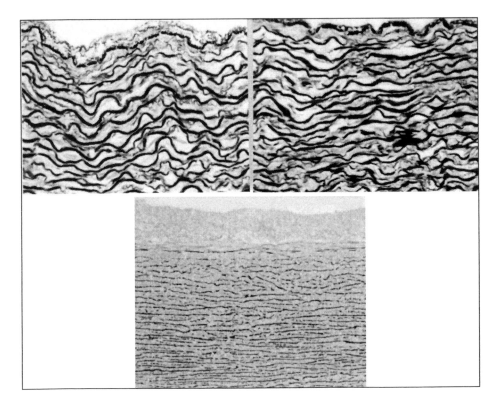

Fig. 6.2. Neonatal aorta showing A moth-eaten appearance of the IEL and in B a thinner IEL with some intimal proliferation containing fibrillary elastica. Both show more fragmentation of medial elastic laminae than in Fig. 6.1 A. C Aorta from a 2 year old infant with diffuse intimal thickening, indistinct intima-media junction, widening of interlamellar spaces and more profound fragmentation of medial elastic tissue. Verhoeff's elastic stain and eosin. A and B X 280, C X 112.

Fig. 6.3. Aorta of 17 year old youth with fragmented IEL similar to that in afferent arteries of AV fistula. Note poorly developed subendothelial elastic lamina in intimal proliferation and extensive fragmentation of medial elastic laminae. Verhoeff's elastic stain and eosin. X 112.

Schlatmann and Becker[31,32] reported that aging changes in the aorta comprise progressive fragmentation of elastic laminae, increasing fibrosis, loss of smooth muscle cells (SMCs), accumulation of proteoglycans and at times large pools of proteoglycans devoid of elastin with but few SMCs. These phenomena, regarded as effects of hemodynamic stress, commence early in life with variation in site and between individuals (Fig. 6.4). They occur in severe form often with dilatation in senile, hypertensive or Marfan patients, variation being one of degree rather than of kind (Fig. 6.5). The fragmentation of medial elastic laminae and disappearance of interlaminar elastic fibers and bands is spectacularly seen by scanning electron microscopy in the inner media more particularly than the outer and especially in dissecting aneurysm subjects.[33] Concomitantly, atherosclerotic changes progressively develop diffusely in the intima with extracellular lipid accumulation and mineralization and though later and to a lesser extent affect the media. These changes represent the diffuse changes of atherogenesis and merge with earlier changes at forks, bends and junctions (reviewed in refs. 2,3 and 30).

Fibromusculoelastic intimal thickening of the fetal aorta (Fig. 6.1B) appears initially about orifices of aortic branches and downstream of the union with the ductus arteriosus.[34] The localization of intimal proliferation is believed to be similar to that in cerebral arteries where it develops over the flow divider of forks, the crescentic ridge at the junction of two branches and at lateral angles beyond the commencement of the curvature of the wall into lateral branches where boundary layer separation occurs.[2,3] It is pronounced in the carotid

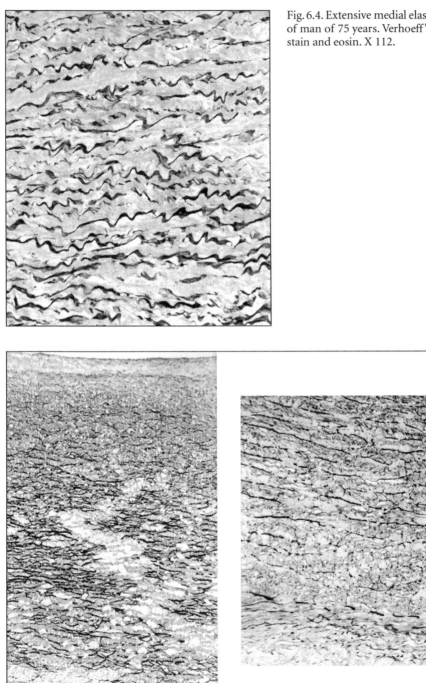

Fig. 6.4. Extensive medial elastic tears of man of 75 years. Verhoeff's elastic stain and eosin. X 112.

Fig. 6.5. Sections of aortae: A from a girl of 8 years and B from a 21 year old male, both with aortic coarctation. Both exhibit extensive fragmentation of medial elastic laminae and B shows cystic medionecrosis. Verhoeff's elastic stain and eosin. A X 50, B X 112.

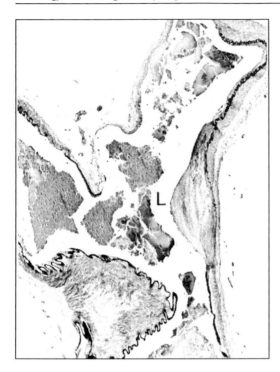

Fig. 6.6. Cerebral arterial fork showing prominent lateral angle thickening (L) with peripheral extension into branch and extensive atrophic lesion on opposite side of branch with loss of IEL and with thinning and attenuation of media and adventitia (incipient aneurysm formation). Flow from below upwards. Verhoeff's elastic stain and eosin. X 22. Reprinted with permission from Stehbens WE. Arch Neurol 1963; 8:272-285.

siphon being found on the distal aspect of the lesser curvatures, which again are sites of boundary layer separation. Intimal proliferation at lateral angles (Fig. 6.6) and bends extends peripherally with age in humans, merges with the diffuse intimal thickening developing later and progresses insidiously to overt atherosclerosis which, remaining most severe at sites of initial development, can involve the entire intimal surface of larger arteries.[2,3] Experimentally similar intimal proliferation occurs in surgically fashioned forks, bends and bypass models.[35-37]

Histologically the intimal proliferation at forks and junctions in human cerebral arteries commences with thinning, fragmentation and loss of the IEL underlying intimal proliferation.[38,39] Ultrastructurally the IEL is depleted and intimal SMCs exhibit granulovesicular degeneration, thickening and multilamination of basement membrane and at times muscle necrosis with redundant basement membrane in the matrix.[40] Hence these intimal thickenings have been classified as early lesions of atherogenesis. They are now accepted as being hemodynamically induced and as sites which ultimately exhibit severe atherosclerosis.[41] Experimental proof of the correlation is provided by enlargement of pre-existing intimal thickenings at branching sites and progression to overt atherosclerosis in rabbit arteries proximal to an AV shunt at serum cholesterol values below those of neonates.[42]

In addition to intimal and medial elastic changes already described, ultrastructural studies of intimal thickenings about arterial forks (human and rabbit), in spontaneous cerebral aneurysms, experimental aneurysms and AV fistulae exhibit other characteristically atherosclerotic features not seen histologically,[40,43-49] viz.

    i. Elongated, attenuated or branched SMCs shed variously sized vesicles (matrix vesicles) and plasma membrane fragments (granulovesicular degeneration). This vesicular debris accumulates in larger aggregates with electron translucent contents

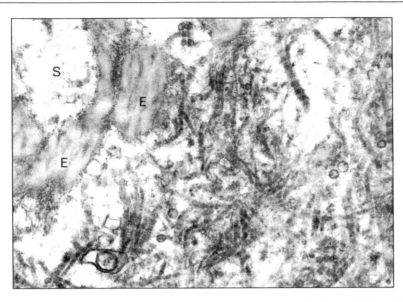

Fig. 6.7. Electron micrograph of haphazardly arranged short, curved collagen fibrils with a few matrix vesicles and elastin (E) in the wall of an experimental aneurysm (rabbit) network. Proteoglycan incompletely bridges the spaces (S). X 60,000.

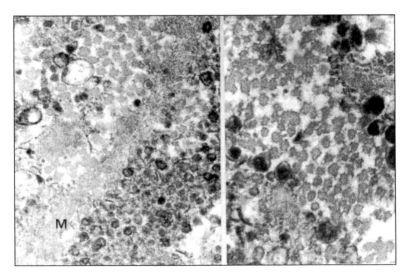

Fig. 6.8. Electron micrographs from experimental rabbit aneurysms. A Matrix vesicles and fragments of SMC plasma membrane in a matrix of redundant basement membrane (M) and misshapen collagen fibrils, seen at larger magnification in B. Some fibrils appear to be fusing and others disintegrating. See bottom right. A X 60,000, B X 90,000.

(lipid accumulation50) or very dense contents due to mineralization.[51] Eventually there is SMC depletion.

ii. Irregular thickening, multiplication, reticulation and redundant basement membranes, at times appearing to be the dominant fibrous matrix component with interspersed collagen and proteoglycans.

iii. Partial separation of endothelium and SMCs from the dystrophic basement membrane and inter-endothelial cell junctions.

iv. Short, curved, haphazardly arranged intimal collagen fibrils (Fig. 6.7) with irregularly shaped cross-sectional profiles similar to those prevalent in genetic disorders associated with connective tissue fragility (Fig. 6.8).[52-54] Some fibrils partially fuse one with another and some appear to disintegrate. Those in the adventitia are larger and more varied in size than usual with occasional misshapen fibrils.

v. Some intimal spaces appear to be partially devoid of a proteoglycan network (Fig. 6.7).

Mild changes seen in intimal thickenings in human infant cerebral arteries[40] nevertheless constitute the beginnings of this train of pathogenetic events of atherosclerosis though generally ignored and considered manifestations of cell fragility,[55] matrix dystrophy and loss of cohesion of mural constituents.[2,3] They are not accelerated experimentally by cholesterol overloading control rabbits[56] or in experimental hemodynamically-induced lesions of animals with cholesterolosis[57,58] but occur in experimental animals on herbivorous diets at cholesterol levels usually below 2 mmol/L.

Poststenotic Aneurysm

The first important clue to the explanation of aneurysmal dilatation was provided by Halsted[59] and Holman[60] who demonstrated poststenotic dilatation beyond an aortic coarctation and subclavian artery aneurysms associated with cervical ribs. The dilatations were attributed to profound turbulence generated beyond the stenosis and sufficient to produce a palpable thrill, harsh blowing murmur, jet lesion, intimomedial tears and transmural rupture in the poststenotic region. The dilatation (fusiform or cylindrical) was associated with a dissecting aneurysm[60] and intimomedial tears. Mural thrombus and some dissection internal to the adventitia have been observed at the commencement of intercostal arteries arising from the poststenotic area in the aorta.[61] Poststenotic aneurysms were reproduced in experimental animals and in rubber tubing with and without flow, the essential feature being the intense vibratory effect on the arterial wall or rubber tubing.[60,62] The poststenotic dilatation occurred with maximal vibrational stress, severe mural degenerative changes, fibrosis, severe fragmentation, loss of elastic tissue, cystic medionecrosis histologically and increased fragility[63-65] as is consistent with dissecting aneurysm development beyond aortic stenosis. The dilatation was considered a nonspecific response to vibrational stress.[66-68]

The extent of the poststenotic dilatation depends on the degree of stenosis and the vibratory effect (frequency and amplitude) produced, the blood pressure and flow velocity.[60] It is not due to hypertension primarily since the mean blood pressure beyond an aortic coarctation is less than proximally where a distal dampening effect on the pulse pressure is effected.[69] Significantly the atherosclerosis is less than proximally or in coronary arteries and abdominal aortic aneurysms have not been reported despite congenital heart disease.

Similar flow disturbances are associated with an AV fistula, the intense vibratory activity producing a bruit (louder with systole), thrill and tortuous aneurysmal dilatation of the artery and vein. There are high amplitude low frequency vibrations as well as lower amplitude vibrations with a frequency up to 2,200 Hz.[70] The arteries are exposed to progressive increases in flow rates (by a factor of 12),[71] velocity (by a factor of 4.4)[72] and pulse pressure (up to 146

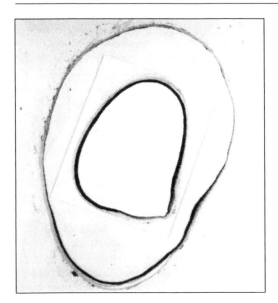

Fig. 6.9. The (perfusion fixed) right external iliac artery in transverse section from a rabbit with a right femoral arteriovenous fistula. The left external iliac artery (same magnification) is the smaller vessel superimposed in the center to show the gross difference in cross-sectional area. Note the mural thinning in the larger artery. Verhoeff's elastic stain and eosin.

mmHg)[73] near the shunt. Arteriovenous communications characteristically produce ectasia, tortuosity and aneurysmal dilatation of both arteries and veins. The arterial walls exhibit severe mural thinning, flaccidity and ectasia perhaps reaching aneurysmal proportions (Figs. 6.9). More localized dilatations may develop on the greater curvature of tortuosities and berry aneurysms develop at forks of cerebral arteries proximal to AV shunts.[74] The arteries do not undergo architectural "venization" nor do veins undergo "arterialization". Overt atherosclerosis develops about the ostia of branches in afferent arteries, on the lesser curvature of tortuosities and more diffusely in the artery nearer the shunt's vibratory activity and in the vein. In rabbits the afferent arteries are thinner and there may be complete loss of the IEL and media with thinning of the adventitia.[75]

In adult survivors of a left-to-right shunt resulting from anomalous origin of the left coronary artery from the pulmonary artery, blood flows from the aorta into the right coronary artery retrograde through myocardial vessels to the left coronary artery and pulmonary artery. The right coronary artery has augmented flow and pulse pressure and exhibits pronounced atherosclerosis possibly with aneurysmal dilatation whereas the left is thin and sometimes ectatic.[76]

The diffuse venous enlargement and tortuosity may produce a large goitrous swelling in the neck of sheep with a carotid-jugular fistula and in humans a large spherical sac can form in the vein of Galen.[77] Veins of experimental herbivorous animals develop atherosclerosis with abrupt intimomedial tears having some dissection and mural thrombosis as in human aortae. Some tears heal as occurs in humans.[42,78] Histological changes in the media display elastic changes similar to cystic medionecrosis in human aortae (Fig. 6.10),[79] indicating that hypertension and arterial hemodynamics are not essential to effect such change because venous pressure in chronic AV fistulae may not be elevated and proximal to the shunt can be negative. Arteries and veins consist of the same cellular and noncellular connective tissues and, though they have different architecture and substantially different hemodynamic stresses, aneurysmal dilatation and overt atherogenesis develop in both arterial and venous limbs of AV fistulae.[44,48,49,75] These experiments provide evidence that the essential changes associated with loss of tensile strength in human arteries can be reproduced hemodynamically in both artery and vein. Aneurysmal dilatation of the afferent artery to an

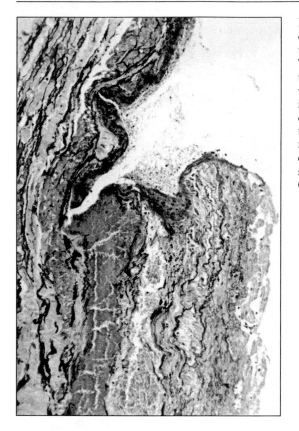

Fig. 6.10. Abrupt tear of a sheep external jugular vein anastomosed to the ipsilateral common carotid artery. The vein wall exhibits much fragmentation of mural elastic laminae with widening of inter-laminar spaces and is dissected with mural thrombus deposition on the exposed floor of the tear following retraction of the tear margins. Verhoeff's elastic stain and eosin. X 58. Reprinted with permission from Stehbens WE. Surg Gynec Obst 1968; 237:327-338.

AV fistula may occur years after surgical closure of the fistula[73,80] and complications associated with coarctation may follow its surgical correction.[21] This evidence, that mural effects of vibrational stress are cumulative and irreversible, is characteristic of fatigue.

Bioengineering Fatigue

Poststenotic dilatations in arteries and aneurysmal dilatation in veins of AV fistulae have been attributed to mechanical or engineering fatigue resulting from prolonged exposure to the associated vibrational stress.[66,73] The phenomenon of mechanical fatigue affects metals which suddenly fracture (failure) possibly preceded by small propagating fractures or cracks at normal operational loads far below the static fracture stress or elastic limit. At a constant level of stress amplitude, fatigue depends on the application of a certain number of load cycles rather than total time under load. Fatigue onset is accelerated by increasing the frequency, stress amplitude or tension and the fracture site is usually smooth and fairly abrupt.[21] It has long been recognized that fatigue affects biological materials such as timber, rubber and bones. March fractures in soldiers date back a century and stress fractures of bones, cartilage, tendons and ligaments are common in joggers, sportspersons and marathon runners.[2,3]

Biological tissues are not immune to the physical laws of nature and no logical reason exists for denying vascular tissue susceptibility to similar long term vibrational stresses. Bioengineering fatigue explains the flow-induced development of poststenotic ectasia, aneurysms, intimomedial tears, dissecting aneurysm and arterial rupture. A similar phenomenon was suggested to explain atherosclerosis and its complications.[81] Even in young subjects with aortic coarctation, aneurysms occur proximal to the coarctation and on collateral vessels. As fusiform and saccular berry aneurysms are unduly prevalent in cerebral arteries, it was reasoned that cerebral arterial aneurysms may also be of similar etiology particularly since hypothetical congenital explanations of their development lack scientific validity.[74,81,82]

The basic mechanism underlying engineering fatigue is that flow-induced tensile stresses cause macromolecules to fragment, each scission occurring at the midpoint and each fragment in turn cleaving centrally and so on.[83,84] In mechanically induced molecular scissions, axial tensile stresses on the chain raise energy levels of bond vibrations within the chain lowering the threshold to bond dissociation and chain cleavage. Each covalent molecular scission produces two free radicals, ordinarily having a short half life and quenched by antioxidants. More complex molecular chains and entangled polymers with highly integrated intermolecular orientations and crosslinks react differently displaying greater resilience to degradation. As a consequence many more stresses are required and midpoint scission is not invariable. Larger stresses involve reorientation of chain segments and opening of voids in the polymer.[85,86] Such cumulative molecular scissions underlie mechanical fatigue development and engineering failure. Molecular defects, voids, inclusions and inhomogenieties in the complex structure of polymers, singly and in variable combinations, predispose to premature onset of fatigue leading to crack formation and stress concentration at such sites with loss of tensile strength and eventual failure (fracture, rupture or tearing) under normal conditions of usage.

The essential difference between engineering and bioengineering fatigue is that vessel walls are viable and comprise many constituents functionally integrated into an architectural design to enable the wall as a whole to fulfill and adapt to functional requirements of the pulsatile flowing blood at high velocity and pressure. Where stresses are greatest, fatigue is likely to occur earlier than elsewhere. Failure of one or more constituents jeopardizes others, the integrity of their function and the whole wall but being viable biological tissue, the wall is capable of repair and to some extent compensation. These responses occur under similar and more severe stresses than those responsible for the initial fatigue effects, so repair is never as satisfactory as original architectural arrangements.[3]

Following examination of serial sections of several hundred arterial forks from humans and lower animals (with and without aneurysms) and flow studies on glass models of arteries with varying geometric configurations in which disturbed flow patterns developed at Reynolds numbers considerably below the critical value for turbulence of 2,000, it was postulated that atherosclerosis was the consequence of hemodynamically induced mechanical fatigue due to mural vibrations evoking compensatory and reparative mural thickening.[81] It was reasoned that fatigue was associated with changes in mural constituents at a molecular level and that the initial lesions involving elastic tissue destruction and intimal thickenings about arterial forks, the latter constituting reparative changes to compensate for the weakness resulting from the elastic tissue destruction. Localization of initial intimal thickenings could only be explained by flow-related stresses. Further experimental work and studies of mural constituents and flow disturbances confirmed the fatigue hypothesis as the unifying thread pervading atherogenesis and its complications which include aneurysms.[2,3,21]

Hemodynamics is a major determinant in the anatomical and architectural mural development of the arterial tree. The initial elastic tissue degeneration and compensatory intimal proliferation commence prior to birth though other degenerative changes including lipid accumulation and calcification may be present in neonates.[34,38] Analysis of hemodynamic stresses explains their transmural effects,[30] the localization of early and severe lesions at forks, junctions and curvatures and lesion progression at variable rates for each individual site.[2,3,21]

In the unloaded state arterial walls are under continual longitudinal tension as evidenced by retraction of the cut edges on transection after their being freed from surrounding tissues. As cut edges also retract when an artery is opened longitudinally, residual stress in both longitudinal and circumferential directions is indicated.[87] The IEL displays longitudinal corrugations in histological sections due to contraction of medial SMCs. These disappear when the artery is distended, the radial compressive stress being 70-80 mm Hg in diastole.

On systolic ejection of blood the aortic pressure rises and initiates transmission of a wide based pulse pressure wave with a velocity of about 4 m/sec in the ascending aorta increasing to about 12 m/sec in the femoral artery[88] while the wavelength is about 4 m.[89] This rapidly moving pulse pressure wave exerts a pressure of at least 40 mm Hg proximally which is higher still in the distal aorta and femoral arteries due to a summation effect of reflected waves. The rapidly moving pulse wave acts like a bougie travelling distally along the lumen accelerating as it travels (Fig. 6.11).[30] Its front (the ascending or anacrotic limb) distends the wall circumferentially and stretching simultaneously endothelium, IEL, each lamellar unit and the adventitia longitudinally till the arrival of the wave peak, then retraction takes place during the descending limb phase with lesser pertubations at the dicrotic notch and during diastole prior to the arrival of each successive wave. Fatigue life is shortened by increases in pulse pressure and rate and when the material is under constant tension the energy required to rupture polymer bonds is reduced.[90]

There is crowding of laminae at the systolic summit of the wave. Attenuation of medial lamellae is greater in the inner than the outer part of the wall. The arterial diameter is reputed to increase 5-10% with each systole,[91] the circumference by 17-25%[92] and the external diameter enlarges only 1-4%.[93] The descending thoracic aorta is said to lengthen by 1% and the abdominal segment to shorten by 1%.[23] The overall measurement does not account for the pulse pressure wave and numerous microwaves and thereby considerable repetitive lengthening of elastic laminae, lamellar units and other mural constituents during pulse wave transmission. Such lengthening would not be apparent in the estimated 1% increase in the overall longitudinal measurements[23] and it has also been suggested that the aorta may be enlarged elliptically in cross section due to nonaxisymetric flow.[94] The differential attenuation of laminae imposes considerable tensile stresses on all mural constituents during pulse wave transmission. Superimposed on these disruptive stresses would be the additional stretching, bending and even torsional stresses on peripheral arteries during body movement and aggravation by physiological and pathological increases in hemodynamic parameters and rapidity of application of the pulse pressure (steep anacrotic limb) and more abrupt decline in aortic valvular incompetence, AV shunts and patent ductus arteriosus.

The considerable force behind these pulse pressure waves is exemplified by the fact it can cause the head or a crossed leg to oscillate with each pulse and a heavy steel bed to shake during ballistocardiography.[95] These forces exert a transmural distending effect and a longitudinal stress that potentiate the development of tortuosities with visible pulsatility at bends. In a water hammer pulse, the pulse pressure may exceed 100 mm Hg (133,300 dynes/cm^2 or 13,330 Pa) which is considerably greater than the estimated endothelial shear stress (15 dynes/cm^2) prevailing in arteries. The pulse pressure in one subject with an AV fistula rose to 146 mm Hg[73] and on auscultation pistol shot sounds may accompany a severe water hammer pulse.

The effect of pulse pressure is more complicated at arterial forks, bends and junctions where secondary effects of disturbed non-axisymmetrical flow patterns demonstrated in fluid mechanical studies[21,96-98] are superimposed. A more slowly developing superimposed effect associated with thinning and disappearance of the IEL with compensatory intimal

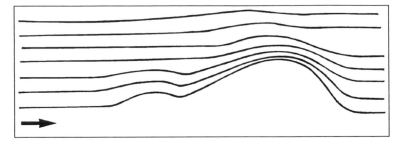

Fig. 6.11. Diagram of elastic artery with the lumen below. The indentation represents the effect of the pulse wave travelling in the direction of the arrow on the elastic laminae (represented by parallel lines) and lamellar units. The effect is like a bougie being driven along the artery but with a much reduced wave length.[30]

proliferation is associated with (i) fluctuating pressures and multidirectional flow in regions of boundary layer separation at lateral angles of bifurcations, the apex of unions and the lesser curvatures of bends and (ii) vortex shedding at the apices of bifurcations, analogous to that of an organ pipe (jet edge effect).[2,3]

The pressure head or kinetic energy associated with the central red cell column of nonaxisymmetric flow impacting on the vessel wall at (i) bifurcations particularly with unexpected imbalance of flow and (ii) the greater curvature of bends complicates the above stresses and correlates with the location of IEL tears and secondarily tears of other elastic elements and loss of medial SMCs possibly culminating in mural atrophy and aneurysms.

Cerebral Berry Aneurysms

IEL tears have also been classed as early lesions of atherosclerosis, the concept stemming from observations that berry aneurysms of cerebral arteries develop at regions of mural thinning (atrophic lesions) adjacent to or involving the apex of bifurcations where the IEL is lost or extensively fragmented (Fig. 6.12) and the underlying media and adventitia are thinned and attenuated (Fig. 6.6). This atrophic thinning progresses and develops into small saccular (Fig. 6.13) or funnel-shaped dilatations with remarkably thin walls. Slight intimal thickening within early sacs may be pre-existing or secondary proliferation at times. The main flux of blood flow at the fork is directed at the fundus where the wall usually remains thin and where rupture most frequently occurs (Fig. 6.14). Intimal proliferation within the sac has a propensity for lateral walls (Fig. 6.14) and resembles that in cerebral arteries. There is a dearth of elastic tissue except in small early aneurysms. The fibromuscular intimal proliferation progresses to overt atherosclerosis indistinguishable from that elsewhere in cerebral arteries except for the absence of the media, a feature erroneously assumed originally to represent a congenitally defective wall.[74]

These aneurysms characteristically exhibit prominent pulsatility with a volumetric increase of $51 \pm 10\%$[99] and high-frequency low amplitude vibrations (460 Hz $\pm$ 130 SD).[100] The bruits being musical suggest sound emanation from vortex shedding[21,96] rather than turbulence. Oscilloscope recordings show systolic accentuation of amplitude as with experimental AV fistulae.[101] This vibratory activity provides incessant repetitive tensile stresses of all mural constituents, and these parameters may be greater for abdominal aortic aneurysms which often display visible pulsatility of the abdominal wall. Whether large extracranial aneurysms in humans also have this high frequency vibrational activity is unknown. Some human aneurysms have been associated with bruits on auscultation but the presence of a related AV shunt cannot be excluded.

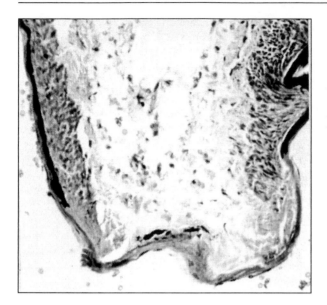

Fig. 6.12. Fragmentation of the IEL at apex of a cerebral arterial fork with early invagination into only part of the medial raphe suggesting that only part of the wall is weak. Verhoeff's elastic stain and eosin. X 225. Reprinted with permission from Stehbens WE. Pathology of the Cerebral Blood Vessels. St. Louis: CV Mosby, 1972:351-470.

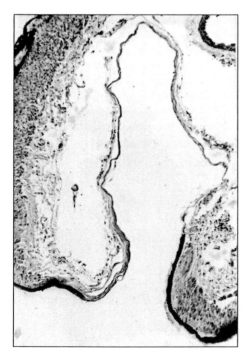

Fig. 6.13. Small cerebral berry aneurysm with flimsy wall without evidence of leakage or rupture. Note irregularity of contour of the IEL and its abrupt termination on each side of the entrance. On the left side the invagination does not commence where the media ceases. Verhoeff's elastic stain and eosin. X 75. Reprinted with permission from Stehbens WE. Pathology of the Cerebral Blood Vessels. St. Louis: CV Mosby, 1972:351-470.

The so-called "medial defects" of Forbus[102] (sites of interruption of the media by a wedge of adventitial fibroelastic tissue) are in reality raphes in relatively acute apical or lateral angles where medial muscle of the adjacent arterial wall is pulled in opposing or disadvantageous directions. Furthermore they increase in prevalence with age and their biological distribution does not correspond to that of aneurysms in cerebral or extracranial

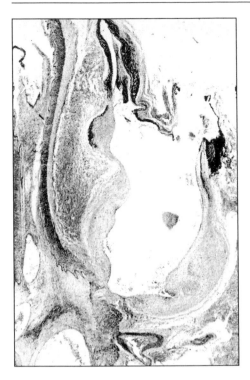

Fig. 6.14. Low magnification of berry aneurysm with rupture of the fundus. Note the intimal proliferation at sides of the sac but thin wall infiltrated with fibrin (staining black) near the rupture site with fibrin lining the wall nearby. Mallory's phosphotungstic acid-hematoxylin. X 18. Reprinted with permission from Stehbens WE. Pathology of the Cerebral Blood Vessels. St. Louis: CV Mosby, 1972:351-470.

circulations or in other animals.[74] Berry aneurysm propensity for cerebral arteries is explained by the architecturally thin media and adventitia of cerebrospinal arteries, human longevity and predisposition to severe atherosclerosis. Other "evidence" used to support the congenital hypothesis was shown to be fallacious.[74]

It has subsequently been demonstrated that atrophic lesions can be produced experimentally in rabbits in afferent arteries of AV fistulae,[42,48,49,57,103] on the greater curvature of bends[36,104] (with their augmented pulsatility) and adjacent to the apex of surgically fashioned bifurcations.[35] The initial manifestation in each instance is the appearance of IEL tears mostly transversely orientated and with a relatively regular periodicity and sharp edges (Figs. 6.15, 6.16). The tears become more numerous and interconnected by longitudinal tears. Medial elastic laminae and even external elastic laminae fragment (Figs. 6.17, 6.19) and the wall thins (Fig. 6.20) as medial muscle degenerates and disappears until eventually only attenuated adventitia remains[36,42] possibly with some elastic tissue remnants. By this mechanism aneurysms develop at the fork, on the stem of arteries afferent to AV shunts both intracranially and extracranially and in arteries opposite the entrance of an extracranial-intracranial bypass graft.[105]

It is well recognized that anatomical variations of the circle of Willis are related to the localization of cerebral berry aneurysms.[74] Some variations correlate better with specific sites of aneurysm development and some indicate aneurysm rarity at other sites.[61] This observation led to the investigation of the effects of carotid artery ligation in the neck on arterial forks within the circle of Willis[106] and the successful hemodynamic production of aneurysms at forks with an imbalance of flow caused by ligating the common carotid artery in hypertensive lathyritic rats.[107] The aneurysms, not always of the berry type, occurred predominantly on arteries associated with collateral flow. Subsequently it was shown that carotid ligation was the essential lesion and hypertension or lathyrism or both could be

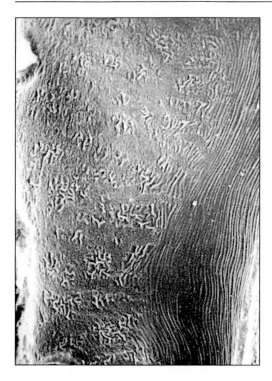

Fig. 6.15. Scanning electron micrograph of intimal surface of rabbit carotid artery. Transversely orientated tears of the IEL with some longitudinally interconnecting tears. Islands of IEL are irregularly folded rather than longitudinally corrugated as at the right of the vessel. X 29.

dispensed with. Early aneurysmal changes were adjacent to the apex in the branch participating in collateral flow and consisted of fragmentation and loss of the IEL, gross medial thinning with eventual aneurysmal bulging of a grossly thinned wall devoid of elastic tissue and most muscle.[107-109]

Abrupt, sustained changes in function or vascular topography are likely to have adverse secondary effects on vessel walls due to their inability to withstand new unaccustomed hemodynamic stresses associated with altered flow.[3,61] This maxim is illustrated in the cerebral circulation for internal carotid ligation proximal to an ipsilateral aneurysm, by lowering the pulse pressure distally, reduces the propensity to hemorrhage from the aneurysm which may progressively enlarge to form a space occupying lesion.[74] Simultaneously it predisposes to berry aneurysms of contralateral arteries participating in the cerebral circulation due to flow imbalance at bifurcations by disturbing the blood flow axis in the parent vessel.[61,107,110]

Increased or decreased run-off from a branch secondarily due to a pathologic or iatrogenic occlusion or an AV shunt produces imbalance of flow with the deflected flux of blood impacting on the vessel wall to the side of the apex to initiate aneurysmal dilatation. In some subjects head turning can interfere with flow through the vertebral or cervical internal carotid artery possibly associated with pathological changes in cervical vertebrae.[111] Increase in the pulse pressure for whatever reason (including hypertension) would potentiate this phenomenon.

More recently a study of experimental production of cerebral berry aneurysms following unilateral carotid ligation in the neck in four strains of rats revealed that genetic propensity to aneurysm production was independent of the predisposition to IEL tears in extracranial arteries and that hypertension but not lathyrism potentiated aneurysm development.[112]

Further support derives from the development of aneurysms on collateral vessels in moyamoya disease, most frequently found in young Japanese and associated with bilateral

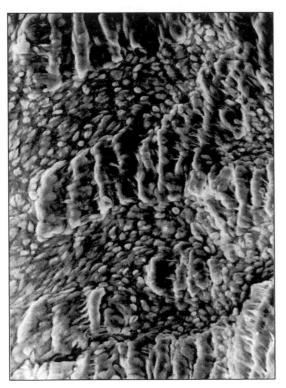

Fig. 6.16. Scanning electron micro-graph of IEL tears showing sharp edges with depressed intervening floor covered by high population density of small endothelial cells. A few interconnecting tears are present. X 180.

occlusion of the internal carotid arteries in the upper cervical segment.[61] Sometimes the occlusion is unilateral and possibly results from arteritis. Angiographically the collateral circulation consists of a large number of dilated and tortuous small vessels from the internal and external carotid, anterior choroidal and posterior communicating arteries and produces an angiomatous vascular network at the base of the brain simulating a rete mirabile or an AV communication. Histologically these small vessels exhibit lipohyalinosis, thrombosed and recanalized vessels, fibrin infiltration of the wall, severe fibrosis and miliary aneurysms which could be responsible for cerebral hemorrhage, although berry aneurysms developing on the large basal arteries produce subarachnoid hemorrhage in young adults. Ultrastructurally moyamoya vessels exhibit severe degenerative changes similar to those in hypertension and atherosclerosis viz. intimal thickening, loss of elastica, vesiculogranular degeneration (matrix vesicles or cell debris) and necrosis of SMCs, medial thinning and much redundant, duplicated basement membrane.[113-116] Tortuosity, ectasia and aneurysms characteristically develop in collateral vessels (arteries or veins) which are prone to hemorrhage following direct rupture or after aneurysmal dilatation. The arterial occlusion appears to be the primary event and hemodynamic stresses associated with augmented flow and pulse pressure are responsible for the mural changes in morphology and loss of tensile strength in the collaterals.

Attributing cerebral aneurysms to undiscovered defects of germ plasma or vestiges of the primordial plexus is a carry-over of the outmoded congenital hypothesis and is based on a lack of understanding of the foregoing pathology and pathogenesis of these aneurysms. Cerebral aneurysms have been attributed to abnormal neural crest development[117] based on the occurrence of cerebral aneurysms or dissecting aneurysms of cerebral or cervical arteries in 14 cases of congenital heart disease in which 10 were likely associated with increased

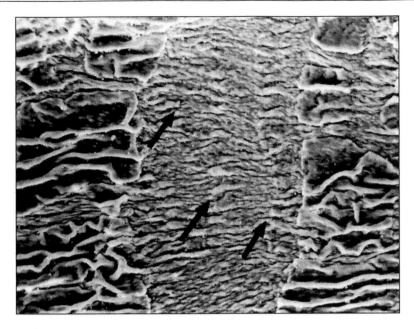

Fig. 6.17. Scanning electron micrograph of a tear of IEL (flow from left to right) which is irregularly folded despite pressure fixation. Transversely orientated ridges (arrows) in the floor of the IEL tear represent similar transversely orientated tears of underlying medial elastic laminae. X 80.

pulse pressure or hypertension. Another two subjects with pulmonary artery stenosis may also have had additional hemodynamic abnormalities (e.g., patent ductus arteriosus). Such weak evidence does not support the hypothesis.

Many authors still believe that some congenital abnormality underlies aneurysm development. Much emphasis is placed on familial occurrence which may be coincidental in view of aneurysm frequency or due to some heritable connective tissue disorder predisposing to or aggravating their development. A deficiency of type III collagen has been invoked[118,119] and whilst this has created controversy,[120,121] it is established fact that inherited disorders associated with connective tissue fragility potentiate mural fatigue and aneurysm development. Some cerebral aneurysms, single or multiple particularly occurring at a young age, could be associated with as yet undisclosed genetic connective tissue disorders but it does not follow that all aneurysms reflect such disorders or that a specific gene for cerebral aneurysm formation exists. Early aneurysm development in the Marfan syndrome is similar to that in other subjects[122] and in experimental lesions. It must be concluded that cerebral aneurysms are acquired degenerative lesions etiologically related to hemodynamic stress and nonspecific complications of biomechanical failure embraced in the widened concept of atherosclerosis. In genuine instances of familial aneurysms, very early development, extreme multiplicity or abnormal location, a possible inherited defect of a mural constituent should be sought but not of berry aneurysms per se.

Experimental Aneurysms

Experimental aneurysms fashioned from external jugular vein autografts by microvascular surgery in rabbits and sheep were designed (i) to produce aneurysm models of viable tissue with walls thinner, probably weaker and more susceptible to arterial hemodynamics than the host artery although with less prior exposure to hemodynamic

Fig. 6.18. External elastic lamina and adventitial elastic fibers are torn abruptly and separated in the absence of foam cells or inflammatory infiltrate in the femoral artery proximal to a femoral AV fistula. Arterial media above. Verhoeff's elastic stain and eosin. X 150.

stress than aneurysms arising spontaneously and (ii) to simulate hemodynamic flow in such configurations in humans though avoiding the gross chemical and physical trauma so often used to produce aneurysms. This was important given that cerebral aneurysms have been alleged to form from defective vascular tissue. Trauma was minimal and healing rapid.

In experimental lateral, berry and fusiform (or spherical) aneurysms fashioned from the external jugular vein of rabbits on herbivorous diets, venous walls developed fibromusculo-elastic phlebosclerotic changes and ultimately overt atherosclerosis.[123,124] However degeneration and loss of elastic tissue was pronounced, correlating with the virtual absence of elastin in human berry aneurysms. Ultrastructurally[47] there was extensive degeneration and necrosis of abnormally shaped SMCs, abundant granulovesicular degeneration with heavy calcium deposits in some vesicles (Fig. 6.8B), lipid accumulation in others and thick and multiple basement membrane material sometimes reticulated and separated from SMCs or lying free in the matrix. A thick layer of reticulated basement membrane developed beneath the endothelium often with patchy separation and partial separation of intercellular spaces. The endothelial disposition suggested a whorling pattern in Haütchen (en face) preparations.[125] Subendothelial connective tissues were also fashioned into whorls with crescentic elevations about depressions suggestive of vortex flow in the jet lesions.[126] Such changes, consistent with flow-related changes in the wall and those of human atherosclerosis, differ markedly from those of dietary cholesterolosis. Fibrin infiltration and slight mural thrombus were also observed in the aneurysms.

Elongated lateral pouch aneurysms fashioned on sheep common carotid arteries progressively enlarged and became spherical exhibiting histological changes similar to those of rabbits although lipid accumulation was less than expected.[127] In advanced stages the walls were extremely fibrotic and almost acellular, only a few elastic remnants remaining near the adventitia. Lipid accumulation, calcification, occasionally ossification, aneurysm growth, rupture, mural tears and laminated thrombus were observed as in human aneurysms. The arterial segment became incorporated within the sac exhibiting fibromuscular intimal

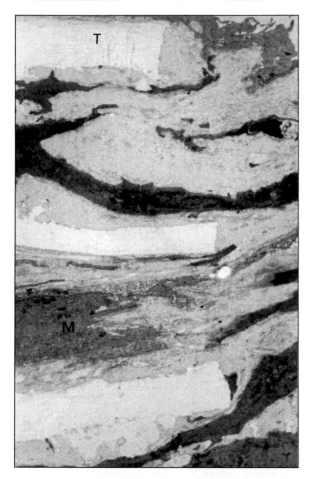

Fig. 6.19. Electron micrograph of a rabbit common carotid artery proximal to a carotid-jugular AV fistula displaying abrupt tears of the IEL and two underlying medial elastic laminae 25 days postoperatively. Many smooth muscle cells exhibit densely staining cytoplasm with shrinkage suggestive of degeneration and necrosis in contrast to the muscle cell (M). Note microfractures (T) mostly on the abluminal surface of the IEL near the tear margin. Such tears do not resemble enzymatic digestion and occur in the absence of monocytes or macrophages. By courtesy of Dr. GT Jones. X 4800.

thickening with considerable elastic tissue loss, lipid accumulation, mural tears, thrombus and fragmentation of medial elastic laminae. Deep tears with dissection developed in either venous or arterial segments (Figs. 6.21, 6.22). Mural tears and rupture did not occur in the presence of foam cell accumulation. Vasa vasorum invaded the thickened wall even penetrating the thrombus but further thrombus organization appeared to be prevented by dissection between the wall and the thrombus indicating some blood may be derived secondarily from small vasa vasorum. Some tears healed and mural laminated thrombus of variable thickness was prevalent. Several aneurysms ruptured. No radially directed tears in the thrombus were found but dissection between the thrombus and sac wall and even of the laminated thrombus was common and attributed to the way laminated thrombus of aneurysms developed with progressive yield of the sac wall proper.[128] It was apparent that the laminated fibrin persisting after dissolution of cellular elements and loss of plasma must be remarkably durable.

Experimental sheep and rabbit aneurysms indicate that augmented hemodynamic stress in aneurysmal sacs is sufficient to initiate degenerative changes in aneurysm walls leading to atherosclerosis and further complicated by mural tears, dissection, rupture and laminated thrombus. Ultrastructural changes in rabbit aneurysms are identical to those occurring in human cerebral atherosclerosis and berry aneurysms though the rate of progression in rabbit

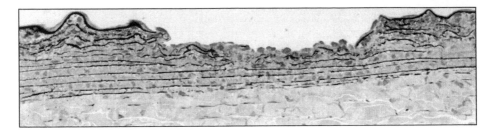

Fig. 6.20. Tear of IEL in rabbit common carotid artery proximal to an AV fistula. Vessel was pressure fixed and medial elastic laminae are straight but IEL is irregularly convoluted. Note disorganized media deep to IEL fragments. Prominent endothelial cells and fragmented medial elastic laminae in floor of tear. Toluidine blue. X 175.

aneurysms was greater than in the sheep model.[127] In rabbit and sheep the collagen fibrils were frayed and misshapen (Fig. 6.8B) as in genetic connective tissue disorders associated with premature arterial fragility and in human atherosclerosis.[52,53] The morphological changes and associated fragility are therefore acquired and hemodynamically induced even in connective tissue disorders.

Internal Elastic Lamina Tears

IEL tears, mostly transversely oriented, are the initial manifestations of progressive atrophy of arterial walls recently classified as atrophic lesions of atherosclerosis. They present early in life on the greater curvature of the carotid siphon and become the nidus for early calcification manifested in the IEL at tear margins with further extension into the lamina. There is also a propensity for lipid accumulation in the floor of the tears[34] where SMCs exhibit degeneration and necrosis experimentally. Similar tears occur on other curvatures e.g. the splenic artery and in common iliac arteries in late fetal life. In infants with a single umbilical artery, tears in the ipsilateral common iliac artery are pronounced and overt atherosclerosis is superimposed on the atrophic lesions within four years postnatally.[34] Analogous tears occurring elsewhere in the aorta and large peripheral arteries during childhood and maturation may be masked by intimal proliferation.[2,3]

In experimentally induced IEL tears, any initial endothelial damage heals rapidly and later tears develop seemingly without endothelial disruption. The initial rupture of trabeculae traversing IEL fenestrae[129] and micro tears on the abluminal surface[130] adjacent to early IEL tears (Fig. 6.19), the persistence of sharp edges of the tears and the absence of detectable elastase though inconsistent with enzymatic elastolysis, are consistent with bioengineering fatigue. Initiation of tears within 5 days in the elastic common carotid artery[103,131] and two days in the muscular femoral artery[104,132] postoperatively reveals the rapidity of lesion development.

With time postoperatively, tears in afferent arteries of AV fistulae increase in frequency and area of involvement such that much of the IEL becomes so fragmented and irregularly convoluted (rather than longitudinally corrugated) that by scanning electron microscopy it appears to be functionally ineffective (Figs. 6.15, 6.20). Underlying elastic laminae are likewise torn, separated or apparently absent (Figs.6.16-6.20) and SMCs undergo degeneration. This type of IEL tear is associated with dissecting aneurysms or occlusive thrombosis in cerebral arteries (Fig. 6.23) and strong experimental evidence indicates hemodynamic induction. They occur preferentially where the flux of blood impinges directly on the vessel wall at

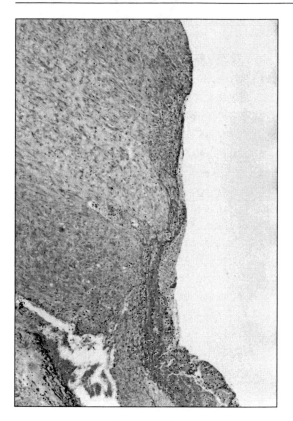

Fig. 6.21. Deep abrupt tear in sheep aneurysm wall with attached thrombus over the torn surface and exposed residual wall. There is some mural dissection and remnants of medial elastic tissue at bottom left. Verhoeff's elastic stain and eosin. X 60. Reprinted with permission from Stehbens WE. Pathology 1997; 29:374-379.

forks and bends and in vessels with unaccustomed, unduly elevated pulse pressure and augmented flow.[2,3]

Hemodynamic stresses have been implicated in IEL tears in caudal and renal arteries of rats: those in caudal arteries are possibly due to bending of the tail. A propensity for stroke in spontaneously hypertensive rats is associated with frequent, early tears in the caudal artery.[133,134] A proclivity to IEL tears, reported in specific breeds of rats and in lathyritic rats,[112,134] suggests that hypertension, genetic and connective tissue disorders predispose to their development. Inability to detect elastolytic activity in afferent arteries to AV fistulae[129] and their ready experimental production merely by altering blood flow supports the concept that IEL tears are due to hemodynamically induced loss of tensile strength as instanced in poststenotic dilatation and dissecting aneurysm in the absence of monocytes or foam cells. The associated elastic tears progressively developing so widely in the arterial circulation throughout life and in dissecting aneurysm are likewise mechanically induced as in poststenotic aneurysm. Any contention that they are physiological adaptive changes to increased blood flow is implausible and inconsistent with the concomitant destructive change in mural architecture of afferent arteries culminating in irregular ectasia, tortuosity and aneurysm formation. More localized aneurysmal dilatation can occur on the greater curvature of bends (as in the carotid siphon) and intimal proliferation over atrophic lesions was considered a compensatory or reparative response to mural atrophy and flaccidity but after a considerable time lag as if SMC replication had been depressed or inhibited. Once dilatation has commenced, augmented hemodynamic stresses accelerate atherogenesis with further dilatation and loss of tensile strength and mural cohesion.

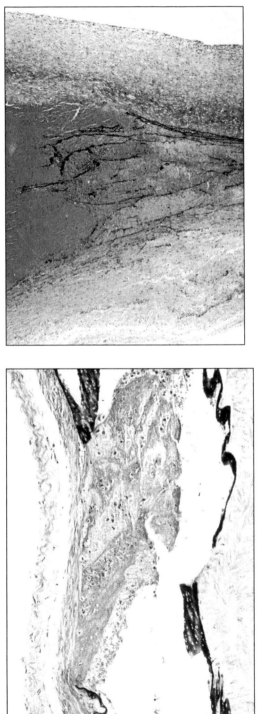

Fig. 6.22. Arterial segment of an experimental sac wall of a sheep displaying intimal thickening, recent mural dissection and extensive medial elastic fragmentation. Lumen is at top. Verhoeff's elastic stain and eosin. X 38. Reprinted with permission from Stehbens WE. Pathology 1997; 29:374-379.

Fig. 6.23. Abrupt tear of IEL in cerebral artery with reflected end of the lamina into the lumen at the left side. Mural thrombus has formed on the exposed medial surface. Verhoeff's elastic stain and eosin.

In anastomosed veins of AV fistulae[46] and in human berry aneurysms[74] accelerated atherosclerosis develops as confirmed experimentally under similar conditions in rabbits and sheep on herbivorous diets.[135] The aneurysmal veins exhibit enhanced permeability to protein-bound Evans blue dye[136] and predilection to lipid accumulation in cholesterol-overfed rabbits.[57] Morphologically the dietary lipid storage disorder (cholesterolosis), rather than accelerating true atherogenesis, is superimposed on the hemodynamically affected wall, nor do aneurysms develop in cholesterol overfed animals.

Similar ectasia, tortuosity and aneurysmal dilatation develop in the abdominal aorta and iliofemoral arteries in humans following above-the-knee amputation.[137] The tortuosity of the aorta and iliofemoral vessels is similarly directed to those vessels in rabbits with femoral AV fistulae depending on the flow imbalance at the aortic bifurcation. In human amputees reflected waves from arteries on the side of the amputation further increase the pulse pressure in the lower aorta thus enhancing aneurysmal development. The higher the amputation, the greater the hemodynamic changes.

Aortic Aneurysms

Noninflammatory and nontraumatic aortic aneurysms, most commonly located in the infrarenal segment of the abdominal aorta, have been considered atherosclerotic in etiology. For more than four centuries there has been argument as to whether they formed by simple dilatation of the aortic coats or resulted from rupture of the inner coats with dilatation of the external coat.[138] The former opinion was seemingly influenced by traumatic aneurysm frequency in older times and probably by the mural dissection about atherosclerotic tears and ulcers with excavation and bulging of the ulcer floor forming a small dome-shaped swelling on the adventitial surface. In 1893 Coates and Auld[139] recognized the phenomenon illustrated in Figures 6.24 and 6.25 where, in the presence of considerable atherosclerotic intimal thickening, medial elastic laminae have torn abruptly like those associated with dissecting aneurysms (Fig. 6.26) in the absence of macrophage accumulation. Retraction of tear margins accompanies variable mural dissection and disruption of mural cellular and other noncellular matrix constituents with many SMCs damaged and dying as occurs with IEL tears experimentally. Fibrosis develops in the region and macrophages or foam cells are not conspicuous. Such abrupt, simultaneous tears of multiple lamellar units are unlikely to derive from enzymatic activity and suggest rather that all the torn elastic laminae and other fibrous matrix constituents simultaneously lost considerable tensile strength as must be the case in aortae exhibiting diffuse elastic fragmentation and cystic medionecrosis in dissecting aneurysms (with or without Marfan syndrome) and poststenotic aneurysms. These are important expressions of mural weakness[140] and Thoma's angiomalacia.[13] Simultaneous tearing of intimal and medial elastic laminae in aneurysmally dilated sheep veins of AV fistulae exhibit dissection and similar elastic changes histologically (Figs. 6.10, 6.27). In experimental sheep aneurysms,[127] tears perpendicular to the endothelial surface and of varying depth penetrate thick fibrous intima with no demonstrable elastin and no local foam cell or monocyte aggregation (Figs. 6.21, 6.22). They are indicative of mural weakness without obvious causal cellular infiltration-inflammatory or otherwise.

Pathologists have always regarded fusiform abdominal aortic aneurysms as atherosclerotic because they develop where atherosclerosis is most severe and in association with generalized advanced atherosclerosis, variable ectasia and tortuosity in conformity with their age. Multiple aneurysms may be present and the severely atherosclerotic common iliac arteries can be displaced, irregularly ectatic or aneurysmal with widening of the bifurcation angle.[2] In early stages they appear as a diffuse slightly expanded lumen with obvious atherosclerosis and variable tortuosity. Anterior curvature and spinal support of the aorta predispose to bulging and mural thinning anteriorly by accentuating pulsatility at the site

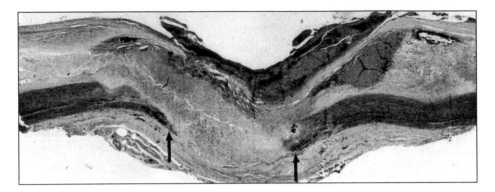

Fig. 6.24. Macrophotograph of an incipient aneurysm underlying intimomedial tear with wide separation of the edges. The arrows indicate the margins of the torn medial elastic laminae and thrombus lies in the floor of mural evagination. Reprinted with permission from Stehbens WE. Progr Cardiovasc Dis 1975; 18:89-102.

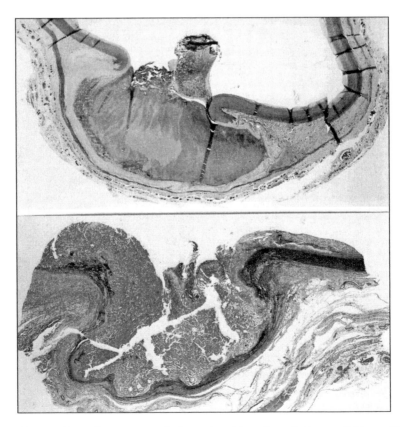

Fig. 6.25. Two small saccular aneurysms in formation. A has abrupt intimomedial tear with extensive mural dissection and bulging on the outer surface of the aorta. B has considerable thinning and tearing of the media beneath an intimal ulcer with external bulging of the aortic wall. Verhoeff's elastic stain and eosin. Reprinted with permission from Stehbens WE. Vascular Pathology. London: Chapman & Hall, 1995:353-414.

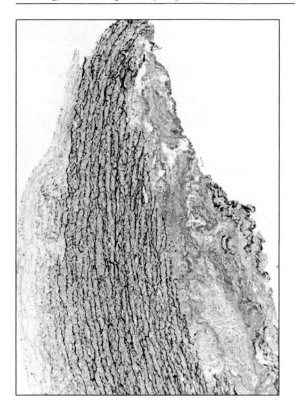

Fig. 6.26. Edge of an aortic intimo-medial tear of a dissecting aneurysm showing numerous medial elastic tears and dissection with thrombus on the disrupted media. Such tears occur on a smaller scale and can undergo healing. Verhoeff's elastic stain and eosin. X 110.

even though marginally.[141,142] Early abdominal aneurysms of this type, by virtue of their mural weakness, disturbed flow patterns,[143] increased mural tension and augmented pulsatility exhibit accelerated atherosclerosis and increasing enlargement as the cumulative effects of biomechanical fatigue gain ascendancy. Reparative compensatory changes are likewise subjected to greatly increased hemodynamic stresses such that they can only retard rather than prevent progression although increase in size is slower than was once believed. Rapidity of growth and dimensions over 6 cm in diameter portend the likelihood of early rupture. They become bulbous or more spherical and large sacs almost invariably contain coarsely laminated, indurated thrombus with occlusion of aortic branches as in severe ulcerative atherosclerosis.[144] With further enlargement the inner surface may contain much black-red thrombus of soft consistency with an irregular channel coursing through the center and occasionally the lumen is occluded. The precise site of rupture is difficult to locate in large aneurysms as surrounding tissues are heavily infiltrated with clotted blood at autopsy. The outermost lamina of thrombus is usually dark suggesting blood dissected between mural thrombus and the sac wall which exhibits variable dissection. As with any aneurysmal rupture, hemorrhage can fortuitously involve adjacent anatomical structures and occasionally the inferior vena cava resulting in an AV shunt.

These aneurysms have not been thoroughly investigated histologically due to their large size. The wall is severely atherosclerotic with much hyaline relatively acellular fibrous tissue and severe medial loss of elastin (Fig. 6.28). Increasing mural tension and augmented pulsatility enhance fatigue, mural fragility and further dilatation. Superficial erosion of the inner surface is common and more severe fibrin infiltration of the intima thickens and

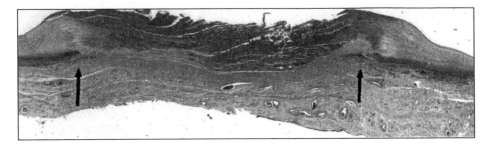

Fig. 6.27. Histological section through an ulcer in the anastomosed phlebosclerotic vein of a sheep AV fistula. The torn elastic laminae of the intimomedial tear are indicated by arrows showing wide separation of the tear margins. Mural thrombus covers the exposed floor of the ulcer so formed and closely resembles the tears and incipient aneurysm of Figure 6.24. Reprinted with permission from Stehbens WE. Progr Cardiovasc Dis 1975; 18:89-102.

possibly reinforces the wall. In larger aneurysms laminated mural thrombus lines the whole sac or much of it. The augmented mural tension and pulsatility accelerate atherosclerotic manifestations of fatigue and continuing sac enlargement culminating in rupture. Menashi et al[145] confirmed the increased collagen and diminished elastin content of the sac wall observing that collagen types were the same as in atherosclerosis.

The contention that in atherogenesis lipid accumulation and foam cell infiltration progressively increase is mistaken. Cellularity in general is severely diminished in endstage disease. The outermost part of the sac wall contains elastic remnants and a variable degree of perivascular accumulation of lymphocytes and plasma cells in the adventitia and perianeurysmal fibrous tissue as is common in severe atherosclerosis.[146] The findings reflect those reported in association with chronic experimental aneurysms[127] and AV fistulae[78] in sheep. Remnants of atherosclerotic aortic media and adventitia often with calcification are found amidst much hematoma and thrombus in ruptured aneurysms. Considerable difficulty is encountered in sampling pure segments of the wall for biochemical or cytological assays with a view to determining the cause of rupture.

Experimental aneurysms in sheep provide valuable etiological information and indicate the wall is prone to deep mural tears with variable dissection (Figs. 6.21, 6.22) which in some instances will prove fatal after premonitory leaks characteristic of aneurysmal rupture. They also demonstrate that foam cells play no demonstrable role in the pathogenesis of aneurysmal wall fragility which can be produced hemodynamically at will.

To suggest that the number of lamellar units in the abdominal aorta predisposes to aneurysm formation is unacceptable when such thin walls devoid of elastic laminae can withstand pulsatile blood pressure for a period of time. Since early in life the aortic IEL is fragmented or lost and a variable number of medial elastic laminae have likewise disappeared and because several lamellar units can be abruptly ruptured in intimomedial tears, the crucial factor has to be the stresses that insidiously over a lifetime cause such tears which are an integral feature of atherogenesis. Invoking a structural weakness specifically of the abdominal aortic wall requires more evidence than assuming a causal role on the basis of the number of medial lamellar units especially as many develop postnatally.

A more likely explanation for the frequent localization of atherosclerotic aortic aneurysms to the abdominal segment is the augmented pulse pressure of the segment due to a reflected wave summation effect. The importance of pulse pressure in producing transmural elastic tissue fragility and fragmentation is seen well in aortic incompetence and

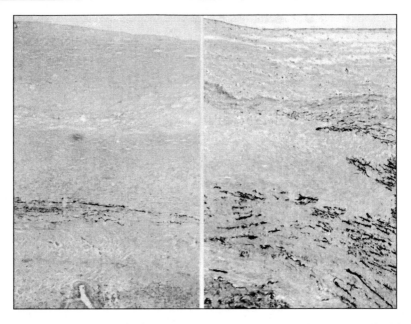

Fig. 6.28. Two sections from a developing fusiform aortic aneurysm showing considerable intimal thickening, much hyaline fibrous tissue with gross loss and fragmentation of medial elastic tissue. Little of the original medial elastin is left in A and B displays very patchy loss of elastin in a thicker segment of the wall. Verhoeff's elastic stain and eosin. X 70.

AV fistulae. Patients with peripheral obstructive vascular disease[147] and those with short legs[148] are alleged to carry an increased risk of these aneurysms.

If atherothromboembolism from the mural thrombus reduces blood flow through one or both lower limbs, the pulse pressure in the abdominal segment is likely to be further augmented by reflected waves thus developing a vicious cycle. Invoking bioengineering fatigue in this scenario of aneurysmal growth is more plausible than attributing growth in size merely to increasing diameter, increasing lateral pressure and greater mural tension with further thinning and yielding of the wall.

Increased collagenase and elastase concentrations have been reported in abdominal aortic aneurysms[149-151] though difficulties arise in sampling walls of ruptured aneurysms and the possibility of confounding results by incorporation of leukocytes and platelets is real. Nor would the presence of these enzymes prove causality because enzymatic digestion is not the sole mechanism of collagen and elastin breakdown. Intimal foam cells, incriminated because of their metalloproteinases, may be involved in breakdown of damaged proteins with reutilization of amino acids and peptides in the process of repair.[3] To suggest foam cells are responsible for aneurysm genesis or rupture is implausible and inconsistent with their presence at other sites without such adverse behavioral and frankly pathogenic activities. Above all they exhibit no such activity in genetic hypercholesterolemia in humans or lower animals nor in experimental cholesterol overfeeding in which aneurysms are notably absent.[2] The latter observation has been used as evidence to suggest that abdominal aortic aneurysms are not due to atherosclerosis but this is a weakness of the lipid hypothesis[152] rather than any proof of nonatherosclerotic origin of these aneurysms. The lack of correlation of aneurysms (aortic and cerebral) with risk factors for CHD is not surprising since CHD can never be synonymous with atherosclerosis, each having a different pathology and pathogenesis.[152]

Enzymatic degradation of arteries to produce aneurysms has been evoked from in vitro studies of canine and human arteries subjected in vitro to higher than physiological pressures whilst being incubated in abnormally high concentrations of elastase or collagenase. The arteries ruptured when treated with collagenase[25] because they had lost tensile strength following enzymatic digestion. The extrapolation to in vivo situations and aneurysms[153,154] is unacceptable. Temporary exposure of arterial segments to high concentrations of enzymes in vivo, like the aforementioned experiments, has limited applicability to spontaneous aortic aneurysms in humans. The experiment does not comply with the logic of Koch's postulates nor explains the localization or precludes atherosclerosis as the cause of aneurysm at any site.

The lower collagen content in atherosclerotic aortic aneurysm walls than in nonaneurysm or occlusive atherosclerotic disease[155] could derive from the thinner aneurysm walls and difficulty in sampling. Enhanced proteolytic degradation is unlikely to be responsible. Tilson[156] postulated that genetic variations of proteases and inhibitors may potentiate aneurysm formation. The alleged pathogenic or causal role of proteases in collagen and elastin degradation has yet to be demonstrated and until substantiating evidence is forthcoming that normal cells lead to their own destruction and the body as a whole, the assumed causal concept has no credence. The possibility of mechanical degradation was never considered and enzymatic degradation is unnecessary in the aneurysmal dilatation of rubber tubing.

A familial occurrence of abdominal aneurysms has been reported by several authors.[156-159] Both an X-linked and autosomal dominant form have been proposed[159] and no pattern of inheritance has been found in a recent investigation.[160] Just how "genetically determined" aortic aneurysms can be differentiated from the "nongenetic" has not been addressed butboth are atherosclerotic. That abdominal aortic aneurysms are distinct from atherosclerosis is an implausible concept when atherosclerosis is ubiquitous. Collagen type III or copper deficiency has been invoked when at most they can be only predisposing factors. While genetic defects of any mural constituent or of an endogenous enzyme or cytokine essential for optimal mural function can conceivably predispose the artery wall to aneurysmal dilatation, in itself it cannot be the sole prerequisite and therefore is not causal.[4] In Marfan syndrome vascular ill-effects of the fibrillin abnormality are not manifest at or before birth. They occur only later in life by predisposing skeletal, pulmonary and vascular tissues to premature fatigue consequent upon normal physiological usage of the tissues. The pathological development of aortic and cerebral berry aneurysms[122] in Marfan syndrome is similar to that in non-Marfan subjects. The fibrillin defect merely increases fatigue vulnerability of the wall when subjected to physiological hemodynamic stresses because the vessel wall is as resilient as its most fatigue-vulnerable mural constituent.[3] Furthermore the age of most subjects with aortic and cerebral aneurysms is well within the age of onset of degenerative diseases.

In Marfan syndrome different types of aneurysm develo at different sites over a period of decades being neither site- nor type-specific. Variation in type is dependent on the rapidity of development of bioengineering fatigue and the site is determined by local hemodynamic stresses peculiar to the site under stress. The stresses vary with the topographical anatomy of the arterial tree, the presence of pathological vascular lesions secondarily affecting flow at the site and the variation in the hemodynamic parameters of pulsatile blood flow. The cause of the aneurysm remains the same as in atherosclerosis, (viz. hemodynamically induced fatigue) the genetic fibrillin defect being a predisposing factor. The extreme vascular fragility in some young adults with Ehlers-Danlos syndrome (type IV) is such that the fetus would be unlikely to survive birth or the neonate to survive infancy. The extraordinary fragility, the misshapen collagen fibrils and other degenerative arterial changes must therefore be acquired manifestations of the disease.

It has been contended that "dilating" disease of the aorta had different pathogenetic determinants from "stenosing" disease and that each had different characteristics and genetics.[161] Reed et al[162] concluded that the determinants were the same in both. The so-called risk factors of atherosclerosis are statistical correlates of CHD[152] which is not pathognomonic of atherosclerosis but is a common cause of morbidity and mortality after surgery for abdominal aortic aneurysms.[163] Both stenosing and aneurysmal lesions are complications of atherogenesis and occur simultaneously in individual patients and therapeutic AV shunts.[46]

A similar genetic concept evolved for the etiology of cerebral aneurysms where the familial occurrence of berry aneurysms and deficiency of type III collagen were also invoked. In a sense the allegation is a carry-over from the outmoded scientific misconception that berry aneurysms resulted from an inherent structural weakness of the arterial wall and it has now been replaced by the current variant—a genetic defect of the mural connective tissue matrix.

Reasons why and how arterial walls become weakened by hemodynamic stresses and experimental hemodynamic reproduction of the essential matrix degradation have been demonstrated. Similarly the initial mural atrophy, berry aneurysm production and the subsequent changes once the aneurysm forms have been explained and reproduced experimentally. For these reasons both types of aneurysm should be considered cmplications of atherosclerosis and thus manifestations of hemodynamically induced ioengineering fatigue. To suggest that abdominal aortic aneurysms are genetically determined is inconsistent with their age incidence because genetic predisposition to mural fatigue becomes manifested generally at a younger age and the difference in incidence in the two sexes diminishes with age as does the severity of atherosclerosis and its complications because of slower progression in females. The rarity of aneurysms in lower animals correlates with their low severity of atherosclerosis: their aneurysms are mostly of infectious or parasitic origin. It is not possible to differentiate ubiquitous atherosclerosis and the histological changes associated with mural weakness from genetically determined changes producing aneurysms. Moreover iatrogenic atherosclerosis and aneurysms are produced at an accelerated rate in the veins of therapeutic bypass grafts and arteriovenous shunts[46,164,165] indicating that only hemodynamic stress can be the factor responsible. If left intact like those elsewhere these veins would have exhibited minimal changes for the remaining years of life.

There can be no specific gene for a nonspecific omplication of many disorders.[3] The wall consists of a heterogeneous mix of disparate constituents each with individualistic susceptibility to fatigue. Any such dilatation must be due to weakening of the wall as a whole and it is logically implausible that a single dominant gene[8] would control the phenotypic expression of all other genes responsible for all mural constituents and their unctions even if only in the aorta and at a specific site. This is also true for cerebral berry aneurysms. The genetic defect in Marfan syndrome is of one mura constituent, viz. fibrillin, and in Ehlers-Danlos syndrome type IV, there is a lack of collagen type III. In both aortic and ceebral berry aneurysms deficiency of type III collagen has been invoked although in the absence of other manifestations of Ehlers-Danlos syndrome. The inecapable conclusion is that both types of aneurysm may be associated with important genetic defects which, while they may predispose the wall to early fatigue onset, are not causes of aneurysms per se and neither hypertension nor hypercholesterolemia are obligatory antecedents. Some subjects with dominant plasma membrane fragility or biochemical disturbances interfering with physiological function of cellular constituents may enhance the "atheromatous" response and those with genetically determined short fatigue endurance of one or more fibrous matrix constituents are more likely to have pronounced and early ectasia, aneurysms and mural dissection accompanied by manifestations of each defect in other similarly affected tissues. These are probably only extremes of a range with many intervening variants, hemodynamic stresses modifying the outcome on a local basis.

With cumulative molecular scissions and simultaneous production of free radicals in the disintegration of matrix constituents, the molecular fragments (even those with crosslinks) may bond with other proteins and lipid oxidation products producing molecular structures in some respects resembling normal mural constituents.[3] Similarly flow-induced dispersion of lipid from low density lipoprotein (LDL) particles or modification of the associated apolipoprotein B100 and other constituents may be suggestive of compositional and structural changes.

Autoxidation products of LDL as well as those of connective tissue proteins can be immunogenic and responsible for the presence of low grade perivascular lymphoplasmacytic infiltration in advanced atherosclerosis. Such a process is not specific for abdominal aortic aneurysm as alleged[147] and seems to be the possible explanation for the so-called molecular mimicry and autoimmune responses invoked in recent years.[166-168]

Miscellaneous Aneurysms

Aneurysms in the proximal aorta develop most often in the ascending aorta where ectasia or a more distinct fusiform aneurysm may occur with a normal aortic valve and hypertension. The prevalence of ectasia and aneurysm of this segment is greatly enhanced by a bicuspid aortic valve and incompetence[169] and by aortic valve stenosis (as a poststenotic dilatation). Medial elastic degenerative changes are severe with elastic laminae diffusely fragmented or patchy as in cystic medionecrosis (Fig. 6.5A). Dilatation of the aortic ring with valvular incompetence and ectasia of the ascending aorta (annuloaortic ectasia) usually indicates Marfan syndrome or a forme fruste. Such lesions may be complicated by dissecting aneurysm or progress to gross aneurysmal dilatation and involve the arch. Hypertension in aortic coarctation, possibly associated with an aortic bicuspid valve and valve incompetence, can potentiate such an aneurysm, a dissecting aneurysm or a cerebral berry aneurysm. The accompanying high pulse pressure has been invoked as a major determinant of early profound elastic tissue destruction at both sites. Syphilitic aortic aneurysms may be of hemodynamic origin secondary to aortic valve incompetence.[61]

The pulmonary artery is a large artery without any propensity to aneurysm formation and with atherosclerosis and hemodynamic stresses much less severe than in the aorta. Aneurysms and dissecting aneurysms involve the pulmonary trunk in the presence of pulmonary valve stenosis, pulmonary hypertension or patent ductus arteriosus thus indicating the importance of hemodynamic stress in these lesions.

While acknowledging that hypertension is an aggravating factor in aneurysmal dilatation of the aorta and cerebral arteries, and though associated periodic pressure increases accompanying exertion or emotional stress can cumulatively contribute, it is not an essential antecedent in aneurysmal development or rupture.

Similar degenerative changes to those of atherosclerosis of large vessels including aneurysms occur in small arteries and are accentuated by pulsatile hemodynamic stress,[170] though the degenerative changes in umbilical arteries including aneurysms, tortuosity and severe elastic loss require more detailed study. Perhaps the biggest enigma is the fusiform dilatation constituting the carotid sinus, there being no obvious physiological benefit to barometric sensors or the carotid body.

Complications

Primary complications of aneurysms are (i) rupture with fatal hemorrhage a strong posibility, (ii) enlargement of the dilatation to cause local pressure effects, erosion of neighboring structures or generalized pressure effects when in a confined space and (iii) embolism of mural thrombi. References 47 and 74 cover these phenomena in more detail. Several general aspects regarding the pathogenesis of these complications and aneurysm etiology require discussion.

The relationship of trauma to aneurysm rupture is pertinent. Little is known of the stresses and strains imposed on large and small arteries (whether superficial or encased within the skull) at the moment of impact. Laceration of arteries can result from blunt indirect trauma even in closed injuries and some sites may be unduly vulnerable. Such stresses may also precipitate rupture of aneurysms in imminent danger of leaking and abdominal aortic aneurysms are vulnerable to cannulae or deep abdominal palpation.

Patel[171] reported the possible tamponade effect of an organizing aortic perianeurysmal hematoma confined by adventitial tissue and mesentery. In the case of cerebral arteries with their thin adventitia and limited arachnoidal or pial fibers, a relatively thin layer of fibrin of variable thickness supported by a most flimsy layer of collagen can breach the gap (Fig. 6.29) withstanding aneurysmal hemodynamics for some time. In contrast thick layers of aortic intima and media apparently readily tear or ulcerate. More than loss of some elastic tissue and collagen is needed for aneurysms to form, and architectural structure or thickness of the vessel wall are poor indicators of mural tensile strength. Spider silk is one of the strongest known filaments for weight and size. It is therefore fitting that fibrin threads so important in clots, thrombi and sealing wounds should be tested for tensile strength and more importantly for fatigue resistance. It is suggested that lining or filling leaking cerebral aneurysmal sacs with fibrin or an autologous fibrin sealant could prove more efficacious modes of therapy than the current introduction of foreign bodies.

The accentuated pulsatility for which large abdominal aneurysms are renowned has the capacity to progressively and insidiously erode adjacent vertebral bodies and less extensively, the fibrous intervertebral discs. Static pressure alone can cause bone thickening though with some architectural rearrangement, erosion is more likely to result from the repetitive tensile stresses associated with aneurysmal pulsatility. Such vertebral erosion has not been thoroughly investigated nor has erosion of skin by aneurysms or even varices in the absence of trauma. The vibrational action of the varix could weaken dermal and subdermal connective tissues instead of a simple pressure effect with blood bursting through the skin by force.

Secondary Infection

Though infrequent, acute secondary hematogenous infection of aneurysms primarily of the abdominal aorta can occur,[146] there being usually a known source of infection. Not widely appreciated is that platelets constitute an important, early line of defense against circulating microorganisms. They tend to adhere to them in the process of clearing the blood forming platelet-leukocyte microthrombi.[172] Consequently platelets with adherent microorganisms may readily attach and become incorporated in microthrombi on cardiac valves or mural thrombi in atherosclerotic aortae, aneurysms or the sites of vascular trauma.[173]

Interest has developed in viral infection of abdominal aortic aneurysms or ulcerated atherosclerotic plaques. Finding serological evidence of endemic Chlamydia and Herpes type viruses in some though not all abdominal aortic aneurysms should not be surprising in an aging population particularly using improved technological laboratory facilities. They may be implicated only secondarily rather than causing aneurysms or their complications. On the other hand these viruses are so common in the population that few may escape

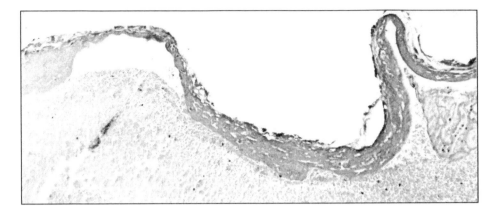

Fig. 6.29. Portion of a false sac at the fundus of a ruptured aneurysm showing a thin layer of fibrin of variable thickness supported externally (above) by only a flimsy layer of connective tissue hardly discernible in places. Below is blood clot and a small thrombus to the right. Verhoeff's elastic stain and eosin. X 112.

infection during a lifetime, little being known of their latency or occult infections. Circulating in the blood their presence in an aneurysm may be enhanced by mural thrombus and the extravasated blood following mural dissection or rupture. Sampling then becomes a confounding factor and rather than assuming causation for a site-specific aneurysm, more thorough investigation is required. However in view of the experimental reproduction of the pathogenetic changes leading to aneurysm formation and similar but less rapid changes in human systemic arteries and atherosclerosis, viruses are unlikely to be causal in atherogenesis or atherosclerotic aneurysms. Secondary infection of these aneurysms will be associated with much more pronounced inflammatory infiltration of the aneurysm than that seen in conventional aneurysms as was the case in the subject reported by Walton et al.[174] This patient died of severe viral respiratory infection. Careful histological assessment is required by experienced vascular pathologists, for any extrapolation from findings in endstage disease to etiology is unwise.

Some abdominal aortic aneurysms are associated with considerable retroperitoneal fibrosis with many chronic inflammatory cell (lymphoplasmacytic) infiltrations, lymphoid germinal centers and giant cells. There may be arteritis of vasa vasorum, other vessels in the neighborhood including the aorta, the aneurysm and visceral vessels including coronary arteries;[175] this is a multivessel involvement reminiscent of the multivessel arteritis common in primates. Inflammatory aortic aneurysms occur in the elderly and, accompanying retroperitoneal fibrosis, may be manifestations of systemic vasculitis or autoimmune disease such as systemic lupus erythematosus.[176] They can accompany ankylosing spondylitis.[175] The fibrous encasement of the aorta, inferior vena cava and sometimes ureter is unlike atherosclerosis at any site so careful histopathological appraisal is mandatory. Confusion between retroperitoneal fibrosis and atherosclerosis has led some authors to mistakenly assume that the former is a variant of the latter and that both are extremes of a spectrum of changes. Atherosclerosis is by far the most common cause of aortic aneurysm and the presence of some immigrant monocytes early in life to phagocytose lipid attractant matrix vesicles (cell debris) and a few chronic inflammatory cells in the intima with larger but variable numbers about adventitial vasa vasorum in endstage disease does not make atherosclerosis an inflammatory disorder.[2,3]

General Comments

In children and young adults, aneurysms (aortic or otherwise) are most likely to be associated with a connective tissue disorder, coarctation of the aorta, trauma, infection and inflammatory diseases, severe hypertension, aortic valve disease or an arteriovenous fistula. Undue multiplicity of aneurysms at a very young age and in the presence of generalized ectasia is suggestive of a connective tissue disorder particularly in the presence of other manifestations and a strong family history of the genetic disorder. Those at unusual sites in the absence of an arteriovenous shunt and not associated with collateral flow require investigation for an infetion or trauma.

In recent years a propensity to cerebral or extracranial arterial aneurysms has been reported accompanying a variety of diseases (type I neurofibromatosis, 3-M syndrome, osteogenesis imperfecta, acid maltase deficiency, sickle cell anemia, fibromuscular dysplasia) with hypertension such that the possibility exists that a connective tissue or other metabolic disorder may be associated with vascular fragility underlying their development. These aneurysms and those of very sall vessels are briefly reviewed in reference 61. They have not been investigated in depth but the generalizations concerning hemodynamics and mural connective tissue failure here expounded are equally applicable.

There is no evidence of inherent mechanical weakness of the wall. If existing, it is unlikely to be genetically determined and even less likely to be site specific. Common misconceptions of atherosclerosis as a lipid metabolic disorder and its alleged restriction to the intima[177,178] rather than recognition of its transmural involvement are partially responsible for the controversy related to aneurysms and atherosclerosis. Aneurysms at specific sites cannot be considered in isolation.

Any causal hypothesis must explain the acquired vascular fragility and loss of mural cohesion that is the unifying thread in the etiology and pathogenesis of atherosclerosis and its primary pathological complications including aneurysms. Hypotheses must be biologically plausible and in the context of vascular pathology. The presence of or even covariance with proteolytic enzymes, inhibitors and other mediators does not warrant assumed causation of pathological lesions. Speculative tissue destruction by proteolytic enzymes resulting in pathological weakness, aneurysmal dilatation and tissue disruption is contrary to the very essence of biological survival and misleads the unwary. Any assumed role of specific constituents (monocytes or enzymes) must be consistent with their presence in other tissues and pathological states and presupposes a sound knowledge of vascular pathology. Repair of proteins equires proteases and deficiency of protease inhibitors (e.g., α_1-antitrypsin) may be detrimental to many diseases without having any specific casual relationship to aneurysms. It follows that benefit from administration f protease inhibitors to aneurysm victims cannot be assumed.

This review indicates the pathological changes in mural architecture, differing experimental models and types of aneurysm, though varying in hemodynamic stresses, are remarkablysimilar. Elastic tissue fragmentation and loss in aneurysms must be considered in the light of similar and progressive changes throughout the arterial circulation and not in isolation. Evidence favoring biomechanical degradation of connective tissue in atherosclerosis and the genesis of aneurysms is strong.[2,3,21,61] The pathological mural changes associated with the types of aneurysm discussed can be reproduced experimentally under conditions analogous to those prevailing in humans. They also comply with the logic of Koch's postulates, the ultimate proof of causality and mirror the histological and ultrastructural features that characterize atheroclerosis. Strong support derives from human vascular pathology and the iatrogenic production of aneurysms by hemodynamic stress, the dominant factor in their etiology, pathogenesis and localization. The cause of these spontaneous aneurysms is, in the scientific and pathological sense, bioengineering mural fatigue as it is in poststenotic aneurysmal dilatation of aortae.

References

1. Sevitt S. Arterial wall lesions after pulmonary embolism, especially ruptures and aneurysms. J Clin Pathol 1976; 29:665-674.
2. Stehbens WE. Atherosclerosis and degenerative diseases of blood vessels. In: Stehbens WE, Lie JT, eds. Vascular Pathology. London: Chapman & Hall, 1995:175-269.
3. Stehbens WE. The pathogenesis of atherosclerosis: A critical evaluation of the evidence. Cardiovasc Pathol 1997; 6:123-153.
4. Stehbens WE. Causality in medical science with reference to coronary heart disease and atherosclerosis. Perspect Biol Med 1992; 36:97-119.
5. Chappell DC, Varner SE, Nerem RM et al. Oscillatory shear stress stimulates adhesion molecule expression in cultured human endothelium. Circ Res 1998; 82:532-539.
6. Gréhant N, Quinquaud H. Mesure de la pression nécessaire déterminer la rupture des vaisseaux sanguins. J Anat Physiol 1885; 21:287-297.
7. Yamada H, Evans FG. Strength of Biological Materials. Huntington NY: RE Kreiger Pub Co., 1973:114-116.
8. Tilson MD. Atherosclerosis and aneurysm disease. J Vasc Surg 1990; 12:371-372.
9. Aschoff L. Introduction In: Cowdry EV, ed. Arteriosclerosis. New York: Macmillan, 1933:1-18.
10. Wilens SL. The postmortem elasticity of the adult human aorta. Its relation to age and to the distribution of intimal atheromas. Am J Pathol 1937; 13:811-834.
11. Hass GM. Elastic tissue. III Relations between the structure of the aging aorta and the properties of the isolated aortic elastic tissue. Arch Pathol 1943; 35:29-45.
12. Ranke O. ‹ber die ƒnderung des elastischen Widerstandes der Aortenintima und ihre Folgen f,r die Entstehung der Atheromatose. Zieglers Beitrager 1923; 121:78-98.
13. Thoma R. Text-Book of General Pathology and Pathological Anatomy. (Transl by Bruce A) London: Adam and Charles Black, 1896:246-247, 384-399.
14. Learoyd BM, Taylor MG. Alterations with age in the viscoelastic properties of human arterial walls. Circ Res 1966; 18:278-292.
15. Johnson WTM, Salenga G, Lee G et al. Arterial intimal embrittlement. A possible factor in atherogenesis. Atherosclerosis 1986; 59:161-171.
16. De Palma RC, Clowes AW. Interventions in atherosclerosis: a review for surgeons. Surgery 1978; 84:175-189.
17. Robertson JS, Smith HV. Analysis of certain factors associated with production of experimental dissection of aortic media, in relation to pathogenesis of dissecting aneurysm. J Pathol Bacteriol 1948; 60:43-49.
18. Hirst AE, Johns VJ. Experimental dissection of media of aorta by pressure. Circ Res 1962; 10:897-903.
19. Lendon CL, Davies MJ, Born GVR et al. Atherosclerotic plaque caps are locally weakened when macrophage density is increased. Atherosclerosis 1991; 87:87-90.
20. Burleigh MC, Briggs AD, Lendon CL et al. Collagen types I and II, collagen content, GAGs and mechanical strength of human atherosclerotic caps: span-wise variations. Atherosclerosis 1992; 96:71-81.
21. Stehbens WE. Hemodynamics and diseases of systemic blood vessels. In: Stehbens WE, ed. Hemodynamics of the Blood Vessel Wall. Springfield: CC Thomas, 1979:294-427.
22. Winternitz MC, Thomas RM, Le Compte PM. The Biology of Arteriosclerosis. Springfield: CC Thomas, 1938:18.
23. Dobrin PB. Biomechanics of arteries and veins. Mechanical properties. In: Abramson DI, Dobrin PB, eds. Blood Vessels and Lymphatics in Organ Systems. Orlands: Academic Press, 1984:64-70.
24. Dobrin PB, Baker NH, Gley WC. Elastolytic and collagenolytic studies of arteries. Arch Surg 1984; 119:405-409.

25. Dobrin PB. Pathophysiology and pathogenesis of aortic aneurysms. Surg Clin Nth Am 1989; 69:687-703.
26. Greenhalgh RM, Taylor GW, Kaye J. A comparison of fasting serum lipid concentrations and lipoprotein patterns in patients with stenosing and dilating forms of peripheral vascular disease. J Cardiovasc Surg 1975; 16:150-151.
27. Lobato AC, Puech-Leão. Predictive factors for rupture of thoraco-abdominal aortic aneurysms. J Vasc Surg 1998; 27: 446-453.
28. Halpern JV, Kline RG, D'Angelo AJ et al. Factors that affect the survival role of patients with ruptured abdominal aortic aneurysms. J Vasc Surg 1997; 26:939-948.
29. Le Veen HH, Diaz C, Christoudias G. The postendarterectomy intimal flap. Arch Surg 1973; 107:664-668.
30. Stehbens WE. Structural and architectural changes during arterial development and the role of hemodynamics. Acta Anat 1996; 157:261-274.
31. Schlatmann TJM, Becker AE. Histologic changes in the normal aging aorta: inmplications for dissecting aortic aneurysm. Am J Cardiol 1977; 39:13-20.
32. Schlatmann TJM, Becker A. Pathogenesis of dissecting aneurysm of the aorta. Comparative histopathologic study of significance of medial changes. Am J Cardiol 1977; 39:21-26.
33. Nakashima Y, Shiokawa Y, Sueishi K. Alterations of elastic architecture in human aortic dissecting aneurysm. Lab Invest 1990; 62:751-760.
34. Meyer WW, Walsh SZ, Lind J. Functional morphology of arteries during fetal and post-natal development. In: Schwartz CJ, Werthessen NJ, Wolf S, eds. Structure and Function of the Circulation. Vol 1. New York: Plenum Press, 1980:95-379.
35. Stehbens WE, Martin BJ, Delahunt B. Light and scanning electron microscopic changes in experimental arterial forks in rabbits. Int J Exp Pathol 1991; 72:183-193.
36. Stehbens WE. Experimental arterial loops and arterial atrophy. Exp Mol Pathol 1986; 44:177-189.
37. Matsuda I, Niimi H, Moritake K et al. The role of hemodynamic factors in arterial wall thickening in the rat. Atherosclerosis 1978; 29:363-371.
38. Stehbens WE. Focal intimal proliferation in the cerebral arteries. Am J Pathol 1960; 36:289-301.
39. Stehbens WE. Histopathology of cerebral aneurysms. Arch Neurol 1963; 8:272-285.
40. Stehbens WE. Cerebral atherosclerotis Arch Pathol 1975; 99:582-591.
41. Stary HC, Blankenhorn DH, Chandler AB et al. A definition of the intima of human arteries and its atherosclerosis-prone regions. Arterioscler Thromb 1992; 12:120-134.
42. Stehbens WE. Experimental induction of atherosclerosis associated with femoral arteriovenous fistulae in rabbits. Atherosclerosis 1992; 95:127-135.
43. Stehbens WE, Ludatscher RM. Ultrastructure of the renal arterial bifurcation of rabbits. Exp Mol Pathol 1973; 18:50-67.
44. Stehbens WE. The ultrastructure of the anastomosed vein of experimental arteriovenous fistulae in sheep. Am J Pathol 1974; 76:377-400.
45. Stehbens WE. Ultrastructure of aneurysms. Arch Neurol 1975; 32:798-807.
46. Stehbens WE, Karmody AM. Venous atherosclerosis associated with arteriovenous fistulas for hemodialysis. Arch Surg 1975; 110:176-180.
47. Stehbens WE. The ultrastructure of experimental aneurysms in rabbits. Pathology 1985; 17:87-95.
48. Jones GT, Stehbens WE. The ultrastructure of arteries proximal to chronic experimental carotid-jugular arteriovenous fistulae in rabbits. Pathology 1995; 27:36-42.
49. Jones GT, Stehbens WE. Ultrastructure of the afferent arteries of experimental femoral arteriovenous fistulae in rabbits. Pathology 1995; 27:333-338.
50. Stehbens WE. The role of lipid in the pathogenesis of atherosclerosis. Lancet 1975; 1:724-727.
51. Greenhill NS, Presland MR, Rogers KM et al. X-ray microanalysis of mineralized matrix vesicles of experimental saccular aneurysms. Exp Mol Pathol 1985; 43:220-232.
52. Stehbens WE, Martin BJ. Ultrastructural alterations of collagen fibrils in blood vessel walls. Connect Tiss Res 1993; 29:319-331.

53. Martin BJ, Leppien B, Stehbens WE. Changes in collagen fibril morphology in experimental aneurysms and arteriovenous fistulae in sheep. Int J Exp Pathol 1993; 74:267-274.

54. Parry DAD, Craig AS. Growth and development of collagen fibrils in connective tissue. In: Ruggeri A, Mota PM, eds. Ultrastructure of the Connective Tissue Matrix. Boston: Martinus Nijhoff Pub Co., 1984:34-64.

55. Rogers KM, Stehbens WE. The morphology of matrix vesicles produced in experimental arterial aneurysms in rabbits. Pathology 1986; 18:64-71.

56. Stehbens WE, Ludatscher RM. The susceptibility of renal arterial forks in rabbits to dietary-induced lipid deposition. Pathology 1983; 15:475-485.

57. Stehbens WE. Experimental arteriovenous fistulae in normal and cholesterol-fed rabbits. Pathology 1973; 5:311-324.

58. Stehbens WE. Predilection of experimental arterial aneurysms for dietary-induced lipid deposition. Pathology 1981; 13:735-747.

59. Halsted WS. An experimental study of circumscribed dilation of an artery immediately distal to a partially occluding band, and its bearing on the dilation of the subclavian artery observed in certain cases of cervical rib. J Exp Med 1916; 24:271-285.

60. Holman E. On circumscribed dilation of an artery immediately distal to a partially occluding band: poststenotic dilation. Surgery 1954; 36:3-24.

61. Stehbens WE. Aneurysms. In: Stehbens WE, Lie JT, eds. Vascular Pathology. London: Chapman & Hall, 1995:353-414.

62. Bruns DL, Connolly JE, Holman E et al. Experimental observations on post-stenotic dilatation. J Thorac Cardiovasc Surg 1959; 38:662-669.

63. Robicsek F, Sanger PW, Taylor FH et al. Pathogenesis and significance of post-stenotic dilatation in great vessels. Ann Surg 1958; 147:835-844.

64. Robicsek F. Post-stenotic dilatation of the great vessels. Acta Med Scand 1955; 61:481-485.

65. Edwards JE, Christensen NA, Clagett OT et al. Pathologic considerations in coarctation of the aorta. Proc Staff Meet Mayo Clinics 1948; 23:324-332.

66. Holman E. The obscure physiology of poststenotic dilatation: its relation to the development of aneurysms. J Thorac Surg 1954; 28:109-133.

67. McKusick VA, Logue RB, Bahnson HT. Association of aortic valvular disease and cystic medial necrosis of the ascending aorta. Circulation 1957; 16:188-194.

68. Heath D, Edwards JE, Smith LA. The rheologic significance of medial necrosis and dissecting aneurysm of the ascending aorta in association with calcific aortic stenosis. Proc Staff Meet Mayo Clin 1958; 33:228-234.

69. Magarey FR, Stehbens WE, Sharp AA. Effects of experimental coarctation of the aorta on the blood pressure of sheep. Circ Res 1959; 7:147-150.

70. Stehbens WE, Liepsch D, Poll A et al. Recording of unexpectedly high frequency vibrations of blood vessel walls in experimental arteriovenous fistulae in rabbits using a laser vibrometer. Biorheology 1995; 32:631-641.

71. Schenk WG, Bahn RA, Cordell AR et al. The regional hemodynamics of experimental acute arteriovenous fistulas. Surg Gynec Obstet 1957; 105:733-740.

72. Ingebrigtsen R, Krog J, Leraand S. Velocity and flow of blood in the femoral artery proximal to an experimental arterio-venous fistula. Acta Chir Scand 1962; 124:45-53.

73. Holman E. Abnormal Arteriovenous communications. 2nd. Edn. Springfield: CC Thomas, 1968.

74. Stehbens WE. Pathology of the Cerebral Blood Vessels. St Louis: CV Mosby, 1972:351-470.

75. Stehbens WE. Abnormal arteriovenous communications and fistulae. In: Stehbens WE, Lie JT, eds. Vascular Pathology. London: Chapman & Hall, 1995:517-552.

76. Burch GE, De Pasquale NP. The anomalous development of left coronary artery. An experiment of nature. Am J Med 1964; 37:159-161.

77. Stehbens WE, Sahgal KK, Nelson L et al. Aneurysm of the vein of Galen and diffuse meningeal angiectasia. Arch Pathol 1973; 5:333-335.

78. Stehbens WE. Haemodynamic production of lipid deposition, intimal tears, mural dissection and thrombosis in the blood vessel wall. Proc Roy Soc (London) Ser B. 1974; 185:357-373.

79. Stehbens WE. The role of hemodynamics in the pathogenesis of atherosclerosis. Progr Cardiovasc Dis 1975; 18:89-103.
80. Eisenbrey AB. Arteriovenous aneurysm of the superficial femoral vessels. JAMA 1913; 61:2155-2157.
81. Stehbens WE. Intracranial Arterial Aneurysms and Atherosclerosis. Thesis, University of Sydney, 1958.
82. Stehbens WE. Etiology and pathogenesis of intracranial arterial aneurysms In: Fox JL ed. Intracranial Aneurysms. Vol 1, New York: Springer-Verlag, 1983,358-395.
83. Odell JA, Keller A, Miles MJ. A method for studying flow-induced polymer degradation: verification of chain halving. Polymer Commun 1983; 24:7-10.
84. Odell JA, Keller A. Flow-induced chain fractures of isolated linear macromolecules in solution. J Polymer Sci 1986; 24:1889-1916.
85. Kausch HH. Polymer Fracture. 2nd. Edn. Berlin: Springer-Verlag, 1987:141-162.
86. Odell JA, Keller A, Muller JA. Thermomechanical degradation of macromolecules. Colloid Polymer Sci 1992; 270:307-324.
87. Chuong CJ, Fung YC. On residual stress in arteries. J Biomech Engin 1986; 108:189-192.
88. O'Rourke MF. Arterial Function in Health and Disease. Edinburgh: Churchill Livingstone, 1982:41-43.
89. Caro CG, Pedley TJ, Schroter RC et al. The Mechanics of the Circulation. Oxford: Oxford University Press, 1978.
90. Nguyen TQ, Kausch H-H. Mechanochemical degradation in transient elongational flow. Advances in Polymer Sci 1992; 100:73-182.
91. Ku DN, Zhu C. The mechanical environment of the artery. In: Sumpio BE, ed. Hemodynamic Forces and Vascular Cell Biology. Austin: RG Landes Co, 1993:1-23.
92. Sumpio BE. Mechanical stress and cell growth. J Vasc Surg 1989; 10:570-571.
93. Cliff WJ. Blood Vessels. Cambridge: Cambridge University Press, 1976:86, 142-144.
94. Moreno MR, Moore JE, Meuli R. Cross-sectional deformation of the aorta as measured with magnetic resonance imaging. J Biomech Eng 1998; 120:18-21.
95. Dock W, Mandelbaum H, Mandelbaum RA. Ballistocardiography. St Louis: CV Mosby, 1953:203.
96. Stehbens WE. Flow in glass models of arterial bifurcations and berry aneurysms at low Reynolds numbers. Quart J Exp Physiol 1975; 60:181-192.
97. Stehbens WE, Fee CJ. Hydrodynamic flow in U-shaped and coiled glass loops simulating carotid arterial configurations. Angiology 1985; 36:442-451.
98. Stehbens WE, Stehbens GR. Flow in glass models simulating vascular junctions under steady flow conditions. Quart J Exp Physiol 1985; 70:515-526.
99. Meyer FB, Huston J. Riederer SS. Pulsatile increases in aneurysm size determined by cine phase-contrast MR angiography. J Neurosurg 1993; 78:879-883.
100. Ferguson GG. Turbulence in human intracranial saccular aneurysms. J Neurosurg 1970; 33: 485-497.
101. Simkins TE, Stehbens WE. Vibrations recorded from the adventitial surface of experimental aneurysms and arteriovenous fistulas. Vasc Surg 1974; 8:153-165.
102. Forbus WD. On the origin of miliary aneurysms of the superficial cerebral arteries. Bull Johns Hopk Hosp 1930; 47:239-284.
103. Greenhill NS, Stehbens WE. Scanning electron-microscopic study of experimentally induced intimal tears in rabbit arteries. Atherosclerosis 1983; 49:119-126.
104. Greenhill NS, Stehbens WE. Scanning electron microscopic investigation of the afferent arteries of experimental femoral arteriovenous fistulae in rabbits. Pathology 1987; 19:22-27.
105. Sasaki T, Kodama N, Itokawa H. Aneurysm formation and rupture at the site of anastomosis following bypass surgery. J Neurosurg 1996; 85:500-502.
106. Hassler O. Experimental carotid ligation followed by aneurysmal formation and other morphological changes in the circle of Willis. J Neurosurg 1963; 20:1-7.
107. Hazama F, Hashimoto N. An animal model of cerebral aneurysms. Neuropathol Appl Neurobiol 1987; 13:77-90.

108. Hashimoto N, Handa H, Hazama F. Experimentally induced cerebral aneurysms in rats: Part III Pathology. Surg Neurol 1979; 11:299-304.
109. Hashimoto N, Kim C, Kikuchi H et al. Experimental induction of cerebral aneurysms in monkeys. J Neurosurg 1987; 67:903-905.
110. Stehbens WE. Etiology of intracranial berry aneurysms—A review. J Neurosurg 1989; 70:823-831.
111. Countee RW, Vijayanathan T, Barrese C. Cervical carotid aneurysm presenting as recurrent cerebral ischemia with head turning. Stroke 1979; 10:144-147.
112. Coutard M, Osborne-Pellegrin M. Genetic susceptibility to experimental cerebral aneurysm formation in the rat. Stroke 1997; 28:1035-1042.
113. Yamashita M, Oka K, Tanaka K. Histopathology of the brain vascular network in moyamoya disease. Stroke 1983; 14:50-58.
114. Yabumoto M, Funahashi K, Fujii T et al. Moyamoya disease associated with intracranial aneurysms. Surg Neurol 1983; 20:20-24.
115. Takebayashi S, Matsuo K, Kaneko M. Ultrastructural studies of cerebral arteries and collateral vessels in moyamoya disease. Stroke 1984; 15:728-732.
116. Grabel JC, Levine M, Hollis P et al. Moyamoya-like disease associated with a lenticulostriate region aneurysm. J Neurosurg 1989; 70:802-803.
117. Schievink WI, Mokri B Piepgras DG et al. Intracranial aneurysms and cervicocephalic arterial dissections associated with congenital heart disease. Neurosurg 1996; 39:685-690.
118. Pope FM, Nicholls AC, Narcisi P et al. Some patients with cerebral aneurysms are deficient in type III collagen. Lancet 1981; 1:973-975.
119. Pope FM, Limburg M, Schievink WI. Familial cerebral aneurysms and type III collagen deficiency. J Neurosurg 1990; 72:156-157.
120. Leblanc R, Lozano AM, van der Rest M et al. Absence of collagen deficiency in familial cerebral aneurysms. J Neurosurg 1989; 70:837-840.
121. Leblanc R, Lozano AM, van der Rest M. Response. J Neurosurg 1990; 72:157-158.
122. Stehbens WE, Delahunt B, Hilles AD. Early berry aneurysm formation in Marfan's syndrome. Surg Neurol 1989; 31:200-202.
123. Stehbens WE. Chronic changes in experimental saccular and fusiform aneurysms in rabbits. Arch Pathol Lab Med 1981; 105:603-607.
124. Stehbens WE. Chronic vascular changes in the walls of experimental berry aneurysms of the aortic bifurcation in rabbits. Stroke 1981; 12: 643-647.
125. Fallon JT, Stehbens WE. The endothelium of experimental saccular aneurysms of the abdominal aorta in rabbits. Brit J Exp Pathol 1973; 54:13-19.
126. Greenhill NS, Stehbens WE. Scanning electron microscopic study of the inner surface of experimental aneurysms in rabbits. Atherosclerosis 1982; 45:319-330.
127. Stehbens WE. Histological changes in chronic experimental aneurysms surgically fashioned in sheep. Pathology 1997; 29:374-379.
128. Stehbens WE. Observations on the development of mural thrombi in chronic experimental aneurysms in sheep. Thromb Haemost 1997; 78:952-957.
129. Martin BJ, Stehbens WE, Davis PF et al. Scanning electron microscopic study of hemodynamically induced tears in the internal elastic lamina of rabbit arteries. Pathology 1989; 21:207-212.
130. Jones GT, Stehbens WE, Martin BJ. Ultrastructural changes in arteries proximal to short term experimental carotid-jugular arteriovenous fistulae in rabbits. Int J Exp Pathol 1994; 75:225-232.
131. Jones GT, Martin BJ, Stehbens WE. Endothelium and elastic tears in afferent arteries of experimental arteriovenous fistulae in rabbits. Int J Exp Pathol 1992; 73:405-416.
132. Jones GT, Martin BJ, Stehbens WE. Endothelium in the aorta and ilio-femoral arteries proximal to femoral arteriovenous fistulae in rabbits. Pathology 1993; 25:277-281.
133. Coutard M. Osborne-Pellegrin MJ. Spontaneous lesions in the rat caudal artery. Atherosclerosis 1982; 44:245-260.
134. Coutard M. Osborne-Pellegrin M. Rupture of the internal elastic lamina and vascular fragility in stroke-prone spontaneously hypertensive rats. Stroke 1991; 22:510-515.

135. Stehbens WE. The experimental and iatrogenic production of atherosclerosis. In: Hosada S, Yaginuma T, Sugawara M et al, eds. Recent Progress in Cardiovascular Mechanics: From Basic Research to Clinical Application. London: Gordon & Breach, 1994:253-268.

136. Stehbens WE. Endothelial permeability in experimental aneurysms and arteriovenous fistulae in rabbits as demonstrated by the uptake of Evans blue. Atherosclerosis 1978; 30:343-349.

137. Vollmar JF, Paes E, Pauschinger P et al. Aortic aneurysms as late sequelae of above-knee amputation. Lancet 1989; 2:834-835.

138. Stehbens WE. History of Aneurysms. Medical History 1958; 11:274-280.

139. Coats J, Auld AG. Preliminary communication on the pathology of aneurysms, with special reference to atheroma as a cause. Brit Med J 1893; 2:456-460.

140. Becker AE, van Mantgem J-P. The coronary arteries in Marfan's syndrome. Am J Cardiol 1975; 36:315-321.

141. Bayle O, Branchereau A, Rosset E et al. Morphological assessment of abdominal aortic aneurysms by spiral computed tomographic scanning. J Vasc Surg 1997; 26:238-246.

142. Vorp DA, Raghavan ML, Webster MW. Mechanical wall stress in abdominal aortic aneurysm: influence of diameter and asymmetry. J Vasc Surg 1998; 27:632-639.

143. Stehbens WE. Flow disturbances in glass models of aneurysms at low Reynolds numbers. Quart J Exp Physiol 1974; 58:167-174.

144. Cluroe AD, Fitzjohn TP, Stehbens WE. Combined pathological and radiological study of the effect of atherosclerosis on the ostia of segmented branches of the abdominal aorta. Pathology 1992; 24:140-145.

145. Menashi S, Campa JS, Greenhalgh RM et al. Collagen in abdominal aortic aneurysm: typing, content and degradation. J Vasc Surg 1987; 6:578-582.

146. Sommerville RL, Allen EV, Edward JE. Bland and infected arteriosclerotic abdominal aortic aneurysms: a clinicopathologic study. Medicine 1959; 38:207-221.

147. van der Vliet JA, Boll APM. Abdominal aortic aneurysm. Lancet 1997; 349:863-866,

148. Smulyan H, Marchais SJ, Pannier B et al. Influence of body height on pulsatile arterial hemodynamic data. J Am Coll Cardiol 1998; 31:1103-1109.

149. Busuttil RW, Abou-Zamzam AM, Machleder HI. Collagenase activity of the human aorta. Arch Surg 1980; 115:1373-1378.

150. Campa JS, Greenhalgh RM, Powell JT. Elastin degradation in abdominal aneurysms. Atherosclerosis 1987; 65:13-21.

151. Herron GS, Unemori E, Wong M et al. Connective tissue proteinases and inhibitors in abdominal aortic aneurysms. Arteriosclerosis Thromb 1991; 11:1667-1677.

152. Stehbens WE. The Lipid Hypothesis of Atherogenesis. Austin: RG Landes Pub Co, 1993.

153. Anidjar S, Salzmann J-L, Gentric D et al. Elastase-induced experimental aneurysms in rats. Circulation 1990; 82:973-981.

154. Nackman GB, Karkowski FJ, Halpern VJ et al. Elastin degradation products induce adventitial angiogenesis in the Anidjar/Dobrin rat aneurysm model. Ann NY Acad Sci 1996; 300:260-262.

155. McGee GS, Baxter BT, Shively VP et al. Aneurysm or occlusive disease—factors determining the clinical course of atherosclerosis of the infrarenal aorta. Surgery 1991; 110:370-375.

156. Tilson MD. Aortic aneurysms and atherosclerosis. Circulation 1992; 85:378-379.

157. Norrgård Ö, Rais O, Ängquist KA. Familial occurrence of abdominal aortic aneurysms. Surgery 1984; 95:650-656.

158. Reilly JM, Tilson MD. Incidence and etiology of abdominal aortic aneurysms. Surg Clin Nth Am 1989; 69:705-711.

159. Tilson MD, Seashore MR. Human genetics of the abdominal aortic aneurysm. Surg Gynec Obstet 1984; 158:129-132.

160. Biddinger A, Rocklin M, Coselli J et al. Familial thoracic aortic dilatations and dissections: a case control study. J Vasc Surg 1997; 25: 506-511.

161. Tilson MD, Stansel HC. Differences in results for aneurysms vs occlusive disease after bifurcation grafts. Arch Surg 1980; 115:1173-1175.

162. Reed D, Reed C, Stemmermann G et al. Are aortic aneurysms caused by atherosclerosis? Circulation 1992; 85:205-211.

163. Emmerich J, Fiessinger J-N. Abdominal aortic aneurysm. Lancet 1997; 349:1699.
164. Liang BT, Antman EM, Tuas R et al. Atherosclerotic aneurysm of aortocoronary vein grafts. Am J Cardiol 1988; 61:185-188.
165. Alexander JT, Liu Y-C. Atherosclerotic aneurysm formation in an in situ saphenous vein graft. J Vasc Surg 1994; 20:660-664.
166. Xia S, Ozsvath K, Hirose H et al. Partial amino-acid sequence of a novel 40-kDa human aortic protein, with Vitromectin-like, fibrinogen-like, and calcium binding domains: aortic aneurysm-associated protein 40 (AAAP-40) [human MAGP-3, proposed]. Biochem Biophys Res Commun 1996; 219:36-39.
167. Tilson MD. Similarities of an autoantigen in aneurysmal disease of the human abdominal aorta to a 36-kDa microfibril-associated bovine aortic glycoprotein. Biochem Biophys Res Commun 1995; 213:40-43.
168. Hirose H, Ozsvath KJ, Xia S et al. Molecular cloning of the complementary DNA for an additional member of the family of aortic aneurysm antigen proteins. J Vasc Surg 1997; 26:313-318.
169. Edwards WD, Leaf DS, Edwards JE. Dissecting aortic aneurysm associated with congenital bicuspid aortic valve. Circulation 1978; 57:1022-1025.
170. Stehbens WE. The elusive local factor in atherosclerosis. Med Hypotheses 1997; 48:503-509.
171. Patel KR. Ruptured abdominal aortic aneurysms: a new perspective. J Vasc Surg. 1992; 16:601-662.
172. Stehbens WE, Sonnenwirth AC, Kotrba C. Microcirculatory changes in experimental bacteremia. Exp Mol Pathol 1969; 10:295-311.
173. Stehbens WE, Manz HJ, Uszinski R et al. Atypical cerebral aneurysm in a young child. Pediat Neurosurg 1995; 23:97-100.
174. Walton LJ, Powell JT, Parums DV. Unrestricted usage of immunoglobulin heavy chain genes in B cells infiltrating the wall of atherosclerotic abdominal aortic aneurysms. Atherosclerosis 1997; 135:65-71.
175. Cohle SD, Lie JT. Inflammatory aneurysm of the aorta, aortitis, and coronary arteritis. Arch Pathol Lab Med 1988; 112:1121-1125.
176. Stehbens WE, Delahunt B, Shirer WC et al. Aortic aneurysm in systemic lupus erythematosus. Histopathology 1993; 22:275-277.
177. O'Brien ERM, de Blois D, Schwartz SM. A critical examination of animal models of restenosis following angioplasty. In: Dobrin PB, ed. Intimal Hyperplasia. Austin: RG Landes Co, 1994:229-256.
178. Tilson MD. Regarding "Atherosclerotic aneurysm formation in an in situ saphenous vein graft". J Vasc Surg 1995; 22:120.

CHAPTER 7

In Vivo Animal Models of Aneurysms

Philip B. Dobrin, Richard R. Keen

Much has been learned regarding the development and progression of arterial aneurysms from histological and biochemical analyses of excised human aneurysms and from in vitro studies utilizing treatment of arteries with degradative enzymes. However, to fully understand the development of aneurysms requires an in vivo animal model. Several do exist. These include aneurysms resulting from 1) atherosclerosis, 2) genetic abnormalities, 3) experimental alterations of metabolism, 4) mechanical and chemical injury, 5) hemodynamic factors, 6) inflammation, and 7) elastase infusion. Most of these methods have not been found to be fully reliable, reproducible and controllable. In addition many of them do not appear to be satisfactorily representative of the lesions in human patients. A recent development is the use of the elastase infusion model in living animals. This model appears to exhibit many of the features of human aneurysms including the inflammatory infiltrate and the development of neovascularization. In the present chapter we shall review a variety of in vivo models with detailed attention given to hemodynamic models and to the elastase infusion models.

Atherosclerosis and Atherosclerosis Regression

Occasionally animals spontaneously develop aneurysms. This has been reported for a cynomolgus monkey which developed a carotid-ophthalmic aneurysm.[1] But this is unusual. In a large colony of squirrel monkeys, only 1.5% of the animals were found to harbor aneurysms. These were dissecting, saccular, or fusiform lesions of the carotid arteries or the aorta.[2] Saccular and fusiform aneurysms were found only in animals that had been fed an atherogenic diet, whereas dissecting aneurysms were found in both normocholesterolemic and hypercholesterolemic animals.[3] Because the number of affected animals was small, it is hard to know if these differences in aneurysm morphology are real.

Zarins, Glagov,[4] and the University of Chicago research group have argued that aneurysms may be caused by the wall degeneration that results from atherosclerotic disease. It was proposed that, under some circumstances, atherosclerotic lesions may lead to vessel occlusion, whereas in other circumstances an atherosclerotic lesion may lead to wall thinning and the formation of an aneurysm.[4] This is in contradistinction to the arguments of Tilson and Stansel[5] who reported that patients with aneurysmal and occlusive disease were of different ages and sex, and had different clinical long term outcomes. Accordingly, these authors suggested that aneurysms result from entirely different processes than those associated with atherosclerotic occlusive disease. In a further examination of this issue, Zarins and co-workers[6] fed cynomolgus and rhesus monkeys a high lipid atherogenic diet. After 16-24 months that diet was discontinued, and the animals were fed an atherosclerosis-regression diet. With regression of disease, 13% of the cynomolgus monkeys and 1% of the rhesus monkeys developed aneurysms suggesting that the atherosclerotic lesions caused degeneration of load-bearing elements in the artery wall. Although this model offers a low yield, it does provide a model that may be representative of the human disease.

Genetic Abnormalities of Metabolism

The blotchy mouse is a genetically-determined in vivo animal model of aneurysms. It results from an abnormality on the X chromosome and manifests as defects in connective tissue, skin color, and neurologic function. Death usually results from aortic rupture. The fundamental biochemical defect in this disorder lies in copper metabolism.[7] Copper is essential for lysyl oxidase activity, an enzyme that facilitates the cross-linking of elastin and collagen.[8] Copper also is essential for tyrosinase activity, an enzyme which initiates the synthesis of melanin. Finally, copper is essential for dopamine-hydrolase activity, an enzyme that plays a role in the synthesis of neurotransmitters. Some aspects of the genetic defect in blotchy mice resemble Menke's Kinky Hair syndrome in man in which there is fragmentation of the internal elastic lamellae,[9] and elongation and tortuosity of the arteries of the brain, viscera and extremities.[9,10] The aneurysms that develop in the mouse usually are fusiform in shape and thoracic in location, but they may also form in the abdominal aorta.[11] Histologically, the connective tissue fibers in the media stain poorly for elastin. The media also possesses pleomorphic smooth muscle cells. These are present even before the aneurysms develop.[12] The skin of these animals exhibits decreased stiffness and decreased breaking strength.[13] Deliberate copper deprivation in chicks produces similar connective tissue defects in those animals.[14] Treating blotchy mice with propranolol delays the appearance of aneurysms[15] and also increases the noncollagenous connective tissues by as much as 150%.[16] Propranolol decreases cardiac inotropism leading to decreased pulse pressure, and decreased dP/dt. It also increases the cross-linking of elastin and collagen.[17]

Copper deficiency[18] or decreased copper tissue levels[19] seemed to be a promising avenue by which to explain the decreased tensile strength of elastin[20] and therefore the pathogenesis of aneurysms, but this has not proven to be a consistent characteristic of the disease in humans.[21-23] In fact, in copper deficient pigs, elastin and collagen actually may undergo increased production.[24]

Experimental Alterations of Metabolism: BAPN, Theophylline, Hormonal Manipulations

Another method of producing aortic aneurysms is by deliberately disturbing the normal synthesis of connective tissues in the arterial wall. A classic example is the production of lathyrism. In 1933, Geiger and co-workers[25] first produced skeletal deformities in animals by feeding them sweet pea meal. The toxic factor in the meal was later identified to be β-aminoproprionitrile (BAPN).[26] Animals fed the meal or BAPN were found to develop aneurysms with spontaneous aortic rupture.[27] Aortic rupture occurred in 5 of 63 rats to 19 of 36 rats, depending upon the amino acid mixture supplement given in the diet.[28] Herniation also occurred in 2 of 48 rats, another measure of connective tissue weakness. Hashimoto and co-workers[29-31] produced saccular aneurysms in the cerebral arteries of rats by treating the animals with BAPN, deoxycorticosterone and salt hypertension with ligation of one carotid artery. This experiment was based on the concept that hypertensive stress applied to the BAPN-weakened arteries might produce cerebral aneurysms. Saccular aneurysms were produced in 14% of the animals with all of the aneurysms located in the anterior cerebral artery-communicating artery complex. Light microscopic examination disclosed histologic changes in the vessels that were similar to aneurysmal lesions observed in the cerebral circulation in man.

Connective tissue defects have been seen with other alterations in amino acid metabolism as well. For example solutions of collagen which contain homocysteine fail to form insoluble collagen fibrils[32] and animals fed cysteamine undergo degeneration of elastic fibers with the formation of aortic aneurysms.[33]

Well-fed Broad Breasted Bronze and American Mammoth Bronze breeds of turkeys undergo spontaneous dissecting aneurysms and rupture of the posterior aorta.[34-36] Often this is associated with a lipid-containing plaque.[34] Feeding the turkeys BAPN increases the risk of rupture of the abdominal aorta.[37] Both momoanime oxidase inhibitors[38] and hydralazine[39] potentiated the lathyrogenic effects of BAPN, whereas addition of propranolol or reserpine decreased the risk of rupture.[39-41] A mortality rate of 24% seen after BAPN rose to 91% when the animals were fed both BAPN and hydralazine. Hydralazine did not alter heart rate or dP/dt, but did lower systolic and diastolic pressures. The mechanism of potentiation by hydralazine may be biochemical. Hydralazine accumulates in the aortic media[42] where it can react with the aldehyde groups on collagen.[43] In vitro, the elastic modulus of collagen is reduced if the protein has been incubated with hydralazine.[44] The mortality of turkeys fed BAPN and hydralazine was lowered to 53% by adding propranolol, or to 61% by adding reserpine to the diet. Propranolol reduced blood pressure and dP/dt,[40] and also appears to increase the cross-linking of elastin.[17] Reserpine reduced heart rate, systolic and diastolic blood pressure, but did not decrease dP/dt.[40]

Gilbert and co-workers produced aortic aneurysms by exposing chick embryos to theophylline.[45-47] The mechanism of aneurysm formation by theophylline appears to be through its action on cyclic adenosine 3', 5'-monophosphate (cyclic AMP). Intracellular theophylline increases cyclic AMP, and this is associated with inhibition of mitosis.[48,49] Thus, exposure to theophylline is thought to lead to arrested development of the arterial wall. Indeed Yokoyama and colleagues[47] found on light and electromicroscopic examination that following exposure of chick embryos to theophylline, there was thinning of the aortic media with dispersion of the elastin and collagen fibers.

Dissecting aortic aneurysms have been reported in hamsters treated with cortisone acetate[50] or anti-ovulatory progesterones,[51] and in blotchy mice treated with hydrocortisone.[52] Aortic rupture also has been seen in turkeys treated with diethylstilbesterol.[53]

Mechanical and Chemical Injury

A number of investigators have produced experimental aneurysms by mechanically or chemically disrupting the integrity of the arterial wall. German and Black[54,55] used a simple vein patch to cover a deliberately-made arteriotomy in the carotid artery in dogs. They described the hemodynamic characteristics of the aneurysm, and described a mathematical relationship between the volume of the resulting aneurysm, the size of its orifice, and its likelihood of remaining patent. Stehbens,[56] Nishikawa and co-workers,[57] and Young and Yasargil[58] used similar surgical methods to create experimental aneurysms. Pappas and Burquist[59] created three types of aortic dissection in dogs. This was accomplished by insufflating carbon dioxide into the thoracic aortic wall with a 25 gauge needle. This created a false channel. An intimal tear was created by partially occluding the aorta, incising the outer aortic layer and excising part of the intima. The outer layer was then closed with a continuous suture. The degree of dissection could be enlarged by producing a brief period of hypertension. This model produced a dissection, not a true aneurysm.

Economou and colleagues[60] performed experiments that combined several features of the above described methods. They exposed the thoracic aorta of dogs, then injected them intramurally with 1-2 ml of 70% acetrizoate, an agent used in the 1960s for contrast angiography. The animals were killed between 1 and 13 weeks after the procedure. Aneurysms were grossly apparent after three weeks. Histological examination demonstrated intramural hemorrhage between the elastic lamellae with destruction of those structures and decreased overall wall thickness. There was persistence of the aneurysms with no evidence of repair even after 30 weeks.

In a second experiment these investigators applied a cryogenic probe to the vessel wall causing full thickness freezing. Following this procedure the vessels histologically demonstrated proliferation of the subendothelial cells leading to repair. This occurred at about six weeks. In a novel experiment subendothelial stripping was performed to create an aneurysm. Half of the length of the vessel was subjected to full thickness freezing. Remarkably, the portion of the vessel that had been stimulated to undergo proliferation by the stripping injury exhibited repair, whereas the portion of the vessel that had not been stripped did not exhibit evidence of repair but instead remained aneurysmally dilated.

Ammirati and co-workers[61] used a microsurgical CO_2 laser to burn a hole in the rat common carotid artery. The previously dissected adventitia was then welded over the hole to repair the arterial wall. Thirty-six percent of the animals died soon after surgery, due mainly to hemorrhage from the operative site. The remaining 64% of animals survived at least one week. These animals developed focal aneurysms at the welded site. Similarly, Quigley and co-workers[62] produced fusiform aneurysms of the carotid artery in rats by making an arteriotomy, then sealing it with a laser. The frequency of aneurysm formation was increased if the rats were hypertensive. Application of the same technique to aortic bifurcations in rats produced saccular aneurysms.[63] Troup and Linne[64] produced lateral aneurysms in rats by partially occluding the carotid artery, making a longitudinal incision in the vessel 2-4 mm in length and then closing it with Kodak methyl-2-cyanocrylate (Eastman 910) cement. The majority of these animals developed aneurysms. Some of these resembled fusiform or cylindrical aneurysms; others were lateral aneurysms that angiographically resembled a diverticulum or a pseudoaneurysm extending from the side of the vessel. McCune, White, and co-workers[65,66] injected a variety of noxious materials into the wall of the aorta or vessels of the cerebral circulation to produce local aneurysms. Injected agents included isotonic saline, hypertonic saline, hyaluronidase, sodium morrhuate, plasmocid (8-[3 diethyl-aminopropylamine]-6-methoxy quinoline) (an agent toxic to muscle), or nitrogen mustard. Hypertonic saline and nitrogen mustard were found to be particularly effective in producing aneurysms.

As summarized above, a number of investigators have created aneurysms by preparing an arteriotomy and patching it with a venous onlay patch, cement or laser repair.[54-66] In a large number of cases the venous patch failed mechanically providing an aneurysm-like outpouching. Although this provides a useful model for testing therapeutic techniques such as the placement of coils, stents, etc., the models are not true aneurysms in the usual clinical sense with their expected progression.

One mechanical model was used to investigate some fundamental histomechanical mechanisms. Zatina and co-workers at the University of Chicago[67] produced experimental aneurysms by applying graded crushing forces to the thoracic aorta in pigs. The arteries normally possess about 75 elastic lamellae. When the number of intact lamellae was reduced by injury to less than 40, the vessels became aneurysmal. This corresponded to an increase in the tension required to achieve equilibrium from 1316 ± 202 dynes per centimeter per lamella in the normal vessel to 4087 ± 871 dynes per centimeter per lamella in the aneurysmal vessel. Thus, there appears to be a critical range of loading above which the wall is unable to maintain equilibrium.

Hemodynamic Factors

Hemodynamic factors can cause arteries to dilate. This is seen in the case of poststenotic dilatation, and in arteriovenous fistulas. Halsted[68] described a patient in whom he had partially narrowed the innominate artery to treat an aneurysm of the subclavian artery.

Four years later he observed that the innominate artery had dilated distal to the region of the partially occluding stenosis. Poststenotic dilatation is commonly seen in the presence of a cervical rib, such as in thoracic outlet syndrome, with coarctation of the aorta, and in many other vessels in which there is a persistent, partial narrowing of an artery. Left untreated, a region of artery that has undergone poststenotic dilatation often progresses to become a true aneurysm. Histologically, arteries that have undergone poststenotic dilatation exhibit fragmentation of the elastic lamellae.[69] Five mechanical factors have been proposed as possible causes for poststenotic dilatation.[70] These are: 1) stasis, 2) increased pressure, 3) cavitation, 4) turbulence, 5) elevated or fluctuating shear stresses. Critical analysis indicates that, of these, only turbulence and increased shear stress are likely candidates.[70] Turbulence occurs at the exit of a stenosis, and this is associated with a bruit and vibration of the vessel wall. Experimental studies by Margot Roach and co-workers at the University of Western Ontario[71] reported that application of 25-30 Hz vibrations to arteries in vitro causes them to dilate; frequencies of 80-1,500 Hz had no effect. This was especially effective when there were resonant frequencies. Vessels obtained from patients younger than 44 years of age responded to low frequencies, whereas vessels obtained from patients older than 60 years of age dilated more readily at higher frequencies. Remarkably, a fluid-filled penrose drain also dilated when subjected to vibrations of 90 Hz producing what was, in effect, a focal aneurysm.[72] Thus, arterial wall vibrations in vivo, possibly those enhanced by reflected pressure waves and resonance,[73] might cause poststenotic dilatations and the formation of an aneurysm. However, other investigators have not been able to elicit vessel dilatation with mechanical vibrations, even with electromechanical vibrators glued directly onto the artery wall.[74]

Another possible mechanism of poststenotic dilatation is exposure of the wall to abnormal shear stresses exerted by the flowing blood. Analysis of flow through stenotic vessels disclosed increased shear stresses at and beyond the outlet.[75,76] Using laser technology it has been possible to separate the flowing stream into seven flow laminae.[77] Analysis demonstrated fluctuating shear stresses that were increased more than eight-fold distal to stenoses.[77] This suggests that high or fluctuating shear stresses may be responsible for the development of vessel dilation. It is noteworthy that, in arteries narrowed by atherosclerotic plaque, the caliber of the vessel is related to the magnitude of shear stress to which the wall is exposed; evidently dilatation occurs until an optimal or normal level of shear stress is obtained.[78] Although the above discussion has focused on the literature concerning poststenotic dilatation, much of it also applies to arteriovenous fistulas where the anastomosed vessels exhibit palpable and audible evidence of vibrations and of markedly increased shear stresses.

These mechanical phenomena can have biochemical consequences. Guzman, Zarins and co-workers at Stanford University[79] examined the effects of chronically increased blood flow and shear stress on the gene expression of the proto-oncogene, c-fos, a nuclear transcription factor involved in the regulation of matrix degrading enzymes. An aortocaval fistula was created in adult mice; this resulted in an immediate four-fold increase in blood flow velocity and shear stress. After 21 days the proximal aorta doubled in diameter, and this was accompanied by a reduction in shear stress to preoperative values. Detectable levels of c-fos in aortic endothelial cells were observed immediately after formation of the fistula, but c-fos levels in endothelial cells were reduced to preoperative values after three days. However there was a large increase in c-fos gene expression in aortic medial smooth muscle cells after six hours, and this was still detectable after 21 days. Similar results were found in experiments by MacIver and co-workers at the University of South Manchester.[80] These investigators produced coarctation of the aorta in rats. Coarctation resulted in increased c-fos, c-myc, and H-ras mRNA levels in the proximal aorta at 72 hours, but not at nine days following production of the coarctation. Thus, in both of these experiments, increased shear stress was associated with a rapid increase in gene expression of several enzyme systems that correlated with or preceded the enlargement of the vessels.

Allografts and Xenografts

Peterson and co-workers[81] working in William Abbott's laboratory at Massachusetts General Hospital developed an experimental aortic allograft in the rat. In this model segments of infrarenal aorta were transplanted into SHR hypertensive and WKY normotensive rats. Graft diameters were measured with magnetic resonance imaging and by direct measurement at the conclusion of the experiments. Control autografts remained at normal dimensions, and also retained their normal histologic structure. Aneurysmal dilatation occurred in the allografts transplanted into spontaneously hypertensive rats (SHR), but did not occur in normotensive WKY rats. All allografts developed an inflammatory cell infiltrate with loss of medial smooth muscle cells. Enhancement of the antigenic response accelerated aneurysmal development in SHR rats, but did not do so in WKY rats. The rates of allograft enlargement were equal and final allograft diameters were similar in SHR rats treated with antihypertensive agents and untreated SHR rats.

The authors concluded that immunologic rejection is more important than hypertension in the formation of aneurysms in this model.

Allaire and coworkers in Alexander Clowes laboratory at the University of Washington[82] studied guinea pig-to-rat xenografts used to replace the aorta in rats. The grafts were decellularized, then seeded with syngeneic Fisher 344 rat smooth muscle cells retrovirally transduced with rat plasminogen activator inhibitor-1 (PAI-1) gene or a control vector without the gene. Other control grafts were not seeded. All of the unseeded grafts, and 40% of the grafts seeded with just vector, ruptured 4-14 days after implantation; those that did not rupture became aneurysmal. By contrast, none of the grafts seeded with plasminogen activator inhibition-1 gene ruptured or were aneurysmal. Moreover, unlike the ruptured and aneurysmal grafts in the control groups, the grafts seeded with PAI-1 gene retained intact elastin. Quantitative zymography demonstrated decreased metalloproteinase-9 and decreased 28kD caseinase. These experiments demonstrate that reducing metalloproteinase-9 by inhibiting tissue plasminogen activator (PAI-1) can protect immunologically-targeted xenografts from aneurysmal degradation. The xenograft method is a unique animal model of aneurysms which elicits a predictable inflammatory response from the host. As shown by this paper, it also is useful for studying perturbations of the host response.

Inflammation

Gertz and co-workers[83] produced aneurysms of the common carotid artery in rabbits by bathing the exposed vessels in vivo in a 0.5 M solution of calcium chloride for 15 minutes. The vessels were excised after 3, 7, 21, 42 or 84 days. The vessels exhibited 60 ± 27% (mean ± SEM) increase in diameter after 21 days. Scanning electronmicroscopic study revealed focal areas of intimal damage with desquamation of groups of contiguous endothelial cells. Leukocytes and platelets were seen attached to the damaged areas of endothelium and to the elastic lamellae. After seven days the adventitia and media showed infiltration of polymorphonuclear cells, lymphocytes and monocytes. By three weeks the internal elastic lamella could not be detected. The wall appeared to be thickened but orcein staining disclosed a deficiency of elastin. There was a calcium-elastin complex which was surrounded by an intense inflammatory reaction consisting of lymphocytes, plasma cells, macrophages and giant cells. After 6 and 12 weeks the inflammatory process was found to be restricted to the media. The smooth muscle cells remained normal in appearance but were reduced in number. The key findings of this model were that calcium precipitation was limited to the internal elastic lamella and the elastic network of the media, and that the lesion was surrounded by inflammatory cells.

Osborne-Pellegrin and co-workers[84] induced aortic aneurysms in rats by combining the chemotactic properties of cotton with mechanical stress of aortic coarctation. A stenosing cotton ligature was placed about the aorta between the renal arteries. After three months 7/12 (58%) of the animals developed an aneurysm of the aorta at the level of the renal arteries. In a series of experiments it was found that aneurysms were not induced by a) a nonstenosing cotton ligature, b) a stenosing cotton ligature with antihypertensive agents, 3) a nylon stenosing ligature, or d) medial damage alone. The investigators concluded that the development of an aneurysm in this model required three simultaneous conditions. These were: 1) a tight stenosis causing injury to the media which damages most of the elastic lamellae, 2) hypertension proximal to the stenosis, and 3) an inflammatory reaction induced by a cotton ligature which stimulates inflammatory cell activity. Inflammatory cells were numerous in the presence of cotton ligatures but were sparse in the presence of nylon sutures. Inflammatory cells seen with the cotton sutures included polymorphonuclear cells, monocyte-macrophages and multinucleated giant cells. As will be discussed in the section below describing the elastase-infusion model, inflammation is probably one of the most important aspects of a realistic animal model of aneurysms. Inflammation has been identified in human aneurysms,[85-87] even in lesions not classified pathologically as "inflammatory aneurysms."

Elastase Infusion Model

In 1990, Samy Anidjar working in Jean-Baptiste Michel's laboratory in Paris,[88] developed an elastase infusion model. Wistar rats, 300-350 grams in weight, are anesthetized with 6% sodium pentobarbital (0.1 ml/100g body weight) intraperitoneally. A midline abdominal incision is made. The intestines are mobilized and brought to the lateral side of the abdomen. The aorta is exposed and, using a binocular surgical microscope, the vena cava and other adherent tissues are dissected free and separated from the aorta. The lumbar arteries and other collateral arteries coming off the posterior aspect of the abdominal aorta are ligated. A PE-10 catheter is inserted into one femoral artery and advanced until its tip is located in the infrarenal aorta. An atraumatic microvascular clamp is placed on the aorta at the level of the renal vein below the renal arteries to occlude the proximal aorta but permit continued perfusion of the renal arteries and permit continued drainage of the renal veins. A ligature clamp or suture loop is placed about the aorta at the level of the bifurcation. This is snugged but not tied. This provides a 1 cm segment of hydraulically isolated aorta. The vessel is filled with 2 mL of fluid which is perfused at a rate of 1 mL per hour. After two hours the catheter is removed and the abdomen is closed with interrupted nonabsorbable sutures. If the perfusate contains elastase, this reliably produces an aneurysm over several days as is discussed in detail below.

However before examining experimental results using this model, a few caveats should be mentioned regarding methods. First, the model is technically challenging and is most readily mastered if performed by a surgeon who uses a binocular microscope or wears loupes. Even then, one must expect a steep learning curve with a substantial number of technical failures. Second, the specific activity of the elastase should be determined. Anidjar and co-workers[88] found that at least six units of elastase were required in order to be effective. We found that, in some experiments 15 units of elastase was sufficient; in others, 30 units were required. In vitro evaluation of enzymatic activity is useful. Moreover, once this has been determined, it is wise to purchase and keep frozen sufficient quantities of enzyme from the same lot, or reserve sufficient quantities of the enzyme from the same lot from the manufacturer. Finally, different methods of perfusion can be used. One way to perfuse the isolated segment of aorta is with an infusion pump, infusing a given volume per unit time, usually one mL per hour for two hours. This assures that the fluid containing the enzyme is driven through the wall. However, it also increases the likelihood that some of the enzyme will be absorbed from the peritoneum damaging unintended tissues causing problems such as respiratory distress or respiratory failure. In fact, histologic examination of lung tissue in some animals dying after perfusion demonstrates destruction of the lung elastin (unpublished observations). Another way to perfuse the aorta is to fill the catheter with saline solution containing a sufficient quantity of the enzyme under a steady pressure. We have used 200-300 mmHg and found this to be effective. This decreases the risk of respiratory failure but tends to be less reliable in producing aneurysms. Of course, both control groups and treatment groups in a given experiment should be perfused using the same methods. Another possibility is to add a few units of plasmin to the elastase. We have not tried this, but as discussed below, Anidjar and co-workers[88] found this to be effective even with low doses of elastase.

In the experiments by Anidjar and co-workers,[88] perfusion of the aorta in some rats consisted of 15 units of hog pancreatic elastase in 2 mL saline. The elastase was type 1; 1 unit = 1 mg elastin hydrolyzed for 20 minutes at pH 8.8, 37C in vitro (Sigma Chemical Co., St. Louis, MO). In other rats the aorta was perfused with 2 mL of saline not containing elastase. The elastase perfused aortas all developed aneurysms (Fig. 7.1). Perfusion with collagenase (clostridium type 7, Sigma Chemical Co.), papain (papaya latex, Sigma Chemical Co.), trypsin (bovine pancreas, Sigma Chemical Co.), chymotrypsin (bovine pancreas, Sigma Chemical Co.), or trypsin plus chymotrypsin did not produce true aneurysms. In other rats perfusion included plasmin (porcine blood, 3-5 units/mg, (Sigma Chemical Co.) or thioglycollate (3% thioglycollate medium), or the combination of thioglycollate plus plasmin. The results of these experiments demonstrated that perfusion with pancreatic elastase uniformly caused the development of aneurysms after several days. Perfusion with saline did not produce aneurysms. Perfusion with the other proteases demonstrated damage or dissolution of the elastic tissue with slight aneurysmal dilatation and fibrin deposition in the lumen. A dose-response analysis of elastase demonstrated that at least six units of elastase were required to develop the formation of aneurysms; lower doses were less effective. Perfusion with plasmin did not result in the destruction of the elastic lamellae, but aortas perfused with small amounts of pancreatic elastase (one unit) plus plasmin (two units) did develop aneurysms in the perfused area, suggesting that the plasmin facilitated diffusion of the elastase through the vessel wall. All of the aortas perfused with thioglycollate plus plasmin (two units) developed aneurysms. The aorta perfused with elastase or with thioglycollate plus plasmin had significantly higher elastolytic activity than did the saline-perfused aortas.

Utilizing these findings, we repeated the experiment at Loyola University[89] with perfusion of elastase (hog pancreatic type 1) E-1250, (Sigma Chemical Co., St. Louis, MO); 15 units of thioglycollate (Difco Laboratories, Detroit, MI) plus plasmin (one unit of porcine plasmin, p8644, Sigma); or perfusion with saline as controls. The vessels were excised between 0 and 12 days. At the time of harvest, vessels were measured before excision in the living animal. Figure 7.2 shows the diameter of the treated aortic region plotted as a function of days. After two hours treatment with elastase the vessels dilated an average of 26%. The diameter remained constant through the second day. Then, between the second and third days, the vessels dilated remarkably to 4 mm, a 300% increase in dimensions. Over the next several days they increased up to 400% dilatation. Histologic examination of the tissues demonstrated that, immediately upon perfusion with elastase, there were numerous disrupted elastic lamellae. Over the next several days the lamellae disappeared completely demonstrating their progressive hydrolysis. A few inflammatory cells were seen in the adventitia. But then, on the third day, there were markedly increased numbers of inflammatory cells in the media.

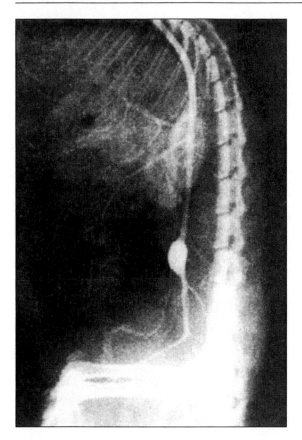

Fig. 7.1. Aortogram of rat showing aneurysm of abdominal aorta following infusion of elastase. Reprinted with permission from: Anidjar S, Salzmann JL, Gentric D et al. Elastase-induced experimental aneurysms in rats. Circulation 1990; 82:973-981.

These persisted for about six days. Somewhere between the sixth and the twelfth days, the inflammatory process regressed. Sections of the tissue also were obtained for immunohistologic analysis. Utilizing mouse antibodies (Table 7.1) these studies demonstrated that there was a large influx of macrophages and T helper cells (Figs. 7.3 and 7.4). These studies show a remarkable temporal correlation between the dramatic secondary enlargement of the vessels and the presence of the inflammatory cells.

In other experiments infusion of thioglycollate plus plasmin was used as a nonspecific stimulant of the immune system. The dimensional and cellular responses of these experiments are summarized in Figure 7.5. Unlike the 26% immediate dilatation following perfusion with elastase, perfusion with thioglycollate plus plasmin produced no immediate increase in diameter. Nor was there the remarkable secondary dilatation that was observed following perfusion with elastase. Instead, there was a gradual and continuous dilatation of the vessels. The arteries were 38% larger than in the control animals after one day, and 213% larger than in the control animals at five days. They were 288% larger than the pretreatment dimensions on days 9 and 16. Histology demonstrated that on the first day after perfusion with thioglycollate plus plasmin there was only minor inflammation. This was localized to the adventitia. But on the fifth day numerous inflammatory cells were found throughout the vessel wall with fractured elastic lamellae. At nine days the elastic lamellae were markedly disrupted with a heavy inflammatory infiltration. At 16 days no elastic lamellae remained to be identified and the cellular infiltrate had regressed to normal. Immunohistologic studies performed on these tissues on days 5 and 9 demonstrated strong evidence for macrophages

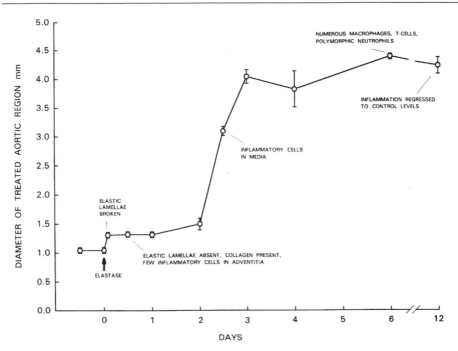

Fig. 7.2. Diameter of 26 rat aortas (means ± SEM) before and after aortic infusion with elastase. Histologic changes also are described. Elastase infusion causes immediate 26% dilatation. After 2-3 days secondary aneurysmal enlargement occurs correlating with inflammation. Reprinted with permission from: Anidjar S, Dobrin PB, Eichorst M et al. Correlation of inflammatory infiltrate with the enlargement of experimental aneurysms. J Vasc Surg 1992; 16:139-147.

and Ia antigen. There was almost no staining for polymorphonuclear leukocytes or T lymphocytes at any time after perfusion with thioglycollate plus plasmin.

These studies again illustrate the temporal relationship between the inflammatory infiltrate and the enlargement of the aorta. Although macrophages and T cells were found following perfusion with elastase, only macrophages were found when vessels were treated with thioglycollate plus plasmin. One may speculate that the foreign protein that comprises the elastase may have served to attract T cells. Because macrophages were found with fractured elastic lamellae in the elastase perfusion experiments, it seems likely that peptide fragments of elastin served to attract the macrophages. Macrophages are known to secrete elastase, collagenase and a variety of cytokines, all of which can contribute to the destruction of the connective tissues. It was shown previously that destruction of collagen is essential for the formation of an aneurysm[90,91] but the initial event probably includes the destruction of elastin.[90,92,93] It would appear in this in vivo model that activity of the inflammatory cells resulted in destruction of the connective tissue. A similar mechanism has been described in the lung.[94]

Halpern and co-workers[95] working in David Tilson's laboratory at St. Luke's Roosevelt Hospital in New York used this model employing a variety of monoclonal antibodies to T cells (CD4, CD5, CD8), to monocyte/macrophages (ED-2), B-cells (LC-A), IgG, and IgM. These investigators perfused the aorta in Wistar rats using either elastase or normal saline. The animals were re-explored and the tissues were harvested at 1, 2, 3, or 6 days after perfusion. In their experiments they did not find that the vessels exhibited the secondary enlargement

Table 7.1 Materials used for immunohistologic preparations

	Monoclonal Antibodies Used	
Mouse Antibody To Rat Determinant	**Specificity**	**Working Dilution**
W3/25	T-helper cells Macrophages	1:500
OX-8	T-suppressor/cytotoxic cells	1:500
W3/13	Thymocytes T-cells Polymorphs	1:500
ED1	Monocytes Macrophages	1:500
OX-6	Immune-associated polymorphic antigen	1:500

until days 3-6, as compared with the earlier findings by Anidjar and Dobrin in which dilatation occurred on days 2-3. This time difference is unexplained, but probably is not critical. Nevertheless, during the period between days 3 and 6 in Halpern's experiments, a variety of endogenous proteinases were observed in the aortic tissues, and immuno-histologic studies revealed various subsets of inflammatory cells. Proteinases with molecular weights between 50 and 90 kDa were found. Pancreatic elastase was not found, demonstrating that the proteinases that were observed were not introduced by the infusate but rather were endogenous enzymes released by the inflammatory cells. ED-2 positive cells were present on day 3, CD-4 positive cells were present on days 1 and 3, CD8 and CD5 positive cells also were present on days 3 and 6. LC-A positive cells were increased on day 6. Specimens stained for IgG and IgM were increased also on day 6. Polymorphonuclear leukocytes appeared in the media early and appeared to enter from both the luminal and adventitial sides of the vessel. Later a large number of polymorphonuclear leukocytes were observed in the adventitia.

Neuman and co-workers[96] working in David Tilson's laboratory performed further studies with the rat elastase infusion model to explore the source of the matrix-degrading proteinases. These included matrix metalloproteinase 1 (MMP-1; interstitial collagenase), matrix metalloproteinase 3 (MMP-3; stromelysin 1) and matrix metalloproteinase 9 (MMP-9; gelatinase B). Immunohistochemical methods were used to identify these substances. These experiments localized MMP-9 to mononuclear cells in the aneurysmal wall. Dual labeling techniques demonstrated that the identity of these cells were macrophages. MMP-3 protein was also detected primarily in macrophage-like mononuclear cells infiltrating the aneurysmal aorta. Immunoreactive material to MMP-1 was identified in mesenchymal cells of the aneurysmal wall suggesting a different source for this enzyme. These findings again suggest that macrophages play a critical role in causing aneurysmal enlargement of the aorta.

In order to test this hypothesis further, experiments were performed in our laboratory at Loyola University using anti-inflammatory agents.[97,98] Rats were treated using the elastase infusion model with 1 mg/kg per day of methylprednisolone, or 5 mg/kg of cyclosporin (5 mg/kg/day), or saline injections as controls, given daily beginning the day before surgery. All three groups of animals underwent perfusion of the aorta with elastase as described previously. As shown in Figure 7.6, the diameter of the aorta of all three groups of rats was

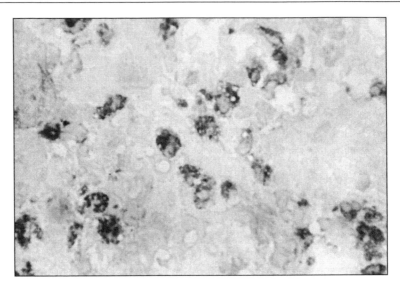

Fig. 7.3. Aorta six days after perfusion with elastase, treated immunohistologically for macrophages. Section stains for macrophages (400x). Reprinted with permission from: Anidjar S, Dobrin PB, Eichorst M et al. Correlation of inflammatory infiltrate with the enlargement of experimental aneurysms. J Vasc Surg 1992; 16:139-147.

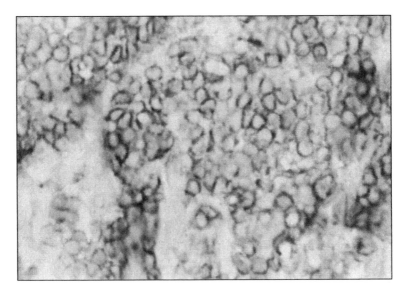

Fig. 7.4. Aorta six days after perfusion with elastase, treated immunohistologically for helper T lymphocytes. Section shows numerous T-cells (400x). Reprinted with permission from: Anidjar S, Dobrin PB, Eichorst M et al. Correlation of inflammatory infiltrate with the enlargement of experimental aneurysms. J Vasc Surg 1992; 16:139-147.

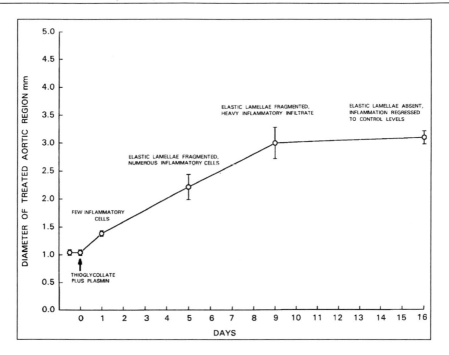

Fig. 7.5. Diameter of the aorta in rats (means SEM) before and after perfusion with thioglycollate plus plasmin. Perfusion caused no immediate change in diameter. Continuous gradual enlargement occurred correlated with the appearance of inflammatory cells. Reprinted with permission from: Anidjar S, Dobrin PB, Eichorst M et al. Correlation of inflammatory infiltrate with the enlargement of experimental aneurysms. J Vasc Surg 1992; 16:139-147.

approximately 1.5 mm before elastase perfusion. This increased to 2.25 mm immediately after two hours of perfusion. Five days later the diameter of the saline-injected control animals was 2.52 mm, the diameter of the methylprednisolone-treated animals was 2.2 mm, and the diameter of the cyclosporin-treated rats was 2.68 mm. These three groups were not statistically different from one another. After nine days, the diameter of the control animals had increased to 4.52 mm whereas the diameter of the methylprednisolone-treated animals was 3.02 mm and the diameter of the aorta of the cyclosporin-treated animals was 3.01 mm. The two immunosuppressed groups were statistically smaller than the saline-injected control group. Histologic examination of the tissues five days after elastase infusion demonstrated that the saline-injected control animals exhibited disruption of the elastic lamellae with moderate edema and marked cellular infiltration throughout all layers of the wall by mononuclear cells including lymphocytes and macrophages (Fig. 7.7A). By contrast, the aorta of the methylprednisolone-treated animals exhibited intact elastic lamellae, no significant edema and no cellular infiltration (Fig.7.7B). The aorta of the cyclosporin-treated animals similarly exhibited intact elastic lamellae. Mild edema was present with a few scattered macrophages (Fig 7.7C). After nine days the saline-injected control animals exhibited disrupted elastic lamellae, but the edema had now regressed. Occasional macrophages were seen. By contrast the aorta of the methylprednisolone-treated animals and the aorta of the cyclosporin-treated animals exhibited intact elastic lamellae with little or no edema and very few macrophages. These data demonstrate that inflammatory cells play an active role in producing the secondary

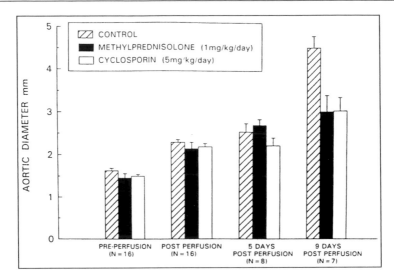

Fig. 7.6. Diameter of aortas in 47 rats (means ± SEM) after perfusion with elastase. At 9 days, the aortas of methylprednisolone- and cyclosporine-treated rats were significantly larger than saline-treated control rats. Reprinted with permission from: Dobrin PB, Baumgartner N, Andijar S et al. Inflammatory aspects of experimental aneurysms. Effect of methylprednisolone and cyclosporine. Ann NY Acad Sci 1996; 800:74-88.

aneurysmal dilatation observed at nine days, and that this was attenuated in animals receiving treatment with anti-inflammatory agents or immunosuppression.

The studies described above with methylprednisolone and cyclosporin utilized a nonspecific form of immunosuppression. In a more specific experiment, Ricci and co-workers at the University of Vermont[99] obtained anti-CD18 monoclonal antibody to inhibit leukocyte CD18 adhesion molecule. The activity of this monoclonal antibody against rat leukocytes was first determined by in vitro immunofluorescence flow cytometry. Experiments then were performed by perfusing the isolated infrarenal abdominal aorta with 15 units of elastase in rats in vivo in normotensive (WKY) and genetically hypertensive (WKHT) rats. Animals of both strains were allocated randomly to control saline injections or to monoclonal antibody treatment. Treatments were begun on the operative day for a total of 4 doses. The initial aortic size in all of the animals was 1.11 mm. The animals were reanesthetized on day 14. At that time, all of the animals had developed aneurysms which were significantly larger than the initial aortic size. For the normotensive WKY rats, treatment with monoclonal antibody produced significantly smaller aneurysms than did the saline-treated controls. Similarly for the hypertensive WKHT animals, treatment with monoclonal antibody also produced significantly smaller aneurysms than did saline treated controls (Fig. 7.8). Histologic examination of the excised tissues demonstrated that neutrophil counts did not differ significantly between the control and treatment groups. Thus, unlike the previously described work utilizing methylprednisolone and cyclosporin, there was no reduction in the number of inflammatory cells, but there was a reduction in inflammatory processes.

Another type of investigation of the role of macrophages was undertaken by Holmes and co-workers working with Jeff Reilly and Rob Thompson's group at the Washington University School of Medicine in St. Louis.[100] It is known that the induction of metalloenzyme expression by macrophages depends upon the production of prostaglandin E_2 (PGE_2) from

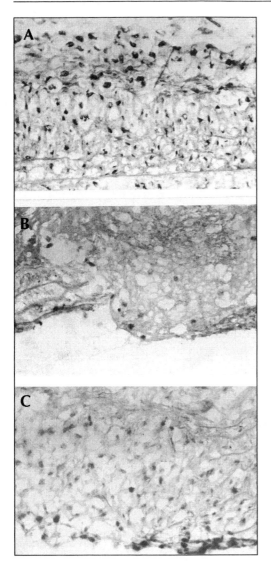

Fig. 7.7. Light histology of aortas of rats treated with (A) saline controls, (B) methylprednisolone, (C) cyclosporine 5 days after elastase infusion. Tissue sections were stained with hematoxylin and eosin. Saline control aortas exhibit a heavy inflammatory infiltrate (A). Aortas of methylprednisolone- and cyclosporine-treated rats (B,C) exhibit much less inflammation. Similar findings were observed after 9 days. Reprinted with permission from: Dobrin PB, Baumgartner N, Andijar S et al. Inflammatory aspects of experimental aneurysms. Effect of methylprednisolone and cyclosporine. Ann NY Acad Sci 1996; 800:74-88.

arachidonic acid, a pathway that is rate limited in regards to prostaglandin synthesis by cellular cyclooxygenase synthase (COX-1 and COX-2) activity. Macrophage activation increases the expression of this enzyme with increased production of PGE_2. The latter has been demonstrated to be increased in the walls of human abdominal aortic aneurysms. If macrophages are responsible for the destruction of elastin through the secretion of macrophage metalloproteinases (MMPs), then blockade of cellular cyclooxygenase synthase should inhibit the progression of aneurysms. Indomethacin is an effective inhibitor of MMP expression by activated macrophages. It acts primarily through its effect on cellular cyclooxygenase synthase. An experiment therefore was undertaken to study the effects of indomethacin on the enlargement of aneurysms.

The in vivo elastase perfusion experiment was used in Wistar rats utilizing 25 units of elastase. Six animals received injections of saline as control subjects, and eight animals received

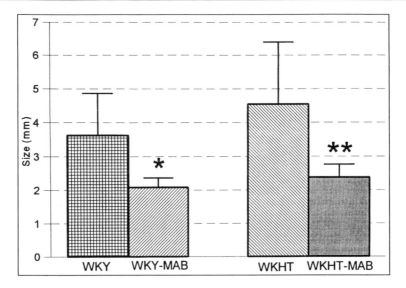

Fig. 7.8. Diameter of the aortas of WKY and WK hypertensive (WKHT) rats 14 days after elastase-infusion. Treatment with monoclonal antibody to CD18 adhesion molecule prevented enlargement in both groups of animals. Asterisks denote significant differences for WKY vs WKY-MAB and for WKHT vs WKHT-MAB. Reprinted with permission from: Ricci MA, Strindberg G, Slaiby JM et al. Anti-CD18 monoclonal antibody slows experimental aortic aneurysm expansion. J Vasc Surg 1996; 23:301-307.

4 mg/kg/day of indomethacin for seven days. As shown in Table 7.2 the mean aortic diameters in the two groups ranged from 1.57-1.59 mm. Immediately after treatment aortic diameters increased to 2.06 mm in the control group and 2.28 mm in the indomethacin treatment group. However, the final diameters seven days after perfusion were 3.54 mm in the saline-treated control group and only 2.46 mm in the indomethacin group. Indeed, five out of six of the saline-treated animals developed aneurysms, whereas *none* of the eight indomethacin-treated animals did so. Histologic examination of the excised tissue demonstrated that the saline-treated animals exhibited marked disruption of the elastic lamellae with a dense inflammatory cell infiltrate. This infiltrate included numerous macrophages. A similar inflammatory infiltrate was noted in the indomethacin-treated animals, but these animals exhibited preserved elastic lamellae (Fig. 7.9). Substrate gel zymography was used to evaluate the effect of indomethacin on aortic wall production of MMP-9 and MMP-2, two of the principal elastolytic metalloproteinases. Aortic extracts from the indomethacin-treated subjects exhibited decreased levels of MMP-9 as compared with saline-treated control subjects. Analysis of MMP-2 production was not conclusive. These data demonstrate the macrophage stabilizing activity of indomethacin and suggest that clinical use of NSAIDs might provide a useful clinical treatment of small aneurysms in patients.

In another experiment from Jeff Reilly and Rob Thompson's group at Washington University in St. Louis,[101] Petrinec and co-workers examined the role of doxycycline in inhibiting aneurysmal degeneration in the elastase perfusion rat model. Tetracycline antibiotics were utilized because they have been demonstrated to be inhibitors of metalloproteinases, acting by mechanisms that are separate from their antimicrobial activity. Tetracyclines inhibit collagenase in vitro and they inhibit MMP-mediated tissue injury in disorders such as arthritis.[102,103] In order to examine the effects of tetracyclines, the elastase

Table 7.2. Diameter of rat aortas (mm)

Treatment	N	Pre Tx	Post Tx	Post Tx (7 days)	AAA
Saline	6	1.57	2.06	3.54	5/6
Indomethacin	8	1.59	2.28	2.46*	0/8*

*Significantly different. From Holmes et al.[99]

infusion model was used in Wistar rats. After aortic perfusion was completed, the animals were treated with subcutaneous injections of doxycycline or saline. The animals were reanesthetized and the aorta was measured and excised after 0, 2, 7, or 14 days. The aorta of the rats was 1.5 mm before perfusion, 2.06 mm immediately after perfusion for both groups, and 2.14 mm at two days for both groups. However, at seven days the saline-treated animals exhibited aortas 3.54 mm in diameter. This was reduced to 2.55 mm in the doxycycline-treated animals. Similarly, at 14 days the aortas were 4.26 mm in the saline-treated group and only 2.73 mm in the doxycycline-treated groups. Thus, at seven and 14 days, doxycycline prevented aneurysmal enlargement of the aorta. Histologic examination showed that both the control and treatment groups had large numbers of inflammatory cells (Fig. 7.10); however, unlike the control group which exhibited destruction of the elastic lamellae, the doxycycline-treated animals possessed intact elastic lamellae (Fig. 7.11). This is comparable to what was observed earlier with methylprednisolone and cyclosporin,[96,97] with monoclonal antibody (mAb),[98] and with indomethacin[99] in that the difference in dimensions between the treatment and control groups both occurred seven days or longer after perfusion with elastase. However, unlike the previously-described experiments which inhibited aneurysmal enlargement by blocking the influx of inflammatory cells,[96,97,99] the doxycycline experiments[100] inhibited aneurysmal enlargement permitting the influx of inflammatory cells, but prevented the production of metalloproteinases. Because 92 kD and 72 kD gelatinases are elastolytic metalloproteinases produced during the course of experimental aneurysms, the effect of doxycycline treatment on the levels of these enzymes in the wall also was examined. Extracts from animals that had been treated with saline solution and killed at 0 and 2 days contained very little 92 kD gelatinase as detected by substrate zymography. Large amounts of this enzyme were present in the seven day group. Ninety-two kD gelatinase declined in the 14-day aortic extracts. By contrast with these data for the control animals, extracts from animals treated with doxycycline and studied on day seven contained levels of 92 kD gelatinase that were almost identical to those seen in the control or two day extracts, and these were markedly less than those present in the seven day extracts from the saline-treated control animals. Thus, as was seen in the two preceding experiments, treatment with MMP-inhibiting doxycycline inhibited the development of aneurysms in the elastase perfusion model. Biochemical analysis suggests that it acts by blocking the expression of MMP in the inflammatory cells.

Aneurysms often exhibit evidence of hypervascularity, consistent with inflammatory processes. Accordingly, the elastase infusion model was used by Nackman and colleagues working in David Tilson's laboratory to examine adventitial angiogenesis.[103] In this experiment the femoral perfusion catheter was used to infuse one of several substances: a) 2 ml of pancreatic elastase, as described in the previous experiments, 25 units per ml; b) 2 ml of normal saline solution; c) a hexamer repeating polypeptide, val/gly/val/ala/pro/gly (VGVAPG). This polypeptide is an elastin degradation product found following elastolysis. This substance was synthesized in the laboratory to permit study of the effects of elastin

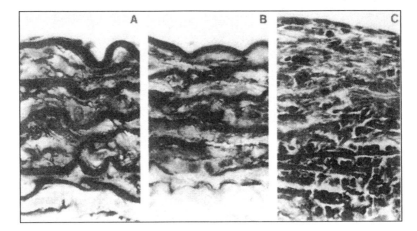

Fig. 7.9. Sections of normal (A), elastase-perfused and indomethacin-treated (B) and elastase-perfused saline-treated aorta, all stained for elastin. The normal (A) and indomethacin-treated vessels (B) exhibit intact elastic lamellae, as compared with elastase-perfused saline-treated vessels (C). Reprinted with permission from: Holmes DR, Petrinec D, Wester W et al. Indomethacin prevents elastase-induced abdominal aortic aneurysms in the rat. J Surg Res 1996; 63:305-309.

degradation products on angiogenesis and the progression of aneurysms. Seven days after perfusion the animals were killed and the aortas were excised. Histologic examination demonstrated a 100-fold increase in the mean number of vessels per high powered field in the aortas perfused with elastase or VGVAPG as compared with saline-infused control animals (Fig. 7.12). In addition, the VGVAPG perfused animals had a 26% increase in diameter, but this was not sufficient to achieve aneurysmal proportions. Nevertheless, this study demonstrated that infusion of this elastin degradation product induced a histologic feature that is seen in human aneurysmal disease. It is unfortunate that an additional treatment group of rats was not followed for longer durations, perhaps 9-12 days, as these aortas may have exhibited true aneurysms.

The elastase infusion model has been useful for evaluating other aspects of the development of aneurysms as well. One might hypothesize that hypertension would increase the rate of enlargement of vessels perfused with elastase. Anidjar and co-workers,[104] working in J-B Michel's laboratory at INSERM, performed an elastase infusion experiment in normotensive rats, in renovascular hypertensive rats, and in spontaneously hypertensive rats. The time course of dimensional changes occurred as described earlier, and there was histologic evidence of destruction of the elastic network in the vessel wall. The enlargement of the aortas in the hypertensive rats was greater than that in the normotensive rats. Moreover, aortic rupture was more frequent in the renovascular hypertensive animals than in the spontaneously hypertensive animals or normotensive animals. As might be expected, the hypertensive rats with aortic rupture were associated with a greater elevation in arterial pressure (220 mmHg vs. 189 mmHg).

In a similar experiment, Gadowski and co-workers working in Michael Ricci's laboratory at the University of Vermont[105,106] examined the extent to which hypertension influences the rate of enlargement of experimental aortic aneurysms. Two strains of Wistar-Kyoto (WKY) rats were used for these experiments. In group I normotensive WKY rats were used and in Group II genetically hypertensive WKHT rats were used. The animals underwent elastase infusion of the abdominal aorta. Aortic diameter was measured with a micrometer

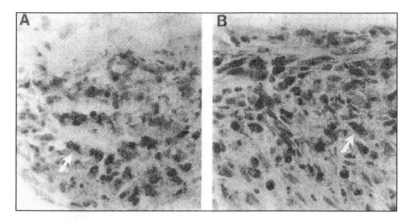

Fig. 7.10. Immunohistology of rat aortas. Sections of elastase-perfused, saline-treated (A), and elastase-perfused (B) indomethacin-treated rats immunostained for macrophages (Arrows). Numerous macrophages are seen in both groups. Reprinted with permission from: Holmes DR, Petrinec D, Wester W et al. Indomethacin prevents elastase-induced abdominal aortic aneurysms in the rat. J Surg Res 1996; 63:305-309.

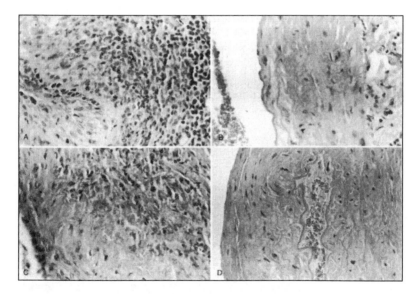

Fig. 7.11. Cellular infiltrate in rat aortas 7 days after elastase-perfusion and treatment with saline solution (A and B) or doxycycline (C and D). A and C are stained with Verhoff-van Giesen stain for elastin, B and D are stained with hematoxyl and eosin for cellular components. Both groups exhibit inflammatory infiltrate (B and D), but doxycycline-treated vessels exhibit intact elastic lamellae (C), unlike the control group which exhibits destroyed elastic lamellae (A). Reprinted with permission from: Petrinec D, Liao S, Holmes DR et al. Doxycycline inhibition of aneurysmal degeneration in an elastase-induced rat model of abdominal aortic aneurysm: Preservation of aortic elastic associated with suppressed production of 92 kD gelatinase. J Vasc Surg 1996; 23:336-346.

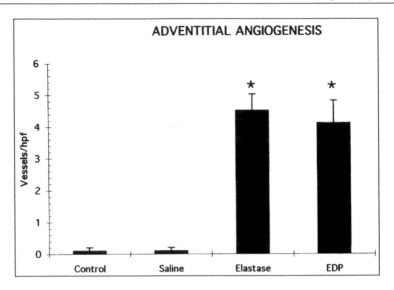

Fig. 7.12. Adventitial vessel count per high-powered field of unperfused aortas (control), perfused with saline, or perfused with elastase or elastin degradation products. Vessel count 7 days after perfusion demonstrated a 100 fold increase in neovascularity. Reprinted with permission from: Nackman GB, Karkowski FJ, Halpern VJ et al. Elastin degradation products induce adventitial angiogenesis in the Anidjar/Dobrin rat aneurysm model. Surgery 1997; 122:39-44.

and systolic blood pressure was measured by tail plethysmography. The rats were killed 7 or 14 days after elastase infusion. Systolic blood pressure was 164 mmHg in the WKHT rats as compared with 119 mmHg in the WKY rats. Initial aortic size was 1.1 mm in both groups of animals. By day seven the aneurysms in the hypertensive animals were 2.54 mm as compared with 2.31 mm in the normotensive animals. However at day 14 the aneurysms were 3.45 mm in the hypertensive animals as compared with 2.36 mm in the normotensive animals. The rate of growth in the hypertensive group was nearly twice that observed in the normotensive group. Growth rates correlated positively with systolic blood pressure. Both groups exhibited elastic lamellar disruption and an inflammatory cell infiltrate in the wall.

The findings of hypertension on aneurysmal expansion is expected based on mechanical considerations. The force distending a blood vessel is given by

$$F_D = P_T \times d_i \times L$$

where F_D is the distending force, P_T is the transmural pressure, i.e., intraluminal pressure minus extravascular pressure, d_i is internal diameter and L is vessel length. A rise in arterial pressure increases the distending force. In addition, because the aneurysmally dilated artery is of larger internal diameter than the normal vessel, the distending force is increased even more.

Slaiby, Ricci and co-workers[107,108] performed a similar experiment using normotensive Wistar-Kyoto rats and genetically hypertensive Wistar-Kyoto (WKHT rats), but this time examined the effects of propranolol. The aorta of these animals was perfused with elastase or saline solution. Some of the animals received saline control injections, while others received propranolol, 10 mg/kg or 30 mg/kg. Results demonstrated that the aortic aneurysms were more than twice the size in the hypertensive rats than in the normotensive animals. Moreover, the dimensions of the aneurysms were smaller in the animals that were treated with propranolol than in the control animals. Propranolol decreases arterial pressure, the rate of rise of arterial pressure (dP/dt), and also has a direct biochemical effect in stabilizing the connective tissues of the arterial wall.[17]

Finally, Boudghene, Anidjar and others working at Jean-Baptiste Michel's INSERM laboratory in Paris[109] utilized the elastase infusion model to produce aortic aneurysms in dogs. These were done in order to provide a model for surgeons to learn to perform endovascular grafting. The dog is a larger animal that requires more elastase but may be technically simpler to perform than with the very small femoral and aortic vessels in the rat.

Acknowledgment

Information given in this Chapter expands on material described in a recent review article: Dobrin DB. Animal models of aneurysms. Ann Vasc Surgery 1999; 13:641-648.

References

1. Espinosa F, Weir B, Noseworthy T. Rupture of an experimentally induced aneurysm in a primate. Can J Neurol Sci 1984; 11:64-68.
2. Boorman GA, Silverman S, Andersen JH. Spontaneous dissecting aortic aneurysm in a squirrel monkey (Saimiri Sciureus). Lab Anim Sci 1976; 26:942-947.
3. Strickland HL, Bond MG. Aneurysms in a large colony of squirrel monkeys (Saimiri Sciureus). Lab Anim Sci 1983; 33:589-592.
4. Zarins CK, Glagov S. Aneurysms and obstructive plaques: Differing local responses to atherosclerosis. In: Bergan JJ, Yao JST eds. Aneurysms: Diagnosis and Treatment. New York: Grune & Stratton 1982; 61-82.
5. Tilson MD, Stansel HC. Differences in results for aneurysm vs. occlusive disease after bifurcation grafts: Results of 100 elective grafts. Arch Surg 1980; 115:1173-1175.
6. Zarins CK, Glagov S, Vesselinovitch D et al. Aneurysm formation in experimental atherosclerosis: relationship to plaque evolution. J Vasc Surg 1990; 12:246-256.
7. Hunt DM. Primary defect in copper transport underlies mottled mutants in the mouse. Nature 1974; 249:852-853.
8. Gacheru S, McGee C, Urin-Houe JY et al. Expression and accumulation of lysyl oxidase, elastin, and type I procollagen in human Menkes and mottled mouse fibroblasts. Arch Biochem Biophys 1993; 301:325-329.
9. Danks DM, Campbell PE, Stevens BJ et al. Menke's kinky hair syndrome: An inherited defect in copper absorption with widespread effects. Pediatrics 1972; 50:186-201.
10. Wesenberg RL, Gwinn JL, Barnes GE. Radiological findings in the kinky hair syndrome. Radiology 1969; 92:500-506.
11. Andrews EJ, White WJ, Bullock LP. Spontaneous aortic aneurysms in blotchy mice. Am J Pathol 1978; 78:199-210.
12. Brophy CM, Tilson JE, Braverman IM, Tilson MD. Age of onset, pattern of distribution and histology of aneurysm development in a genetically predisposed mouse model. J Vasc Surg 1988; 8:45-48.
13. Elefteriades J, Panjabi MJ, Tilson MD. Reduced tensile strength of skin from the spontaneously aneurysm-prone Blotchy mouse. Surg Forum 1982; 33:58-60.
14. O'Dell BL, Hardwick BC, Reynolds G et al. Connective tissue defect in the chick resulting from copper deficiency. Proc Soc Exp Biol Med 1961; 108:402-405.
15. Brophy C, Tilson J, Tilson MD. Propranolol delays the formation of aneurysms in the male blotchy mouse. J Surg Res 1988; 44:687-689.
16. Brophy CM, Tilson JE, Tilson MD. Propranolol stimulates the crosslinking of matrix components in skin from the aneurysm-prone Blotchy mouse. J Surg Res 1989; 46:330-332.
17. Boucek RJ, Gunia-Smith Z, Nobel NL et al. Modulation by propranolol of the lysyl cross-links in aortic elastin and collagen of the aneurysm-prone turkey. Biochem Pharmacol 1983; 32:275-280.

18. Tilson MD, David G. Deficiencies of copper and a compound with ion-exchange characteristics of pyridinoline in skin from patients with abdominal aortic aneurysms. Surgery 1983; 94:134-141.

19. Tilson MD. Decreased hepatic copper levels. A possible chemical marker for the pathogenesis of aortic aneurysms in man. Arch Surg 1982; 1212-1213.

20. Kimball DA, Coulson WF, Carnes WH. Cardiovascular studies on copper-deficient swine III. Properties of isolated aortic elastin. Exp Mol Pathol 1964; 3:10-18.

21. Read R. Discussion of paper by Tilson MD, Davis G. Deficiencies of copper and a compound with ion-exchange characteristics of pyridinoline in skin from patients with abdominal aortic aneurysms. Surgery 1983; 94:134-141.

22. Alston J, Fody E, Couch L et al. A prospective study of hepatic and skin copper levels in patients with abdominal aortic aneurysms. Surg Forum 1983; 24:466-468.

23. Senapati A, Carlsson L, Fletcher C et al. Is tissue copper deficiency associated with abdominal aortic aneurysms? Brit J Surg 1985; 72:352-353.

24. Hill KE, Davidson JM. Induction of increased collagen and elastin biosynthesis in copper-deficient pig aorta. Arteriosclerosis 1986; 6:98-104.

25. Geiger BJ, Steenback H, Parsons HT. Lathyrism in the rat. J Nutrition 1933; 6:427-442.

26. Schilling ED, Strong FM. Isolation, structure and synthesis of a lathyrus factor from L. Odoratus. J Am Chem Soc 1954; 76:2848.

27. Ponseti IV, Baird WA. Scoliosis and dissecting aneurysm of the aorta in rats fed with lathyrus odoratus seeds. Am J Pathol 1952; 28:1059-1077.

28. Lalich JJ. Production of aortic rupture in rats fed purified diets and beta-amino-proprionitrile. AMA Arch Pathol 1956; 61:520-524.

29. Hashimoto N, Handa H, Hazama F. Experimentally induced cerebral aneurysms in rats. Surg Neurol 1978; 10:3-8.

30. Hashimoto N, Handa H, Hazama F. Experimentally induced cerebral aneurysms in rats. Part II. Surg Neurol 1979; 11:243-246.

31. Hashimoto N, Handa H, Hazama F. Experimentally induced cerebral aneurysms in rats. Part III. Surg Neurol 1979; 11:299-304.

32. Kang AH, Trelstad RL. A collagen defect in homocystinuria. J Clin Invest 1973; 52:2571-2578.

33. Jayaraj AP. Dissecting aneurysm of aorta in rats fed with cysteamine. Br J Exp Pathol 1983; 64:548-552.

34. Gresham GA, Howard AN. Aortic rupture in the turkey. J Atheroscler Res 1961; 1:75-80.

35. Ball RA, Sautter JH, Pomeroy BS et al. Natural and experimental dissecting aneurysms in turkeys. Pathol Bacteriol 1965; 89:599-606.

36. Middleton CC. Naturally occurring atherosclerosis in turkeys. Proc Soc Exp Biol Med 1969; 130:638-642.

37. Barnett BD, Bird HR, Lalich JJ et al. Toxicity of β -aminoproprionitrile for turkey poults. Proc Soc Exp Biol Med 1957; 94:67-70.

38. McDonald BE, Bird HR, Strong FM. Production of aortic aneurysms in turkeys and effect of various compounds in potentiating induced aortic ruptures. Proc Soc Exp Biol Med 1963; 113:728-732.

39. Simpson CF, Kling JM, Robbins RC et al. β-aminoproprionitrile-induced aortic ruptures in turkeys. Inhibition by reserpine and enhancement by monoamine oxidase inhibitors. Toxicol Appl Pharmacol 1968; 12:48-59.

40. Simpson CF, Taylor WJ. Effect of hydralazine on aortic rupture induced by β-aminoproprionitrile in turkeys. Circulation 1982; 65:704-708.

41. Simpson CF, Kling JM, Palmer RF. BAPN-induced dissecting aneurysm of turkeys: Treatment with propranolol. Toxicol Appl Pharmacol 1970; 16:143-153.

42. Moore-Jones D, Perry HM. Radioautographic localization of hydralazine-l-C14 in arterial walls. Proc Soc Exp Biol Med 1966; 122:576-579.

43. Paz MA, Seifter S. Immunological studies of collagen modified by reaction with hydralazine. Am J Med Sci 1972; 263:281-290.

44. Kraemer HP, Nemetschek T, Gross F. Effects of hydralazine on the elasticity of n collagen. Experimentia 1979; 35:527-528.
45. Gilbert EF, Bruyere HJ Jr, Ishikawa S et al. The effects of methylxanthines on catecholamine-stimulated and normal chick embryos. Tetrology 1977; 16:47-52.
46. Ishikawa S, Gilbert EF, Bruyere HJ, Jr et al. Aortic aneurysm associated with cardiac defects in theophylline stimulated chick embryos. Tetrology 1978; 18:23-30.
47. Yokoyama H, Matsuoka R, Bruyere HJ Jr et al. Light and electron-microscopic observations of theophylline-induced aortic aneurysms in embryonic chicks. Am J Pathol 1983; 112:285-266.
48. DeWys WD, Bathina SH. Synergistic antileukemic effect of theophylline and 1,3-bis-(2-chloroethyl)-1-nitrosurea. Cancer Res 1980; 40:2202-2208.
49. Kolb CA, Mansfield JM. Effects of theophylline treatment on mouse B-16 melanoma cells in vitro. Oncology 1980; 37:343-352.
50. Stefee CH, Snell KC. Dissecting aortic aneurysm in hamsters treated with cortisone acetate. Proc Soc Exp Biol Med 1955; 90:712-714.
51. Cobb LM, Bloom HJG, Roe FJC et al. Rupture of the aorta produced in the hamster by anti-ovulatory progestens. Nature 1971; 229:50-51.
52. Reilly JM, Savage EB, Brophy C et al. Hydrocortisone rapidly induces aortic rupture in a genetically susceptible mouse. Arch Surg 1990; 125:707-709.
53. Beall CW, Simpson CF, Pritchard WR et al. Aortic rupture in turkeys induced by diethylstilbesterol. Proc Soc Exp Biol Med 1963; 113:442-443.
54. German WJ, Black SPW. Experimental production of carotid aneurysms. N Eng J Med 1954; 250:104-106.
55. Black SPW, German WJ. Observations on the relationship between the volume and the size of the orifice of experimental aneurysms. J Neurosurg 1960; 17:984-990.
56. Stehbens WE. Experimental production of aneurysms by microvascular surgery in rabbits. Vasc Surg 1973; 7:165-175.
57. Nishikawa M, Yonekawa Y, Matsuda I. Experimental aneurysms. Surg Neurol 1976; 5:15-18.
58. Young PH, Yasargil MG. Experimental carotid artery aneurysms in rats: A new model for microsurgical practice. J Microsurg 1982; 3:135-146.
59. Pappas G, Burquist J. Creation of thoracic aortic aneurysms in dogs. J Surg Res 1970; 10:333-336.
60. Economou SG, Taylor CB, Beattie EJ, Davis CB. Persistent experimental aortic aneurysms in dogs. Surgery 1960; 47:21-28.
61. Ammirati M, Ciric I, Robin E. Induction of experimental aneurysms on the rat common carotid artery using a microsurgical CO_2 laser. Microsurgery 1988; 9:78-81.
62. Quigley MR, Heiferman K, Kwann HC et al. Laser-sealed arteriotomy: A reliable aneurysm model. J Neurosurg 1987; 67:284-287.
63. Quigley MR, Tuite GF, Cozzens JW. Histology and angiography in a bifurcation aneurysm model. Surg Neurol 1988; 30:445-451.
64. Troupp H, Torbjorn R. Methyl-2-cyanoacrylate (Eastman 910) in experimental vascular surgery. With a note on experimental aneurysms. J Neurosurg 1964; 21:1067-1069.
65. McCune WS, Samadi A, Blades B. Experimental aneurysms. Ann Surg 1953; 138:216-218.
66. White JC, Sayre GP, Whisnant JP. Experimental destruction of the media for the production of intracranial arterial aneurysms. J Neurosurgery 1961; 18:741-745.
67. Zatina MA, Zarins CK, Gewertz BL et al. Role of medial lamellar architecture in the pathogenesis of aortic aneurysms. J Vasc Surg 1984; 1:442-448.
68. Halsted WS. Cylindrical dilation of the common carotid artery following partial occlusion of the innominate and ligation of the subclavian. Surg Gynecol Obstet 1918; 27:547-554.
69. Trillo A, Haust MD. Arterial elastic tissue and collagen in experimental poststenotic dilatation in dogs. Exp Mol Pathol 1975; 23:473-490.
70. Dobrin PB. Poststenotic dilatation. Surg Gynecol Obstet 1991; 172:503-508.
71. Roach MR, Melech E. Effect of sonic vibration on isolated human iliac arteries. Can J Physiol Pharm 1971; 49:288-291.

72. Bruns DL, Connolly JE, Homan E, Stofer RC. Experimental observations on poststenotic dilation. J Thorac Cardiovasc Surg 1959; 38:662-669.
73. McDonald DA. Blood flow in arteries. Baltimore: Williams and Wilkins, 1974.
74. Grow BS, Legg MG, Yu W. Does vibration cause poststenotic dilatation in vivo and influence atherogenesis in cholesterol-fed rabbits? J Biomech Eng 1992; 114:20-25.
75. Rodbard S, Ikeda K, Montes M. An analysis of mechanisms of poststenotic dilatation. Angiology 1967; 18:349-367.
76. Kawaguti M, Hamano A. Numerical study on poststenotic dilation. Biorhealogy 1983; 20:507-516.
77. Ojha M, Johnston KW, Cobbold RSC. Evidence of a possible link between poststenotic dilatation and wall shear stress. J Vasc Surg 1990; 11:127-135.
78. Zarins CK, Zatina MA, Giddens DP. Shear stress regulation of artery lumen diameter in experimental atherogenesis. J Vasc Surg 1987; 5:413-420.
79. Guzman RJ, Krystowiak AJ, Zarins RK. Smooth muscle cell, C-fos gene expression precedes shear stress-induced aortic enlargement. J Vasc Surg, in press.
80. MacIver DH, Green NK, Gammage MD, Durkin H, Izzard AS, Franklyn A, Heagerty AM. Effect of experimental hypertension on phosphoinositide hydrolysis and proto-oncogene expression in cardiovascular tissues. J Vasc Res 1993; 30:13-22.
81. Petersen MJ, Abbott WM, H'Doubler PB, Jr, L'Italien GJ, Hoppel BE, Rosen BR, Fallon JT, Orkin RW. Hemodynamics and aneurysm development in vascular allografts. J Vasc Surg 1993; 18(6):955-964.
82. Gertz SD, Kurgan A, Eisenberg D. Aneurysm of the rabbit common carotid artery induced by periarterial application of calcium chloride in vivo. J Clin Invest 1988; 81:649-656.
83. Osborne-Pellegrin MJ, Coutard M, Poitevin P et al. Induction of aneurysms in the rat by a stenosing cotton ligature around the inter-renal aorta. Int J Pathol 1994; 75:179-190.
84. Koch AE, Hanes GK, Rizzo R et al. Human abdominal aortic aneurysms: Immuno-phenotypic analysis suggesting an immune mediated response. Am J Pathol 1990; 137:1199-1213.
85. Brophy CM, Reilly JM, Smith GJW et al. The role of inflammation in nonspecific abdominal aortic aneurysm disease. Ann Vasc Surg 1991; 5:229-233.
86. Pearce WH, Koch AE. Cellular components and features of immune response in abdominal aortic aneurysms. Ann N Y Acad Sci 1996; 800:175-185.
87. Anidjar S, Salzmann JL, Gentric D et al. Elastase-induced experimental aneurysms in rats. Circulation 1990; 82:973-981.
88. Anidjar S, Dobrin PB, Eichorst M et al. Correlation of inflammatory infiltrate with the enlargement of experimental aortic aneurysms. J Vasc Surg 1992; 16:139-147.
89. Dobrin PB, Baker WH, Gley WC. Elastolytic and collagenolytic studies of arteries. Arch Surg 1984; 119:405-409.
90. Dobrin PB, Mrkvicka R. Failure of elastin or collagen as possible critical connective tissue alterations underlying aneurysmal dilatation. Cardiovasc Surg 1994; 2:484-488.
91. White JV, Haas K, Phillips SJ et al. Adventitial elastolysis is a primary event in aneurysm formation. 1993; 17:371-381.
92. White JV, Mazzacco SL. Formation and growth of aortic aneurysms induced by adventitial elastolysis. Ann N Y Acad Sci 1996; 300:97-120.
93. Senior RM, Connolly NL, Cury JD et al. Elastin degradation by human alveolar macrophages: A prominent role of metalloproteinase activity. Am Rev Respir Dis 1989; 139:1251-1256.
94. Halpern VJ, Nockman GB, Gandhi RH et al. The elastase infusion model of experimental aortic aneurysms: Synchrony of induction of endogenous proteinases with matrix destruction and inflammatory cell responses. J Vasc Surg 1994; 20:51-60.
95. Newman KM, Jean-Claude J, Li H et al. Cytokines that activate proteolysis are increased in abdominal aortic aneurysms. Circulation 1994; 90:II224-II227.
96. Dobrin PB. In vivo model of aortic aneurysms. J Vasc Surg 1994; 20:150-152.
97. Dobrin PB, Baumgartner N, Anidjar S et al. Inflammatory aspects of experimental aneurysms. Effect of methylprednisolone and cyclosporine. Ann NY Acad Sci 1996; 800:74-88.

98. Ricci MA, Strindberg G, Slaiby JM et al. Anti-CD 18 monoclonal antibody slows experimental aortic aneurysm expansion. J Vasc Surg 1996; 23:301-307.
99. Holmes DR, Petrinec D, Wester W et al. Indomethacin prevents elastase-induced abdominal aortic aneurysms in the rat. J Surg Res 1996; 63:305-309.
100. Petrinec D, Liao S, Holmes DR et al. Doxycycline inhibition of aneurysmal degeneration in an elastase-induced rat model of abdominal aortic aneurysm: Preservation of aortic elastic associated with suppressed production of 92 kD gelatinase. J Vasc Surg 1996; 23:336-346.
101. Greenwald RA. Treatment of destructive arthritic disorders with MMP inhibitors: potential role of tetracyclines. Ann N Y Acad Sci 1994; 732:181-198.
102. Greenwald RA, Moak SA, Ramamirthy NS et al. Tetracyclines suppress matrix metalloproteinase activity in adjuvant arthritis and in combination with flurbiprofen ameliorate bone damage. J Rheumatol 1992; 19:927-938.
103. Nackman GB, Karkowski FJ, Halpern VJ et al. Elastin degradation products induce adventitial angiogenesis in the Anidjar/Dobrin rat aneurysm model. Surgery 1997; 122:39-44.
104. Anidjar S, Osborne-Pellegrin M, Coutard M et al. Arterial hypertension and aneurysmal dilatation. Kidney International 1992; 41(Suppl. 37):S61-S66.
105. Gadowski GR, Ricci MA, Hendley ED et al. Hypertension accelerates the growth of experimental aortic aneurysms. J Surg Res 1993; 54:431-436.
106. Ricci MA, Slaiby JM, Hendley ED et al. Hemodynamic and biochemical characteristics of the aorta in the WKY, SHR, WKHT and WKHA rat strains. Ann NY Acad Sci 1996; 800:121-130.
107. Slaiby JM, Ricci MA, Gadowski GR et al. Expansion of aortic aneurysms is reduced by propranolol in a hypertensive rat model. J Vasc Surg 1994; 20:178-183.
108. Ricci MA, Slaiby JM, Gadowski GR et al. Effects of hypertension and propranolol upon aneurysm expansion in the Anidjar/Dobrin aneurysm model. Ann NY Acad Sci 1996; 800:89-96.
109. Boudghene F, Anidjar S, Allaire E et al. Endovascular grafting in elastase-induced experimental aortic aneurysms in dogs: Feasibility and preliminary results. J Vasc Interven Radiol 1993; 4:497-504.

CHAPTER 8

Cytokine-Mediated Inflammation and Aortic Aneurysm Formation:
Molecular Studies Supporting the Inflammatory Model

Andy C. Chiou, William H. Pearce

Introduction

Early investigators interested in the mechanical properties of the aortic wall recognized that aneurysms must result from degradation of important structural proteins, including both collagen and elastin.[1-4] These investigators hypothesized that aneurysm formation was the result of an imbalance between synthesis and destruction of collagen and or elastin. Busuttil's original work investigating collagenolytic activity in aneurysm formation[5] provided the original experimental evidence for this theory and, in so doing, ushered in two decades of basic science research into aortic aneurysms. The original hypothesis, that aneurysmal degeneration of the aorta is associated with enzymatic degradation of structural matrix proteins by enzymes such as collagenases, elastases and other proteinases, has been significantly advanced in recent years.[6-8] From these investigations, it is clear that a complex remodeling process is present throughout the aortic wall and that elastin fragmentation, medial attenuation, and adventitial collagen turnover are part of the pathogenesis and progression of abdominal aortic aneurysmal disease.

Presently, four different classes of proteinases (endopeptidases) that degrade vascular extracellular matrix components are distinguishable. These are the serine-proteinases, cysteine-proteinases, aspartic-proteinases, and metalloproteinases.[9]

Matrix metalloproteinases (MMPs) have emerged as the principle class of destructive enzyme within the aneurysm wall.[10-12] Metalloproteinases belong to a family of zinc-dependent enzymes that are capable of degrading aortic wall substrates such as collagen, elastin, and gelatin. They require intrinsic zinc and extrinsic calcium for full enzymatic activity, and are generally inhibited by chelating agents.[9] MMPs are initially synthesized as proenzymes from a variety of cell types and are activated after secretion into the pericellular or extracellular space. Activation of the proenzyme occurs with amino-terminal propeptide cleavage, resulting in exposure of the active site zinc-binding domain. Early studies of aortic proteins detailed the presence of elevated levels of several MMPs, and each suggested a role for that particular enzyme in aneurysm formation. While these early descriptive studies provided a basic foundation, their fragmented data often left the casual reader confused as

---

to the role of each particular metalloproteinase in the overall disease process. More recent investigators have employed increasingly advanced molecular and biochemical techniques to precisely study the interplay between inflammation, cellular degradation and MMP expression.

The bulk of resulting experimental data point to inflammatory cytokines as the mediators of metalloproteinase expression within the aneurysm wall.[13-21] To date, several MMP's and their tissue inhibitors have been well documented within the wall of human abdominal aortic aneurysms. Table 8.1 summarizes these findings, including the cellular sources of these metalloproteinases. Studies employing in vitro cell culture techniques and in situ histologic analysis have slowly painted a clearer picture of the complex interplay between inflammatory cells, native aortic smooth muscle cells and cytokine-mediated MMP expression within the aneurysm wall. This chapter reviews the data from tissue, in vitro and in situ cytokine studies in an effort to paint for its reader a clearer picture of the complex interplay at work within the wall of aortic aneurysms.

Tissue Studies

Although cytokines stimulate metalloproteinase production in a variety of disease states, interest in the role of cytokines in aneurysm formation originated in early histologic studies demonstrating increased numbers of inflammatory cells within the diseased aortic wall.[14] Soon after the first inflammatory cytokines were discovered, several authors identified corresponding increases in cytokine levels in aneurysm tissue.

Aortic aneurysms have long been known to be characterized by periadventitial inflammatory infiltrates. Though recognized as potentially important for decades, recent histologic studies quantifying increases in the amount of inflammation have focused attention upon the role in the disease state. In fact, these studies demonstrate that the degree of aortic inflammation, as measured by the number of periadventitial macrophages and lymphocytes, steadily increases from normal aorta to atherosclerotic aorta to aneurysms.[29] In addition, when occlusive tissue specimens were compared to aneurysm biopsies, the bulk of the inflammatory infiltrate was localized to the aneurysm adventitia as compared to the intimal and medial occlusive infiltrates.[29] When studied with immunohistochemistry, over 80% of the cells comprising the aneurysm infiltrate were T and B lymphocytes, with the remainder evenly split between macrophages and vascular smooth muscle cells.[29,30] Finally, aortic vascular smooth muscle cells surrounding the inflammatory infiltrate were seen to express the HLA-DR antigen, suggesting that soluble inflammatory factors had precipitated activation of the normally quiescent smooth muscle cell.[30]

Following the lead of investigators in other fields, especially those studying rheumatoid arthritis and wound healing, several groups sought to identify cytokines within aneurysm tissue. TNF and IL-1β were the first two cytokines identified as increased in aneurysm tissue extracts.[18,20] Subsequent studies have shown similar increases in IL-6 and IFN-γ protein levels.[31,32] Recently, several studies have demonstrated expression of multiple cytokines within the aneurysm wall; findings which serve to confirm protein data but which also demonstrate the difficulty inherent in a descriptive study of a complex system. Table 8.2 provides a listing of the cytokines identified within the aneurysm wall to date.

In Vitro Studies

In an effort to obtain some mechanistic information about the role of the dizzying array of cytokines identified in aneurysm tissue, a number of investigators turned to cell culture experiments. Although admittedly limited by the artificial environments, such experiments do allow implementation of controls necessary for true hypothesis testing.

Table 8.1. Matrix metalloproteinases expressed in abdominal aortic aneurysms

Enzyme	Cellular Source	kDa	References
MMP-1 (Interstitial Collagenase)	Endothelial Cells	52	9
	Macrophages	22	
	Monocytes	23	
MMP-2 (Gelatinase A)	Macrophages	72	24
	Smooth muscle cells	22	
MMP-3 (Stromelysin-1)	Macrophages	57	36
MMP-7 (Matrilysin)	Unknown	28	25
MMP-8 (Collagenase-2)	Macrophage	75	26
MMP-9 (Gelatinase-B)	Macrophage	92	22,27
MMP-12 (Macrophage metalloelastase)	Macrophage	57	28
MMP-14 (Membrane-Type 1 MMP)	Smooth muscle cell	66	25
	Macrophage	25	

Table 8.2. Cytokines identified in aortic aneurysms

Cytokine	Tissue Studied	Measured As	References
TNF-α	Aneurysm Extract	Protein	18,20
IL-1β	Aneurysm Extract	Protein	18,20
IL-1α	Aneurysm Tissue	mRNA	45
IL-2	Aneurysm Tissue	mRNA	45
IL-4	Aneurysm Tissue	mRNA	45
IL-6	Aneurysm Cell Culture	Protein	32
IFN-γ	Aneurysm Cell Culture	Protein	32

Stimulation experiments are the most widely applied cell culture technique in the study of aortic aneurysms. Keen and coworkers were the first to identify differential patterns of MMP-1 expression by vascular smooth muscle cells cultured from aneurysmal and normal donors.[33] This experiment suggested that phenotypic differences in aneurysm cells may lead to an exaggerated response to inflammation and eventual aneurysm formation. Nagasi and coworkers demonstrated that platelet derived growth factor stimulated MMP-1 production by cultured aortic smooth muscle cells and suggested that cytokines released as part of an inflammatory response to an as yet unidentified stimulus result in MMP-1 based collagenolysis and aneurysm formation.[15]

Later studies identified particular patterns of inducible metalloproteinase expression by proinflammatory cytokines. In one study, interleukin-1 (IL-1) and platelet derived growth factor (PDGF) synergistically increased MMP-9 expression by cultured vascular smooth muscle cells while having no effect on tissue-inhibitor of metalloproteinase (TIMP) or other MMP expression.[34] These results suggest that proinflammatory cytokines selectively induce expression of degratory enzymes by vascular smooth muscle cells and, in so doing, may cause matrix destruction and resultant aneurysm formation. Finally a recent study by Koch et al demonstrated that IL-8 and TNF were responsible for endothelial cell migration in response to aneurysm supernatants,[45] suggesting that cytokines may also be responsible for the marked neovascularization of the aneurysm adventitia.

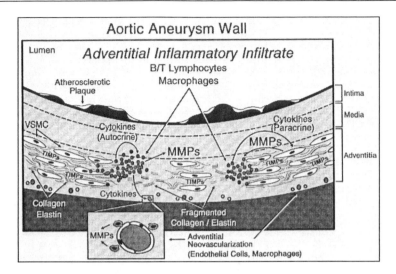

Fig. 8.1. Schematic drawing illustrating localization of inflammation and proposed mechanisms of cytokine-mediated connective tissue destruction within the aneurysm wall.

Despite the large number of cytokines identified with aneurysm tissue and mechanistic suggestions supported by in vitro data, the relative importance of cytokines in aneurysm formation is unknown. Although the concept (Fig. 8.1) of inflammatory cell cytokine-mediated stimulation of metalloproteinase expression leading to AAA expansion is appealing, future studies are needed to identify which, if any, cytokines initiate the inflammatory cascade or prolong metalloproteinase expression within the aortic wall.

In Situ Studies

Recently there has been renewed interest in histologic study of the aneurysm wall. Authors from several centers have recognized that the adventitia is the site of metalloproteinase production, structural protein remodeling, and inflammatory infiltrate formation. This is in contrast to studies of peripheral vascular atherosclerotic disease and occlusive aortic specimens, which show that atherosclerotic plaque is the most active site of inflammatory cell-associated tissue remodeling.[33,36] Advances in molecular techniques allowing for in situ hybridization of mRNA, coupled with immunohistochemistry and zymography have allowed for more precise characterization of inflammatory and metalloproteinase activity within the aneurysm wall. A summary of the known and hypothesized activity is discussed below.

Early studies of aneurysm histology identified large numbers of inflammatory infiltrates within the adventitia.[14] Later, several structural studies of aortic aneurysms suggested that it was the loss of adventitial elastin and collagen which occurred early in aneurysm formation.[37-39] These two observations prompted a number of recent investigators to hypothesize that the adventitia is the active site of ongoing tissue remodeling within aortic aneurysms.

Currently, a number of metalloproteinases and TIMPs have been localized to cells within the adventitia. Immunohistochemical studies have identified broad, low level production of MMP-1 by adventitial vascular smooth muscle cells.[40] On the other hand, a combination of immunohistochemical and in situ hybridization studies have localized MMP-9, MMP-2 and MMP-3 expression to inflammatory cells within periadventitial infiltrates, suggesting an active role for inflammatory cells in the ongoing aneurysm expansion.[33,39,41,42] Other studies have localized metalloproteinase production to endothelial cells surrounding adventitial vasa vasorum, suggesting that metalloproteinases may facilitate the marked neovascularization which characterizes aortic aneurysms.[33,43] In limited studies, TIMPs appear to colocalize with MMPs suggesting that the enzymes are expressed concurrently.[33,40,41] As previously mentioned, work has concentrated on the identification of specific MMPs in AAA. In our laboratory, we compared relative levels of expression of particular MMPs to one another and their TIMPs.[44] We found increases in expression of MMP-1, MMP-9, and MMP/TIMP ratios. MMP-9 appeared to be the predominant metalloproteinase expressed in AAA, with its mRNA levels more than 20 times and two times higher than those of MMP-1 and MMP-2, respectively. Other laboratories also have applied reverse transcriptase-PCR techniques with similar findings of a significant increase in MMP-9 levels compared to the other MMPs. There were small increases in MMP-2 levels in the aortic specimens.[27] However, there are no data regarding the relative tissue concentrations of TIMPs and MMPs at various sites within the aortic wall; it may be that a relative imbalance of metalloproteinase production at a particular site is crucial to aneurysm formation.

Recently, work by Allaire et al[46] has also focused on the prevention of aneurysm development by specific inhibition of MMP activation. They described the local overexpression of plasminogen activator inhibitor-1 (PAI-1) in a rat model of aortic aneurysm. In this study, aortic xenografts were seeded with smooth muscle cells retrovirally transduced with rat PAI-1 gene or the vector alone. Quantitative zymography showed decreased levels of MMPs in the PAI-1 group, thus linking MMP activation to the plasminogen activator/plasmin pathway. Such animal models may provide unique opportunities to further our understanding of inflammatory, cytokine-mediated pathways leading to aortic aneurysm formation.

Although MMP and TIMP expression have been localized to the adventitia of aneurysms, a number of questions remain unanswered. Specifically, what is the role of cytokines in metalloproteinase production? Do the inflammatory cytokines act in an autocrine fashion within the infiltrates, or do they act in a paracrine fashion stimulating adjacent vascular smooth muscle cells to produce metalloproteinase? Is it an imbalance in a cellular response to cytokine stimulation, an absence of particular cytokine receptors, or a defect in autoregulation which leads to ongoing aneurysm expansion. These basic unanswered questions are crucial to a clearer understanding of the complex interplay between inflammatory cytokines, vascular smooth muscle cells, and inflammatory cells which eventually leads to elevated metalloproteinase production and ongoing destruction of structural protein within the adventitia of aortic aneurysms.

Conclusion

It is an exciting time to the study of aortic aneurysms. The data from a number of sources is rapidly converging and it appears that cytokine-mediated metalloproteinase expression within the adventitia of aneurysms is, at least, in part responsible for aortic expansion. Future studies which more precisely define the interplay of cytokines and metalloproteinases within the aortic wall would allow for the identification of a basic defect leading to aneurysm formation. Immunologic, hemodynamic, infectious or genetic causes could all initiate the inflammatory cascade detailed within this review. Thus, while recent advances in scientific technique have allowed for better molecular understanding of ongoing aortic expansion, the basic etiologic question remains unanswered.

Acknowledgment

Supported in part by grants from: Alyce F. Salerno Foundation and the Violet Baldwin Family Research Fund.

References

1. Glagov S. Morphology of collagen and elastin fibers in atherosclerotic lesions. Adv Exp Med Biol 1975; 82:767-773.
2. Sumner DS, Hokansen DE, Strandness DE. Stress-strain characteristics and collagen-elastin content of abdominal aortic aneurysms. Surg Gynecol Obstet 1970; 130:459-466.
3. Campa JS, Greenhalgh RM, Powell JT. Elastin degradation in abdominal aortic aneurysms. Atherosclerosis 1987; 65:13-21.
4. Brown SL, Backstrom B, Busuttil RW. A new serum proteolytic enzyme in aneurysm pathogenesis. J Vasc Surg 1982; 2:393-399.
5. Busuttil RW, Abou-Zamzam AM, Machleder HI. Collagenase activity of the human aorta: a comparison of patients with and without aortic aneurysms. Arch Surg 1980; 115: 1373-1378.
6. Ghorpade A, Baxter BT. Biochemistry and molecular regulation of matrix macromolecules in abdominal aortic aneurysms. Ann N Y Acad Sci 1996; 800:138-50.
7. Gandhi RH, Irizarry E, Cantor JO et al. Analysis of elastin cross-linking and the connective tissue matrix of abdominal aortic aneurysms. Surgery 1994; 115:617-620.
8. Minion DJ, Davis VA, Mejezchleb PA et al. Elastin is increased in abdominal aortic aneurysms. J Surg Res1994; 57:443-446.
9. Emonard H, Grimaud JA. Matrix metalloproteinases. A review. Cell Mol Biol 1990; 36(2):131-153.
10. Vine N and Powell JT. Metalloproteinases in degenerative aortic disease. Clinical Science 1991; 81:233-239.
11. Newman KM, Ogata Y, Malon AM et al. Identification of matrix metalloproteinases 3 (stromelysin-1) and 9 (gelatinase B) in abdominal aortic aneurysm. Arterioscler Thromb 1994; 14:1315-1320.
12. Irizarry E, Newman KM, Gandhi RH et al. Demonstration of interstitial collagenase in abdominal aortic aneurysm disease. J Surg Res 1993; 54:571-574.
13. Tilson MD, Elefriades J, Brophy CM. Tensile strength and collagen in abdominal aortic aneurysm disease. In: Greenhalgh RM, Mannick JA, Powell JT, eds. The Cause and Management of Aneurysms. London: WB Saunders, 1990; 97-104.
14. Koch AE, Haines GK, Rizzo RJ et al. Human abdominal aortic aneurysms: Immuno-phenotypic analysis suggesting an immune-mediated response. Am J Pathol 1990; 137:1199-1213.
15. Yanagi H, Sasagur Y, Sugama K et al. Production of tissue collagenase (matrix metalloproteinase 1) by human aortic smooth muscle cells in response to platelet-derived growth factor. Atherosclerosis 1992; 91:207-216.
16. Faggioli GL, Gargiulo M, Bertoni F et al. Parietal inflammatory infiltrate in peripheral aneurysms of atherosclerotic origin. J Cardiovasc Surg 1992; 33:331-336.
17. Okada Y, Katsuda S, Okada Y et al. An elastolytic enzyme detected in the culture medium of human arterial smooth muscle cells. Cell Biology International 1993; 17(9)863-869.
18. Pearce WH, Sweis I, Yao JST et al. Interleukin-1β and tumor necrosis factor-a release in normal and diseased human infrarenal aortas. J Vasc Surg 1992; 16:784-789.
19. Evans CH, Georgescu HI, Lin CW et al. Inducible synthesis of collagenase and other neutral metalloproteinases by cells of aortic origin. J Surg Res 1991; 51:339-404.
20. Lee E, Vaughan DE, Parikh SH et al. Regulation of matrix metalloproteinases and plasminogen activator inhibitor-1 synthesis by plasminogen in cultured human vascular smooth muscle cells. Circ Res 1996; 78:44-9.
21. Newman KM, Malon AM, Shin RD et al. Matrix metalloproteinases in abdominal aortic aneurysm: Characterization, purification, and their possible sources. Connective Tissue Research 1994; 30:265-276.
22. Welgus HG, Campbell EJ, Bar-Shavit Z et al. Human alveolar macrophages produce a fibroblast-like collagenase and collagenase inhibitor. J Clin Invest 1985; 76:219-224.
23. Campbell EJ, Cury JD, Lazarus CJ et al. Monocyte procollagenase and tissue inhibitor of metalloproteinases. Identification, characterization and regulation of secretion. J Biol Chem 1987; 262:15862-15868.

24. Hibbs MS, Hoidal JR, Kang AH. Expression of a metalloproteinase that degrades native type V collagen and denatured collagens by cultured human alveolar macrophages. J Clin Invest 1987; 80:1644-1650.

25. Thompson RW, Parks WC. Role of matrix metalloproteinases in abdominal aortic aneurysms. Ann NY Acad Sci 1996; 800:138-50.

26. Shipley JM, Wesselschmidt RL, Kobayashi DK et al. Metalloelastase is required for macrophage-mediated proteolysis and matrix invasion in mice. Proc Natl Acad Sci USA 1996; 93(9):3942-6.

27. Elmore JR, Keister BF, Franklin DP et al. Expression of matrix metalloproteinases and TIMPs in human abdominal aortic aneurysms. Ann Vasc Surg 1998; 12:221-8.

28. Matsumoto S, Kobayashi T, Katoh M et al. Expression and localization of matrix metalloproteinase-12 in the aorta of cholesterol-fed rabbits: relationship to lesion development. Am J Pathol 1998; 153:109-19.

29. Koch AE, Haines GK, Rizzo RJ et al. Human abdominal aortic aneurysms: Immuno-phenotypic analysis suggesting an immune mediated response. Am J Pathol 1990; 137;1199-1213.

30. Pasquinelli G, Gargiulo PM, Vici M et al. An immunohistochemical study of inflammatory abdominal aortic aneurysms. J. Submicrosc Cytol Pathol 1993; 5:103-112.

31. Ramshaw AL, Roskell DE, Parums DV. Cytokine gene expression in aortic adventitial inflammation associated with advanced atherosclerosis (chronic periaortitis). J Clin Pathol 1994; 47:721-727.

32. Szekanecz Z, Shah MR, Pearce WH et al. Human atherosclerotic abdominal aortic aneurysms produce interleukin (IL)-6 and interferon-gamma but not IL-2 and IL-4: The possible role for IL-6 and interferon-gamma in vascular inflammation. Agents, Actions 1994; 41:1-4.

33. McMillan WD, Patterson BK, Keen RR et al. In situ localization and quantification of seventy-two kilodalton type IV collagenase in aneurysmal, occlusive, and normal aorta. J Vasc Surg 1995; 22:295-305.

34. Fabunmi RP, Baker AH, Murray EJ et al. Divergent regulation by growth factors and cytokines of 95kDa and 72 kDa gelatinases and tissue inhibitors or metalloproteinases-1, -2, -3 in rabbit aortic smooth muscle cells. Biochem J 1996; 315:335-42.

35. Pearce WH, Koch AE. Cellular components and features of immune response in abdominal aortic aneurysms. Ann NY Acad Sci 1996; 800:175-83.

36. White JV, Haas K, Phillips S et al. Adventitial elastolysis is a primary event in aneurysm formation. J Vasc Surg 1993; 17:371-381.

37. Stehbens WE, Martin BJ. Ultrastructural alterations of collagen fibrils in blood vessel walls. Conn Tiss Res 1993; 29:319-331.

38. Dobrin PB, Baker WH, Gley WC. Elastolytic and collagenolytic studies of arteries. Arch Surg 1984; 119:405-409.

39. Newman KM, Jean-Claude J, Li H et al. Cellular localization of matrix metalloproteinases in the abdominal aortic aneurysm wall. J Vasc Surg 1994; 20:814-820.

40. McMillan WD, Patterson BK, Keen RR et al. In situ localization and quantification of mRNA for 92kD Type IV collagenase and its inhibitor in aneurysmal, occlusive, and normal aorta. Thomb Vasc Biol 1995; 15:1139-1144.

41. Freestone T, Turner RJ, Coady A et al. Inflammation and matrix metalloproteinases in the enlarging abdominal aortic aneurysm. Arterioscler Thromb Vasc Biol 1995; 15:1145-1151.

42. Newman K, Ogata Y, Malon A, et al. Identification of matrix metalloproteinases 3 (Stromelysin-1) and 9 (Gelatinase B) in abdominal aortic aneurysm. Arterioscler Thromb 1994; 14:1315-1320.

43. Herron GS, Unemori E, Wong M et al. Connective tissue proteinases and inhibitors in abdominal aortic aneurysms. Arteriosclerosis Thromb. 1991; 11:1667-77.

44. Tamarina NA, McMIllan WD, Shively VP et al. Expression of matrix metalloproteinases and their inhibitors in aneurysms and normal aorta. Surg 1997; 122:264-71.

45. Szekanecz, Shah, Harlow et al. Interleukin-8 and tumor necrosis factor-alpha are involved in human aortic endothelial cell migration. Pathobiology, 1994; 62:134-139.

46. Allaire E, Hasenstab D, Kenagy RD et al. Prevention of aneurysm development and rupture by local overexpression of plasminogen activator inhibitor-1. Circulation 1998; 98:249-255.

CHAPTER 9

Pharmacologic Treatment of Aneurysms

Richard R. Keen

Introduction

The pathophysiology of arterial aneurysms consists of two stages. The first stage is the initiation or formation of the aneurysm. Damage to the elastic fibers within the internal elastic lamina and the subsequent loss of the integrity of this elastin network is a consistent histologic finding in the initial stages of aneurysm development. The hemodynamic forces of shear stress, hypertension, and pulse wave reflection at arterial bifurcations and branch points each appear to contribute to the degeneration of the arterial wall elastic layer that is required for aneurysm formation to occur. However, thinning of the arterial wall medial layer through the loss of elastin and smooth muscle cells alone is associated only with the earliest stages of arterial expansion in the course of aneurysm development.[1,2] The initial formation of aneurysms is characterized histologically by the *lack* of an inflammatory cell infiltrate.[3]

The second stage of aneurysm pathophysiology is the growth or progression of the aneurysm. During aneurysm growth, more extensive changes occur within the architecture of the arterial wall, as the adventitia and intima also experience significant changes, in addition to further changes found in the arterial media. Aneurysm growth appears to be related closely to changes in arterial wall collagen metabolism as much as to changes in elastin synthesis and degradation. As the aneurysm grows, the medial layer of the artery becomes more attenuated, but the total arterial wall thickness does not decrease because both the adventitia and intima undergo increases in thickness. Remodelling of the arterial wall collagen in the adventitia and intima results in thickening of the adventitial and intimal layers. The total content of arterial wall collagen and elastin increases.[4] However, further increased synthesis of other arterial wall matrix proteins besides collagen in these regions of the arterial wall results in a net decrease in arterial wall collagen concentration. The new synthesis of collagen and elastin does not appear to compensate for the original loss of organized elastin from the media, as the artery never regains its original dimensions. In fact, the total collagen content continues to increase and its concentration decrease with increasing aneurysm diameter. Aneurysm rupture is the "terminal step" in aneurysm growth.

Even though the initial stages of aneurysm development are characterized by the lack a of significant inflammatory reaction within the arterial wall,[5] the subsequent stages of aneurysm growth and progression are remarkable for histologic evidence of an inflammatory process within the arterial wall. This inflammatory infiltrate of macrophages and lymphocytes

Development of Aneurysms, edited by Richard R. Keen and Philip B. Dobrin. ©2000 Eurekah.com.

produces cytokines.[6] Through transcellular metabolism,[7] these inflammatory cells appear to modulate changes in arterial wall architecture by influencing collagen and elastin metabolism. Angiogenesis is a component of the inflammatory reaction that occurs during aneurysm growth, and this angiogenesis is most marked in the arterial wall adventital layer.[8]

Pharmacologic agents may prove useful in the prevention or medical treatment of arterial aneurysms. Drug treatments that could block either the formation or the growth of aneurysms would play a real role in the management of this disease. This chapter discusses the potential for using pharmacologic agents to either prevent the formation or postpone the progression of arterial aneurysms. The mechanisms by which these medications may alter collagen and elastin synthesis and collagen and elastin degradation within the arterial wall are discussed. In addition, pharmacologic agents that affect matrix protein metabolism through their effects on inflammation also are reviewed.

The Etiology of Aneurysms: Is Prevention Possible?

The evidence supporting an etiologic relationship between atherosclerosis and aneurysm formation is difficult to refute.[9] Atherosclerosis is a degenerative process of the arterial wall that begins before birth.[10] Thickening of the arterial intima is the initial manifestation of atherosclerosis. This intimal proliferation develops at predictable locations, which include the lateral walls of arteries at bifurcations and at arterial bends.[11]

Initmal plaque forms in response to hemodynamic injury.[12] The finding that plaque forms at locations within the arterial tree where hemodynamic stress is greatest is consistent with this observation. This hemodynamic stress causes tears and microfractures within the internal elastic lamellae[13,14] and causes reduced tensile strength within the arterial wall. These degenerative changes are visible on electron microscopy at the earliest stages of the development of atherosclerosis.[3]

Areas of the arterial wall where hemodynamic stress is greatest and that initially only demonstrate intimal thickening progress over many years into areas of developed atherosclerotic plaque. Accompanying the maturation and further thickening of the atherosclerotic plaque are other changes within the entire arterial wall, including atrophy of both the medial layer and the medial smooth muscle cells, lipid accumulation, ground substance deposition, inflammatory responses, and dystrophic calcification.[9]

But not all arteries that develop atherosclerotic plaques become stenotic. Arterial dilation occurs in nearly all arteries with advancing age, and arterial enlargement is a ubiquitious finding in atherosclerosis.[15] Compensatory arterial enlargement accompanies plaque deposition and aging in the coronary arteries,[16] the carotid arteries,[17] the aorta,[18] and in the superficial femoral artery.[19] But while arterial enlargement is a consistent finding with atherosclerotic plaque formation, arterial enlargement is not equivalent to aneurysm formation.

Stehbens proposed that the same degenerative process that is responsible for the formation of aneurysms, that is mural degeneration and the loss of tensile strength, is the same process that leads to atherosclerosis.[20] Stehbens contends that arterial aneurysms are a continuation, or late complication, of atherosclerosis.[3] Stehbens suggests that the degeneration of the atherosclerotic plaque leads to the development of the complications of athersclerosis and the clinical manifestations of this disease. These complications of atherosclerosis include arterial stenosis, plaque ulceration, thrombosis, atheroembolization, and aneurysm formation (See Fig. 10.12). Patients with aneurysms tend to be significantly older than patients with occlusive arterial disease. This theory helps explain the approximate 10 year lag between the time that patients present with aortic aneurysms as opposed to aortic occlusive disease.[9]

Similar gross morphologic changes occur in the arterial wall with both occlusive disease and aneurysms, but the histologic findings are different.[20] Accompanying the plaque atrophy found in aneurysms is mural thinning and the loss of the medial elastin lamellar architecture.[21] This loss of the medial elastic structure has been shown to be the histologic finding common to both animal models of aneurysms and human aneurysm specimens. The evolution of the atherosclerotic plaque to the stage of plaque regression, which is accompanied by mural medial thinning, also helps to explain the approximately 10 year greater age difference between patients with aortic aneurysms and those with occlusive disease. The processes of plaque evolution and degeneration also help to explain why a large percentage of patients with arterial aneurysms often present with several of the other late complications of atherosclerosis in other vascular beds.[22]

Plaque formation is accompanied by medial atrophy.[5] Stable atherosclerotic fibrous plaque that overlies a thinned medial elastic layer may provide increased tensile strength to an otherwise weakened arterial wall.[17] Plaque progression, when manifest as plaque regression or thinning, may render the artery with an inability to sustain the mechanical pressures it could with a stronger, more fibrous plaque. Plaque calcification may act as buttress to increase arterial tensile strength.[24] The question arises as to whether the current cholesterol lowering agents accelerate plaque thinning and involution.[25] If so, then the plaque thinning that would accompany the use of these agents would have the potential to accelerate the loss of tensile strength that occurred with mural thinning. The ensuing result could be further arterial dilation.

The necessary change that must occur within in the arterial wall in order for an arterial aneurysm to form is the loss of mural tensile strength. The pathogenesis of aneurysms[26] should not be confused with their etiology,[27] and Stehbens has proposed that the ultimate cause of the the loss of arterial wall tensile strength is hemodynamic stress, whereby the cumulative effects of tensile pressure, vibration, and oscillating flow result in a biomechanical fatigue of the arterial wall that permits arterial dilation to occur.[12]

Dobrin has proposed that the activation of the elastolytic and collagenolytic enzymes that accompanies an inflammatory event within the arterial wall is the predominant requirement for arterial enlargement to occur.[28] The exact role of the proteolytic enzymes in the development of aneurysms has yet to be determined, but this role may be significant. Inflammation within the arterial wall activates enzymes that degrade elastin and collagen. If these enzymes are essential for aneurysms to grow, then one could block interfere with aneurysm progression by blocking the function of these proteolytic enzymes.[29]

Matrix Metalloproteinases and Their Pharmacologic Blockade

The matrix metalloproteinases (MMPs) are a family of zinc-dependent enzymes with a broad range of activity for the structural proteins within the extracellular connective tissue matrix. The primary role of the matrix metalloproteinases appears to be connective tissue remodelling.[30] The proteins that serve as substrates for the active matrix metalloproteinases include collagen, gelatin (also known as fibrillar collagen), and elastin. Metalloproteinases are secreted from their respective synthesizing cells into the extracellular matrix as inactive zymogens. This proenzyme form of inactive matrix metalloproteinase may serve as the substrate of another active metalloproteinase. Experimental evidence to date has suggested that four of the matrix metalloproteinases, MMP-1, MMP-2, MMP-3, and MMP-9, may play a significant role in the pathogenesis of aneurysms.[26]

A varied nomenclature is used to denote these four particular matrix metalloproteinases that may be involved in the pathogenesis of aneurysms. These matrix metalloproteinases can be referred to either by a number, such as MMP-1, or by description. The functional or descriptive term for MMPs often incorporates into its name both the substrate of the enzyme and the molecular weight of the active form of the enzyme that is found on gel enzymography. For example, MMP-1 is most commonly called interstitial collagenase and it degrades type I collagen, the most prevalent type of collagen found within the walls of large arteries.[4] MMP-2 is called gelatinase A or 72 kDa type IV collagenase and it degrades both gelatin (fibrillar collagen) and elastin. MMP-3 is called stromelysin 1. Activated stromelysin is able to cleave the proenzyme form of interstitial collagenase (pro-MMP-1) into the active form of MMP-1. MMP-9 is referred to as gelatinase B or 92 kDa type IV collagenase and it also degrades both gelatin (fibrillar collagen) and elastin.

The arterial wall medial smooth muscle cells are the synthetic source for three of these matrix metalloproteinases:[31] MMP-1 (intersititial collagenase),[32] MMP-2 (gelatinase A),[33] and MMP-3 (stromelysin)[34-36] are produced by the medial smooth muscle cells and are found in the walls of both normal and aneurysmal arteries.[37]

The fourth matrix metalloproteinase, MMP-9 (gelatinase B) is not produced by arterial wall smooth muscle cells.[38] Instead, infiltrating macrophages are the source of MMP-9 that is found in atherosclerotic arteries, including aneurysms.[39,40]

The capacity of both activated MMP-9[29] and MMP-2[41] to degrade elastin potentially could be regulated by pharmacologic intervention. Since MMP-9 is produced by macrophages and MMP-2 is synthesized by smooth muscle cells, two distinct cell types could be the target of interventions designed to block elastin degradation.

Matrix metalloproteinase activity within the arterial wall is tightly regulated at several levels (Fig. 9.1).[42] The MMP cascade within the arterial wall is controlled initially at the level of transcription. Processing of the MMP mRNA and the synthesis of the MMP proenzyme are forms of posttranscriptional regulation.[43] Metalloproteinases are secreted from the cell as zymogens, where they are subject to activation in the extracellular environment. This system provides another potential mechanism for controlling MMP activity.

Plasminogen activators and inhibitors comprise the fibrinolytic cascade and control the cleavage of plasmin from plasminogen. The plasmin that is formed is a serine protease that controls matrix protein levels by converting the metalloproteinase zymogens into active, functioning MMPs.[44-47] The conversion of proenzyme forms of MMP-1, MMP-2, MMP-3, and MMP-9 into active metalloproteinases is dependent on active plasmin. Active forms of the metalloproteinases have been found in the human aneurysms.[48,49]

The activation of the plasmin from serum plasminogen is dependent on either of the two plasminogen activators: tissue plasminogen activator (tPA) or urokinase (uPA).[50] The presence of tissue plasminogen activator (tPA) or urokinase (uPA) could be important in the regulation of MMP activity in aneurysms, so both of these plasminogen activators could to be important in aortic aneurysm pathogenesis.[51] Urokinase (uPA) is produced by inflammatory cells within the aneurysm wall.[52] Tissue plasminogen activator (tPA) is produced by smooth muscle cells.[53]

The plasminogen activators tPA and uPA are themselves inhibited by the plasminogen activator inhibitors, PAI-1 and PAI-2. The plasminogen activator inhibitors provide another level for the control of MMP activity by preventing the plasminogen activators tPA and uPA from converting plasminogen into plasmin.[54-57]

The active forms of the MMP enzymes are directly inhibited by the tissue inhibitor of metalloproteinases, or TIMPs.[43] Activated MMPs form dimer complexes with the appropriate TIMP that blocks all MMP enzyme activity: MMP-1, MMP-3, and MMP-9 form dimers with TIMP-1,[58] and MMP-2 forms a dimer complex with TIMP-2.[59]

Pharmacologic strategies have been proposed for inhibiting the degradation of arterial wall matrix proteins by blocking the function of active matrix metalloproteinases. Tetracycline derivatives[60] and the tissue inhibitor of metalloproteinases (TIMPs)[61] block the active metalloproteinase enzyme. Alternatively, the activation of the MMP zymogens could be

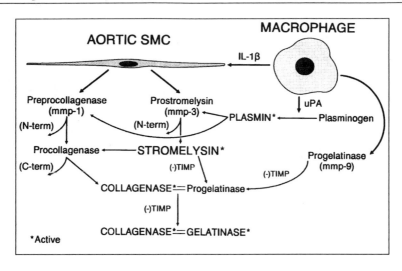

Fig. 9.1. The interaction between macrophages and smooth muscle cells (SMC) in the aortic wall. Macrophages directly affect matrix protein degradation through the secretion of MMP-9 and indirectly through the production of urokinase (uPA) and interleukin-1β (IL-1β). Reprinted with permission from: Pearce WH, Koch AE. Cellular components and fractures of immune response in abdominal aortic aneurysms. Ann NY Acad Sci 1996; 800:175-185.

inhibited by using the plasminogen activator inhibitor (PAI) to block the formation of plasmin that is required to activate the MMP proenzymes (Fig. 9.2).[57]

Other pharmacologic interventions that may inhibit aneurysm progression could work by blocking the inflammatory response found in the aneurysm wall. A large part of the intracellular and intercellular messaging characteristic of inflammation is dependent on the activity of the enzyme cyclooxygenase (COX).[62] Cyclooxygenase plays an important role in inflammation, as cyclooxygenase forms prostaglandins, thromboxanes, and leukotrienes (eicosanoids) from free fatty acids. Cyclooxygenase is not the rate-limiting step in prostaglandin biosynthesis but it is the committed step in the formation of prostaglandins.[63] Cyclooxygenase is activated by lipid hydroperoxides, an oxygen radical that is a powerful intracellular and intercellular messenger.[64] The cyclooxygenase reaction also produces more lipid hydroperoxides as a byproduct, resulting in the formation of greater levels of hydrperoxides. Accelerated cyclooxygease activity causes further activation of cyclooxygenase and a higher "peroxide tone".[62] The nonsteroidal anti-inflammatory drugs act by either irreversibly (e.g., aspirin) or reversibly (e.g., ibuprofen) blocking cyclooxygenase. These nonsteroidal anti-inflammatory medications may have a role in the treatment of aneurysms.

Finally, antihypertensive medications may have a surprisingly crucial role in the drug therapy of aneurysms.

Tetracycline Derivatives

Tetracyclines achieve their broad-spectrum antibiotic properties by blocking mitochondrial protein synthesis. Tetracycline derivatives appear to block matrix metalloproteinase activity by a different mechanism, whereby they bind to Zn or Ca in the enzyme molecule. One theory proposes that tetracycline derivatives may bind to a secondary

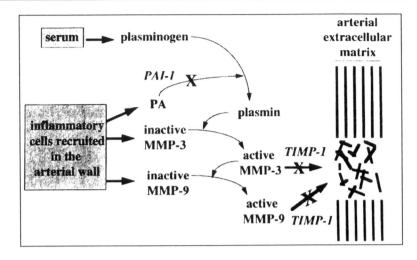

Fig. 9.2. The interaction between plasminogen and matrix metalloproteinases (MMP) in the extracellular matrix. The plasminogen activators that participate in MMP activation are blocked by the plasminogen activator inhibitors (PAI-1). The active MMP enzymes are blocked by tissue inhibition of metalloproteinase (TIMP-1). Reprinted with permission from Allaire E, Masanstab D, Kenegy RD et al. Prevention of aneurysm development and rupture by suppression of plasminogen activator inhibitor-1. Circulation 1998; 98:249-255.

Zn or Ca that is not at the active site. The result is a conformational change in the MMP proenzyme. When the MMP proenzyme interacts with the serine protease plasmin, the conformational change in the proenzyme induced by the tetracycline derivative may cause the proenzyme to be cleaved by the plasmin into smaller, inactive MMP fragments.[65] The new application of the tetracycline derivative doxycycline to inhibit MMP activity in aortic aneurysm disease originated from studies of tetracycline derivatives in peridontal disease.[66] Doxycycline has been shown to have a direct inhibitory effect on both MMP-2 and MMP-9.[67] One appeal of the use of tetracycline derivatives is that their long-term safety has been demonstrated.[68]

Petrinec et al investigated the possible role of tetracycline derivatives in the suppression of aneurysmal growth in animals where aneurysmal degeneration was initiated with the elastase infusion model.[69] Using the Andijar-Dobrin model of elastase infusion in rats, Petrinec et al observed that doxycycline inhibited aneurysm formation compared to animals not treated with the tetracycline derivative (Fig. 9.3).[60] Structural deterioration of aortic elastin was decreased by doxycycline treatment even while aortic wall inflammatory cell infiltration was not decreased with doxycycline. Compared to the control animals who formed aneurysms but did not receive doxycycline, the levels of active MMP-9 were decreased in the aortic wall of animals who were treated with doxycycline. One important finding was that doxycycline did not completely arrest the progressive increase in arterial diameter (Fig. 9.3).

The capability of doxycycline to reduce matrix protein degradation in arterial organ cultures was examined by Boyle et al.[69] Doxycycline at a dose of 10 mg/mL reduced aortic tissue elastin loss (Fig. 9.4).[69] The production of MMP-9 usually is increased in the aortic wall in this model. Doxycycline decreased aortic wall MMP-9 activity (Fig. 9.5),[69] but once again the doxycycline did not completely stop the loss of aortic elastin.[68,69] Tissue levels of metalloproteinase inhibitor TIMP-1 and TIMP-2 were not decreased in the doxycycline-treated group.[69] More recently, doxycycline treatment initiated preoperatively in patients

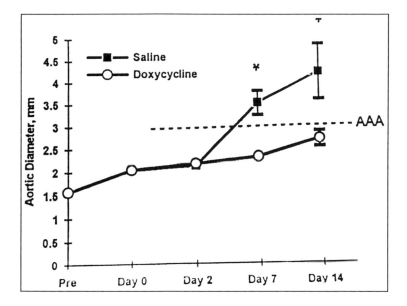

Fig. 9.3. The effect of doxycycline on aortic diameter in the elastase infusion model. Comparison of rat aortic diameter after elastase perfusion and treatment with either doxyclyine or saline solution. Open circles, doxycycline-treated rats; closed squares, saline solution-treated control group. Abdominal aortic aneurysm, defined as increase in aortic diameter to at least twice that of preperfusion control, is indicated by dashed line. Data shown are mean = SEM of aortic diameter for animals in each group (n=6). * Significant difference between doxycycline and saline solution-treated groups (p(0.01, Student's t -test). Reprinted with permission. From: Petrinec D, Liao S, Holmes DR et al. Doxycycline inhibition of aneurysmal degeneration in an elastase-induced rat model of abdominal aortic aneurysm: Preservation of aortic elastin associated with suppressed production of 92 kD gelatinase. J Vasc Surg 1996; 23:336-46.

undergoing open repair of abdominal aortic aneurysms has been associated with the inhibition of MMP production and activity in the aneurysm wall.[70]

Tissue Inhibitor of Metalloproteinases (TIMP) and Plasminogen Activator Inhibitors (PAI)

Allaire et al have evaluated another approach using gene therapy to block MMP activity in the extracellular matrix. Allaire et al developed a xenograft model of aneurysm formation (see Chapter 7) whereby guinea pig abdominal aorta was transplanted into rats.[71,72] Rejection of the aortic xenograft was manifest by infiltration of the guinea pig aortic wall with macrophages, monocytes, and lymphocytes from the host rat.[72] Elastin degradation occurred in this setting of inflammation and rejection. Aortic xenograft dilation and aneurysm formation occurred within one month of graft placement. The inflammatory process that accompanied the degradation of the elastin was associated with increased tPA activity. Rat recipients of guinea pig aortic grafts that were preimmunized before graft implantation with guinea pig aortic extracellular matrix experienced accelerated rejection and aortic xenograft rupture. MMP-9 was upregulated in this model. Gene therapy was used to overexpress TIMP-1, and MMP-9 was blocked with TIMP-1 overexpression.[61]

Since MMP activity also can be inhibited by interrupting the plasminogen/plasmin cascade, another strategy would be to overexpress PAI-1.[57] PAI-1, which is produced by

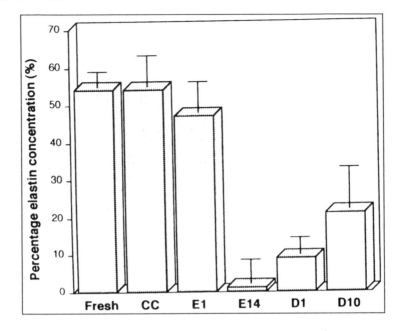

Fig. 9.4. Doxycycline inhibition decreased elastin loss in aortic organ cultures. Graph plots percentage elastin concentration (median values) for fresh aorta (Fresh), cultured control (CC), elastase exposed after 24 hours (E1), elastase exposed after 14 days (E14), doxycycline-treated 10 mg/L (D10). Reprinted with permission from Boyle J, McDermott E, Crowther M et al. Doxycycline inhibits elastin degradation and reduces metalloproteinase activity in a model of aneurysmal disease. J Vasc Surg 1998; 27:354-361.

endothelial cells, smooth muscle cells, and macrophages, interacts with the plasminogen activators tPA and uPA by binding with the catalytic domain of the plasminogen activator. Plasminogen activators convert plasminogen, which is synthesized in the liver, into active plasmin. Plasmin can be inhibited by alpha-2 macroglobulin, alpha-1 antitrypsin, and TIMP, but this step is further along in the cascade. Active plasmin can convert proMMP-3 into active MMP-3, which can then activate the proenzyme forms of MMP-1 and MMP-9 into active enzymes (Fig. 9.2).

Blocking the formation of plasmin through the actions of PAI-1 on tPA and uPA was a second approach used by Allaire et al to inhibit the MMP cascade in their xenograft model of aneurysm formation.[57] Syngenic rat smooth muscle cells were retrovirally transduced with rat PAI-1 gene (LPSN) or vector alone (LXSN), and transfected into the guinea pig aorta just after aortic graft transplantation. PAI-1 activity was overexpressed in the aortic graft LPSN cells (Fig. 9.6). MMP-9 gelatinase activity was greater in the aortic grafts seeded with control (LXSN) cells than in the aortic grafts seeded with cells that overexpressed PAI-1 (LPSN), suggesting that the increased PAI-1 was efficacious in inhibiting the plasminogen cascade and the sequential activation of MMP-9 by plasmin. MMP-2 activity was not unchanged in this model (Fig. 9.7).[57] In aortic cells seeded with PAI-1, neither MMP-9 nor MMP-2 was completely blocked (Fig. 9.7), but the conversion of proMMP-3 into active stromelysin (MMP-3) appeared to be blocked.

If MMP activity is a significant factor in aneurysm growth and progression in humans, these experiments by Allaire et al suggest that gene therapy with overexpressed TIMP to

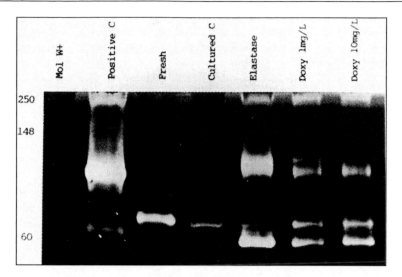

Fig. 9.5. Gelatin zymogram shows metalloproteinase activity for one tissue culture aorta and demonstrates reduction in MMP-9 activity in doxycycline-treated segments. Upper band Gelatinase B (92 kDa type IV collagenase) Lower band: Gelatinase A (72 kDa type IV collagenase) Positive C, HT 1080 fibrosarcoma cell line, which produces large quantities of MMP-2 and MMP-9. Fresh, freshly harvested, noncultured aortic sample. Cultured C Aorta cultured without exposure to elastase for 14 days. Elastase, aorta exposed to elastase for 24 hours and then cultured in standard conditions for a further 13 days. Doxy 1 mg/l and Doxy 10 mg/l, Aortas exposed to elastase for 24 hours and then cultured with doxycycline for 13 days. Reprinted with permission from Boyle J, McDermott E, Crowther M et al. Doxycycline inhibits elastin degradation and reduces metalloproteinase activity in a model of aneurysmal disease. J Vasc Surg 1998; 27:354-61.

block the metalloproteinases or gene therapy with PAI to block plasminogen activators could be future therapeutic options.

Anti-Inflammatory Agents

If inflammation is a critical event in the pathogenesis of aortic aneurysms, then the macrophage plays a key role in this process.[73] The macrophage assists in the regulation of extracellular matrix turnover in both normal and diseased states. Macrophage products act synergistically in the arterial wall. Macrophages release metalloproteinases and their inhibitors and cytokines. Furthermore, the macrophage produces urokinase which interacts with plasmin to convert pro-MMP-9 into the active gelatinase B.

The synthesis of metalloproteinases by the macrophage appears to be dependent on the production of prostaglandin E_2 (PGE_2).[74] The synthesis of PGE_2 from arachadonic acid is determined by the action of the enzyme cyclooxygenase (COX). Two distinct isozymes of cyclooxygenase encoded by two different genes are referred to as COX-1 and COX-2. Transcripts of the COX-2 gene produce the cyclooxygenase enzyme that is found in macrophages, and this COX-2 gene is regulated by proinflammatroy cytokines.[75] Macrophage activation and the release of MMP-9 is associated with the induction of COX-2.[75]

The nonsteroidal anti-inflammatory drugs (NSAIDS) are nonspecific inhibitors of COX-1 and COX-2. Aspirin and indomethacin are irreversible inactivators of cyclooxygenase, while ibuprofen is a reversible inhibitor. In the elastase infusion model of aneurysm formation, COX-2 induction occurs in the inflammatory infiltrate of the aneurysm.

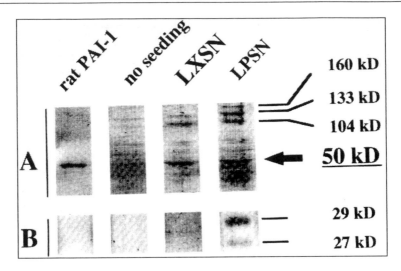

Fig. 9.6. Western blot of PAI-1 protein in guinea pig aortas that were transplanted into rats and then seeded with either smooth muscle cells transfected with the PAI-1 gene (LPSN) or control smooth muscle cells without the excess PAI-1 gene (LXSN). The xenograft overexpressed the PAI-1 gene. A. PAI-1 (50 kD) and PAI-1 complexes (104 kD) B. PAI degradation products (27 kD and 29 kD). Reprinted with permission from: Allaire E, Masanstab D, Kenegy RD et al. Prevention of aneurysm development and rupture by suppression of plasminogen activator inhibitor-1. Circulation 98; 98:249-255.

Holmes et al proposed that blocking COX-2 activity would decrease MMP expression in aneurysms.[77] Aneurysm progression in the elastase infusion model was retarded with indomethacin,[78] an NSAID that is an inhibitor of both COX-1 and COX-2. The advantages of selective COX-2 inhibitors such as celecoxib (Celebrex) and rofecoxib (Vioxx) are the lack of gastrointestinal and renal complications that have been attributed to deleterious effects caused by the inhibition of the COX-1 enzyme. If inflammation and MMP-9 production is found to be associated with the progression of small aneurysms in humans, then COX-2 inhibitors could be a logical therapeutic agent in the treatment of small aneurysms because COX-2 inhibitors potentially could decrease MMP-9 production and release from macrophages in aortic aneurysms.

In experimental aortic aneurysms created using the elastase infusion model, a close correlation was observed between the inflammatory cell response, the induction of proteinases, and an increase in aortic diameter,[79] suggesting that inflammation is an integral component for aneurysm growth in this model. Dobrin et al observed that methyl-prednisolone and cyclosporine were effective in inhibiting inflammation in experimental aneurysms, and observed decreased aneurysms growth with these anti-inflammatory agents.[80] Steroids do not inhibit cyclooxygenase activity but do appear to depress synthesis of cyclooxygenase protein.[63] CD18 monoclonal antibody, which acts by blocking neutrophil-endothelial cell adhesions, also was effective in decreasing aneurysm growth in the elastin infusion model.[81]

In human aortic aneurysm pathologic specimens, the loss of elastin precedes the inflammatory response.[5] The initial intimal and medial fractures of the elastic laminae are not associated with macrophages when the elastin fragmentation is induced by hemodynamic stress.[10] This may be the case in human aortic aneurysms.[10] Brophy et al observed that the mononuclear inflammatory infiltrate found in the human aortic aneurysm wall was located

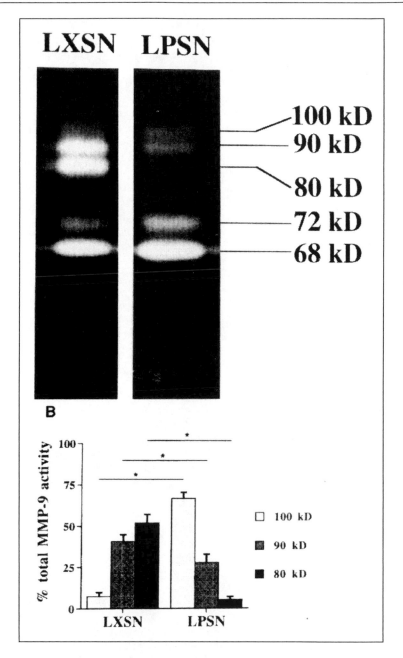

Fig. 9.7. Gelatinase activities in aortic xenograft extracts. Extracts of grafts seeded with control (LXSN) or PAI-1-expressing (LPSN) cells were made 3 days after implantation and analyzed by gelatin zymography. The 80 kD band is the activated form of MMP-9 and this is decreased in the LPSN cells. No changes in MMP-2 activity were detected (72 and 68 kD bands). Reprinted with permission from: Allaire E, Masanstab D, Kenegy RD et al. Prevention of aneurysm development and rupture by suppression of plasminogen activator inhibitor-1. Circulation 1998; 98:249-255.

at the junction of the media and adventitia.[83] This inflammatory infiltrate did not codistribute with the loss of elastin that occurred in the intima and media of the aneurysm wall.[83] Stehbens proposed that the presence of macrophages and metalloproteinases in aneurysms does not indicate their pathogenic role in the genesis of elastin loss and elastin fragmentation any more than the presence of osteoclasts in bone growth means that osteoclasts are responsible for pathologic fractures in bones or tendons.[84] The abrupt edges and tears seen in fractures of the internal elastic laminae in small aneurysms are more consistent with biomechanical fatigue than enzyme-induced destruction by metalloproteinases. The presence of macrophages and metalloproteinases in aneurysms may in fact reflect an attempt by the aneurysm to remodel its matrix and attempt to restore lost tensile strength.

Freestone et al examined both small and large human aortic aneurysms and found evidence to support the hypothesis that inflammation in large aneurysms may be an effect of attempts at healing rather than a cause of aneurysm growth.[85] In small human aortic aneurysms, the primary gelatinase found was MMP-2,[85] which is produced by smooth muscle cells.[38] This metalloproteinase MMP-2 was found in fibrous tissue of the media and in the atherosclerotic plaque. Larger aneurysms had a higher density of inflammatory cells in the adventitia, and larger aneurysms were associated with increasing activity of MMP-9 that was produced by macrophages.[85] In humans, it appears that the existence of small aneurysms can predate any significant inflammatory response.[5]

In the elastase infusion model, this temporal window is brief where small aneurysms exist without evidence of significant inflammation before developing into larger aneurysms. This short window is probably due to the markedly accelerated time course of aneurysm formation in this model, compared to the decades that it takes for human aneurysms to form. In the elastase model, one must consider that the importance of inflammation in aneurysm development may be an artifact of this model.

The trigger for the inflammatory response that is seen in human aneurysms is the subject of much debate (and is further discussed in Chapter 11). Human elastin peptide fragments have been found to be chemotactic for monocytes.[86] Arterial wall elastin fragments may develop in response to cumulative hemodynamic stress and biomechanical fatigue. Macrophage infiltration in large aneurysms may be secondary to this elastin fragmentation.

Since inflammation within the aneurysm wall is associated with the enlargement of small aneurysms, anti-inflammatory drugs that decrease the action of the metalloproteinases, such as doxycycline and the cyclooxygenase (COX) inhibitors, may have a role in inhibiting the growth of small aortic aneurysms in humans. Since smoking also is associated with the growth of small aortic aneurysms in humans, smoking cessation would be important in any drug program designed to prevent the growth of small aneurysms.[87]

Antihypertensive Agents

Antihypertensive agents have been shown to delay the presentation of other cardiovascular complications of atherosclerosis, such as stroke, heart attack, and peripheral vascular disease. Antihypertensive drugs also may prove to be useful in extending the time course before aneurysmal disease presents itself, since antihypertensive agents can reduce the hemodynamic stress on the arterial wall and therefore delay the onset of the biomechanical fatigue characteristic of aneurysms.

Propanolol both delayed the formation and decreased the incidence of aneurysms in male blotchy mice,[88] an animal genetically determined to form aneurysms. In a study of 114 mice over four months, Brophy et al found that 86% of control animals and only 33% of propanolol-treated animals developed aneurysms.[88]

Experimental hypertension accelerated the aneurysm growth in several animal models which induced inflammation and aneurysm formation in rats.[89,90] Propanolol has been found to decrease aneurysm growth in the elastase infusion model.[91]

Beta-blockers may decrease the growth rate of aneurysms in humans. Leach et al reported that the growth rate in human aortic aneurysms was only 0.17 cm/ year in patients receiving beta-blockers, but 0.44 cm/ year in patients not receiving beta-blockers. [92] However, Lindholt et al in a retrospective review found that the use of beta-blockers did not correlate with aneurysm expansion in humans.[93] A prospective, randomized multi-center trial involving over 1000 patients in 30 centers is currently underway to evaluate the efficacy of propanolol in the decreasing the risk of aneurysm rupture.[94]

Conclusion

The prevention of aneurysms is a challenging problem. Aneurysms form as a result of a loss of tensile strength of the arterial wall, probably due to cumulative hemodynamic stresses. Preventing this biomechanical fatigue can probably only be accomplished by decreasing the hemodynamic load that the arterial wall sustains over a lifetime. In order to prevent aneurysms, the optimal pharmacologic agent for preserving the tensile strength of the arterial wall would have to decrease the hemodynamic stress imparted on the arterial wall. Since pulse pressure appears to be a critical factor in the loss of aortic wall tensile strength, antihypertensive agents that would decrease the pulse pressure may be the best pharmacologic agents for preventing the formation of small aneurysms.

At first glance, it appears reasonable to determine, by means of prospective randomized studies, the efficacy of medications such as the teracycline derivative doxycycline and the COX inhibitors such as aspirin and celecoxib in treating small aortic aneurysms and preventing their progression. However, there are logistic problems in performing these studies in a prospective manner. Since the vast majority of patients with aneurysms are already taking the COX inhibitor aspirin on a daily basis for co-existing coronary artery, cerebrovascular, or peripheral vascular disease, enrolling control patients who would not take COX inhibitors, on a randomized basis, would be virtually impossible. There have not been any data reported to suggest that aspirin has decreased the growth of aneurysms in this patient population who are taking aspirin for other reasons. In light of the apparent low risk profile,[68] some vascular specialists have begun prescribing doxycycline to patients with small aneurysms.

References

1. Anidjar S, Dobrin P, Eichorst M et al. Correlation of inflammatory infiltrate with the enlargement of experimental aortic aneurysms. J Vasc Surg 1992; 16:139-47.
2. Dobrin PB, Baker WH, Gley WC. Elastolytic and collangenolytic studies of arteries. Arch Surg 1984; 119:405-509.
3. Stehbens WE. The pathology of atherosclerosis and experimental atherosclerosis. In: Stehbens WE, ed. The Lipid Hypothesis of Atherogenesis. Austin: RG Landes Co., 1993; 10-47.
4. Halloran BG, Baxter BT. Pathogenesis of aneurysms. Sem Vasc Surg. 1995; 2:85-92
5. Stehbens WE. Atherosclerosis and degenerative diseases of blood vessels. In: Stehbens WE, Lie JT, eds. Vascular Pathology. London: Chapman & Hall, 1995; 175-269.
6. Pearce WH, Sweis I, Yao JST et al. Interleukin-1 beta and tumor necrosis factor—alpha release in normal and diseased human infrarenal aortas. J Vasc Surg 1992; 16:784-789.
7. Keen RR, Flanigan DP, Lands WEM. Differences between arterial and mixed venous levels of plasma hydroperoxides following major thoracic and abdominal operations.. J Free Rad Biol Med 1991; 9:485-494.
8. Thompson MM, Jones L, Nasim A et al. Angiogenesis in abdominal aortic aneurysms. Eur J Vasc Endovasc Surg 1996; 11: 464-469.

9. Zarins CK, Glagov S. Atherosclerotic process and aneurysm formation. In: Aneurysms New Findings and Treatments, Yao JST and Pearce W. eds. Norwalk: Appleton and Lange, 1994:35-46.

10. Stehbens WE: Structural and architectural changes during arterial development and the role of hemodynamics. Acta Anat 1996; 157:261-274.

11. Glagov S, Zarins C, Giddens DP et al. Hemodynamics and atherosclerosis: Insights and perspectives gained from studies of human arteries. Arch Path Lab Med, 1988; 112:1018-1031.

12. Stehbens WE. The pathogenesis of atherosclerosis: A critical evaluation of the evidence. Cardiovas Pathol 1997; 6:123-153.

13. Greenhill NS, Stehbens WE. Hemodynamically-induced intimal tears in experimentally U-shaped arterial loops as seen by scanning electron microscopy. Br J Exp Pathol 1985; 66:577-584.

14. Jones GT, Stebhens WE, Martin BJ. Ultrastructural changes in arteries proximal to short term experimental carotid-jugular arteriovenous fistulae in rabbits. Pathology 1995; 75:225-232.

15. Zarins CK, Glagov S, Vesselinovitch et al. Aneurysm formation in experimental atherosclerosis: Relationship to plaque evolution. J Vasc Surg 1990; 12:246-256.

16. Glagov S, Weisenberg E, Kolettis et al. Compensatory enlargement of human atherosclerotic coronary arteries. N Engl J Med 1987; 316:1371-1375.

17. Masawa N, Glagov S, Bassiouny et al. Intimal thickness normalizes neural tensile stress in regions of increased intimal area and artery size at the carotid bifurcation. Arteriosclerosis. 1988; 8:621a.

18. Pearce WH. Slaughter MS, LeMaire S et al. Aortic diameter as a function of age, gender, and body surface area. Surgery 1993; 114:691-697.

19. Blair JM, Glagov S and Zarins CK. Mechanism of superficial femoral artery abductor canal stenosis. Surg Forum. 1990; 41:359-360.

20. Stehbens WE. Pathology and pathogenesis of degenerative atherosclerotic aneurysms. In: Development of Aneurysms. Keen RR, Dobrin PD, Eds. Austin, Eurekah.com, 2000.

21. Zarins CK, Weisenberg E, Kolettis G et al. Differential enlargement of artery segments in response to atherosclerotic plaques. J Vasc Surg 1988; 7:386-394.

22. Keen RR, McCarthy WM, Shireman P et al. Surgical management of atheroembolization. J Vasc Surg 1995; 21:773-781.

23. He CM, Roach MR. The composition and mechanical properties of abdominal aortic aneurysms. J Vasc Surg 1994; 20:6-13.

24. Stehbens WE. Vascular complications in experimental atherosclerosis. Prog Cardiovas Dis 1986; 29:221-237.

25. Brown G, Albers JJ, Fisher LD et al. Regression of coronary artery disease as a result of intensive lipid lowering therapy in men with high levels of apolipoprotein B. N Engl J Med 1990; 323:1289-1298.

26. Wills A, Thompson MM, Crowther m et al. Pathogenesis of abdominal aortic aneurysms—Cellular and biochemical mechanisms. Eur J Vasc Endovasc Surg 1996; 12:391-400.

27. Stehbens WE. Casuality in medical science with reference to coronary heart disease and atherosclerosis. Perspect Biol Med 1992; 36:97-119.

28. Dobrin PB. Elastin, collagen and the pathophysiology of arterial aneurysms. In: Development of Aneurysms. Keen RR, Dobrin PB, eds. Austin; Eurekah.com, 2000.

29. Thompson R, Parks W. Role of matrix metalloproteinases in abdominal aortic aneurysms. Ann NY Acad Sci 1996; 800:157-174.

30. Woesnner J. Matrix metalloproteinases and their inhibitors in connective tissue remodeling. FASEB J 1991; 5:2145-2154.

31. Wills A, Thompson M, Crowther M et al. Elastase-induced matrix degradation in arterial organ cultures: An in vitro model of aneurysmal disease. J Vasc Surg 1996; 24:667-679.

32. Irizarry E, Newman K, Gandhi R et al. Demonstration of interstitial collagenase in abdominal aortic aneurysm disease. J Surg Res 1993; 54:571-574.

33. McMillan W, Patterson B, Keen R et al. In situ localization and quantification of seventy-two-kilodalton type IV collagenase in aneurysmal, occlusive, and normal aorta. J Vasc Surg 1995; 22:295-305.
34. Keen RR, Murphy T, Cipollone M et al. Upregulation of prostromelysin in aortic aneurysms is mediated by interleukin-1B 1993. In Proceedings of the Association for Academic Surgery, 27th Annual Meeting, Nov. 10-13, Hershey, PA.
35. Newman K, Ogata Y, Malon A, et al. Identification of matrix metalloproteinases 3 (Stromelysin-1) and 9 (Gelatinase B) in abdominal aortic aneurysm. Arterioscler Thromb 1994; 14:1315-1320.
36. Newman K, Malon A, Shin R et al. Matrix metalloproteinases in abdominal aortic aneurysm: Characterization, purification, and their possible sources. Conn Tiss Res 1994; 30:265-276.
37. Vine N, Powell JT. Metalloproteinases in degenerative aortic disease. Clin Sci 1991; 81:233-239.
38. Keen RR, Nolan KB, Cippollone M et al. Interleukin-1B induces differential gene expression in aortic smooth muscle cells. J Vasc Surg 1994; 20:774-786.
39. McMillan W, Patterson B, Keen R et al. In situ localization and quantification of mRNA for 92-kD type IV collagenase and its inhibitor in aneurysmal, occlusive, and normal aorta. Arterioscler Thromb Vasc Biol 1995; 15:1139-1144.
40. Thompson R, Holmes D, Mertens R et al. Production and localization of 92-kilodalton gelatinase in abdominal aortic aneurysms. J Clin Invest 1995; 96:318-326.
41. Senior R, Griffin G, Fliszar C et al. Human 92- and 72-kilodalton type IV collagenases are elastases. J Biol Chem 1991; 266:7870-7875.
42. Pearce WH, Koch AE. Cellular components and features of immune response in abdominal aortic aneurysms. Ann NY Acad Sci 1996; 800:175-185.
43. Dollery CM, McEwan JR, Henney AM. Matrix metalloproteinases and cardiovascular disease. Circ Res 1995; 77:863-868.
44. Baramova EN, Bajou K, Remacle A et al. Involvement of PA/plasmin system in the processing of pro-MMP-9 and in the second step of pro-MMP-2 activation. FEBS Letters 1997; 405:157-162.
45. He S, Wilhelm S, Pentland A et al. Tissue cooperation in a proteolytic cascade activating human interstitial collagenase. Proc Natl Acad Sci, USA, 1989; 86:2632-2636.
46. Jean-Claude J, Newman KM, Li H et al. Possible key role for plasmin in the pathogenesis of abdominal aortic aneurysms. Surgery 1994; 116:472-478.
47. Peuhkurinen KJ, Risteli L, Melkko JT et al. Thrombolytic therapy with streptokinase stimulates collagen breakdown. Circulation 1991; 83:1969-1975.
48. Sakalihasan N, Delvenne P, Nusgens V et al. Activated forms of MMP-2 and MMP-9 in abdominal aortic aneurysm. J Vasc Surg 1996; 24:127-33.
49. Patel M, Melrose J, Ghosh P et al. Increased synthesis of matrix metalloproteinases by aortic smooth muscle cells is implicated in the etiopathogenesis of abdominal aortic aneurysms. J Vasc Surg 1996; 24:82-92.
50. Reilly JM, Sicard GA, and Lucore CL. Abnormal expression of plasminogen activators in aortic aneurysmal and occlusive disease. J Vasc Surg 1994; 19:865-872.
51. Anidjar S, Salzmann JL, Gentric D et al. Elastase-induced experimental aneurysms in rats. Circulation 1990; 82:973-981.
52. Newman KM, Jean-Claude J, Li H et al. Cellular localization of matrix metalloproteinases in the abdominal aortic aneurysm wall. J Vasc Surg 1994; 20:814-820.
53. Clowes A, Clowes M, Kirkman T et al. Heparin inhibits the expression of tissue-type plasminogen activator by smooth muscle cells in injured rat carotid artery. Circ Res 1992; 70:1128-1136.
54. Simpson AJ, Booth NA, Moore NR et al. Distribution of plasminogen activator inhibitor (PAI-1) in tissues. J Clin Path 1991; 44:139-143.
55. Schneiderman J, Bordin G, Engelberg I et al. Expression of fibrinolytic genes in atherosclerotic abdominal aortic aneurysm wall. J Clin Invest 1995; 96: 639-645.
56. Kenagy RD, Vergel S, Mattsson E et al. The role of plasminogen, plasminogen activators, and matrix metalloproteinases in primate arterial smooth muscle cell migration. Arterioscler

Thromb Vasc Biol 1996; 16:1373-1382.

57. Allaire E, Hasenstab D, Kenagy R et al. Prevention of aneurysm development and rupture by local overexpression of plasminogen activator inhibitor-1. Circulation 1998; 98:249-255.

58. Docherty AJP, Lyons A, Smith BJ et al. Sequence of human tissue inhibitor of metalloproteinases and its identity to erythroid-potentiating activity. Nature 1985; 318:66-69.

59. Stetler-Stevenson WG, Brown PD, Onisto M et al. Tissue inhibitor of metalloproteinases-2 (TIMP-2) mRNA expression in tumor cells and human tumor tissues. J Biol Chem 1990; 265:13933-13938.

60. Petrinec D, Liao S, Holmes D et al. Doxycycline inhibition of aneurysmal degeneration in an elastase-induced rat model of abdominal aortic aneurysm: Preservation of aortic elastin associated with suppressed production of 92 kD gelatinase. J Vasc Surg 1996; 23:336-346.

61. Allaire E, Forough R, Wang T et al. Tissue inhibitor of metalloproteinase inhibitor-1(TIMP-1) overexpression prevents arterial rupture and dilation in a xenograft model. FASEB J 1996; 10;A1134

62. Lands WEM and Keen RR. Peroxide tone and its consequences. In: Biological Oxidations Systems. Eds: Reddy CC, Hamilton GA, Madyastha KM. San Diego; Academic Press. 1990; 657-666.

63. Marnett LJ. Prostglandin H Synthase In: Biological Oxidations Systems. Reddy CC, Hamilton GA, Madyasthma KN, eds. San Diego; Academic Press 1990; 637-656.

64. Keen RR, Stella LA, Lands WEM: Differential detection of plasma hydroperoxides in sepsis. Crit Care Med 1991; 19:1114-1119.

65. Golub L, Evans RT, McNamara TF et al. A nonantimicrobial tetracycline inhibits gingival matrix metalloproteinases and bone loss in *porphyromonas gingivalis*-induced periodontitis in rats. Ann NY Acad Sci 1994; 732:96-111.

66. Sorsa T, Ding Y, Salo T et al. Effects of tetracyclines on neutrophil, gingival, and salivary collagenases. Ann NY Acad Sci 1994; 732:112-130.

67. Uitto V, Firth J, Nip L et al. Doxycycline and chemically modified tetracyclines inhibit gelatinase A (MMP-2) gene expression in human skin keratinocytes. Ann NY Acad Sci 1994; 732:140-151.

68 Sauer GC. Safety of long-term tetracycline therapy for acne. Arch Dermatol 1976; 112:1603-1605.

69 Boyle JR, McDermott E, Crowther M et al. Doxycycline inhibits elastin degradation and reduces metalloproteinase activity in a model of aneurysmal disease. J Vasc Surg 1998; 27:354-361.

70. Curci J, Mao D, Bohner DB et al. Preoperative treatment with doxycycline inhibits MMP production and activation in patients with abdominal aortic aneurysms. Presented at The Society for Vascular Surgery 53rd Annual Meeting, Washington, DC, June 7, 1999.

71. Allaire E, Guettier C, Bruneval P et al. Cell-free arterial grafts: morphologic characteristics of aortic isografts, allografts and xenografts in rats. J Vasc Surg 1994; 19:446-456.

72. Allaire E, Bruneval P, Mandet C et al. The immunogenicity of the arterial extracellular matrix in arterial xenografts. Surgery 1997; 122:73-81.

73. Louwrens H, Pearce WH. Role of inflammatory cells in aortic aneurysms. In: Aneurysms: New Findings and Treatments. Yao JST and Pearce WH. eds. Norwalk. Appleton and Lange, 1994; 35-46.

74. Corcoran M, Stetler-Stevenson W, Brown P et al. Interleukin-4 inhibition of prostaglandin E_2 synthesis blocks interstitial collagenase and 92-kDa type IV collagenase/gelatinase production by human monocytes. J Biol Chem 1992; 267:515-519.

75. Arias-Negrete S, Keller K and Chadee K. Proinflammatory cytokines regulate cyclooxygenase-2 mRNA expression in human macrophages. Biochem Biophys Res Comm 1995; 208:582-589.

76. Newman K, Jean-Claude J, Li H et al. Cytokines that activate proteolysis are increased in abdominal aortic aneurysms. Circulation 1994; 90:224-227.

77. Holmes D, Wester W, Thompson R et al. Prostaglandin E2 synthesis and cyclooxygenase expression in abdominal aortic aneurysms. J Vasc Surg 1997; 25:810-815.

78. Holmes D, Petrinec D, Wester W et al. Indomethacin prevents elastase-induced abdominal aortic aneurysms in the rat. J Surg Res 1996; 63:305-309.
79. Halpern V, Nackman G, Gandhi R et al. The elastase infusion model of experimental aortic aneurysms: Synchrony of induction of endogenous proteinases with matrix destruction and inflammatory cell response. J Vasc Surg 1994; 20:51-60.
80. Dobrin P, Baugmartner N, Andijar S et al. Inflammatory aspects of experimental aneurysms. Ann NY Acad Sci 1996; 800:74-88.
81. Ricci MA, Strindberg G, Slaiby JM et al. Anti-CD 18 monoclonal antibody slows experimental aortic aneruysms. J Vasc Surg 1996; 23:301-307.
82. Stebhens W. The lipid hypothesis and the role of hemodynamics in atherogenesis. Prog Cardiovasc Dis 1990; 33:119-136.
83. Brophy C, Reilly J, Smith W et al. The role of inflammation in nonspecific abdominal aortic aneurysm disease. Ann Vasc Surg 1991; 5:229-233.
84. Stehbens WE. Apoptosis and matrix vesicles in the genesis of arterial aneurysms of cerebral arteries. Storke 1998; 29:1478-1479.
85. Freestone T, Turner RJ, Coady A et al. Inflammation and matrix metalloproteinases in enlarging abdominal aortic aneurysms. Arterioscler Thromb Vasc Biol 1995; 15:1145-1151.
86. Senior R, Griffin G and Mecham R. Chemotactic activity of elastin-derived peptides. J Clin Invest 1980; 66:859-862.
87. Powell J, Worrell P, MacSweeney STR et al. Smoking as a risk factor for abdominal aortic aneurysm. Ann NY Acad Sci 1996; 800:246-248.
88. Brophy C, Tilson J and Tilson MD. Propranolol delays the formation of aneurysms in the male blotchy mouse. J Surg Res 1988; 44:687-689.
89. Anidjar S, Dobrin P, Chejfec G et al. Experimental study of determinants of aneurysmal expansion of the abdominal aorta. Ann Vasc Surg 1994; 8:127-136.
90. Gadowski G, Ricci M, Hendley E et al. Hypertension accelerates the growth of experimental aortic aneurysms. J Surg Res 1993; 54:431-436.
91. Ricci M, Slaiby J, Gadowski G et al. Effects of hypertension and propranolol upon aneurysm expansion in the Anidjar/Dobrin aneurysm model. Ann NY Acad Sci 1996; 800:89-96.
92. Leach S, Toole A, Stern H et al. Effect of B-adrenergic blockade on the growth rate of abdominal aortic aneurysms. Arch Surg 1988; 123:606-609.
93. Lindholt JS, Heickendorff L, Henneberg EW et al. Serum-elastin-peptides as a predictor of expansion of small abdominal aortic aneurysms. Eur J Vasc Endovasc Surg 1997; 14:12-16.
94. Ricci M, Pilcher D, cBride R. Design of a trial to evaluate the effect of propranolol upon abdominal aortic aneurysm expansion. Ann NY Acad Sci 1996;800:252-253.

CHAPTER 10

Is Blood Flow the Cause of Aneurysms?
The Roles of Shear Stress and Vibration

Richard R. Keen

Introduction

Greater than 95% of all noncerebral artery aneurysms are located in the infrarenal aorta or in the iliac, femoral or popliteal arteries that perfuse the lower extremities. Aneurysmal dilation in all of these locations usually starts in the most distal aspect the artery, immediately proximal to points of bifurcation or major branching (Fig. 10.1).

All of these arteries where aneurysms tend to form are located distal to the level of the renal arteries. The nature of the blood flow pulse wave changes dramatically between the suprarenal aorta and the infrarenal aorta: in the suprarenal aorta, pulsed-wave Doppler spectral analysis shows that blood flow is biphasic, with no flow reversal at the end of systole with significant forward flow during diastole (Fig. 10.2).[1] In the infrarenal aorta, Doppler spectral analysis reveals that blood flow is triphasic with flow reversal in late systole (Fig 10.3).[1] Do aneurysms develop in these characteristic, parallel locations because arteries below the level of the renal arteries and immediately above arterial bifurcations share common hemodynamic features that cause aneurysms to form? The consistent localization of arterial aneurysms in the human infrarenal aorta and its runoff vessels may be a reflection of a specific altered hemodynamic environment that is common to these locations.[2]

Alterations in the hemodynamic environment, through the creation of coarctations or arteriovenous fistulas, can lead to the development over time of aneurysms in other arteries that do not characteristically form aneurysms. For example, aneurysms of the common carotid and innominate arteries are unusual, but aneurysms can be induced to form if the hemodynamic environment is altered. William Halsted treated a right subclavian artery aneurysm by placing of a constricting aluminum band around the innominate artery and ligating the right subclavian artery.[3] Unexpectedly, this surgically-created coarctation resulted 12 years later in the formation of an innominate and right carotid artery aneurysm distal to this constricting band.[3] Aneurysms of the external iliac artery and proximal superficial femoral artery also are very unusual. However, Osler[4] reported a true aneurysm of the external iliac artery in a 30 year old patient who sustained a knife wound above his knee 20 years earlier. The stab wound resulted in an arteriovenous fistula between the distal superficial femoral artery and vein.[5] Aneurysms of the superficial femoral artery have been observed

Development of Aneurysms, edited by Richard R. Keen and Philip B. Dobrin.
©2000 Eurekah.com.

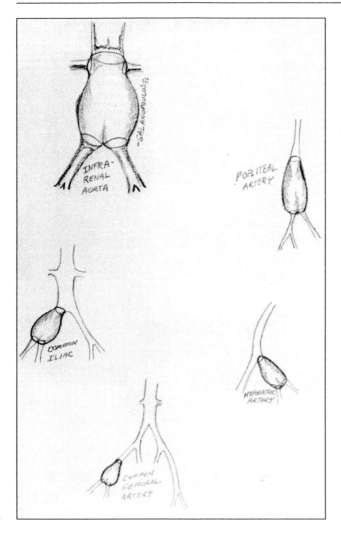

Fig. 10.1. Aneurysms develop proximal to major branches and bifurcations of the infrarenal aorta and the common iliac, popliteal, hypogastric, and common femoral arteries. The external iliac and superficial femoral arteries that have no major branches almost never form aneurysms.

20 years after the formation of a traumatic popliteal fossa arteriovenous fistula.[6] These pathologic changes of arterial dilation and aneurysm formation in arteries proximal to an arteriovenous fistula are the rule rather the exception (Fig. 10.4)[7] since William Hunter first reported this finding in 1757.[4]

These specific cases from the literature served as human experiments whereby the hemodynamic factors that appear to cause the development of aneurysms were accentuated. This chapter will examine the hemodynamics of blood flow: pulse pressure, shear stress, and vibration to help explain the "where aneurysms form" and the "why aneurysms form".

Pulse and Pressure

The mean blood pressure drops as blood travels in normal arteries from the level of the ascending aorta to the level of the ankle, but this drop in mean pressure is only a few millimeters of mercury. This mean pressure differential is a requirement for blood to flow in a distal direction, and is governed by the laws of fluid mechanics. More interesting is the

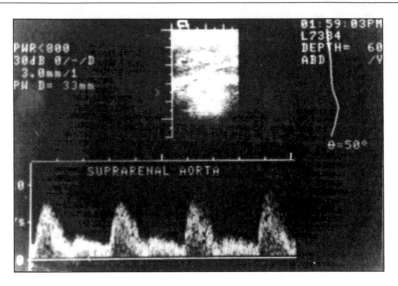

Fig. 10.2. The velocity profile of the suprarenal abdominal aorta demonstrates flow that is biphasic: No flow reversal is seen at the end of systole and significant forward flow occurs during diastole. Reprinted with permission from: Bassiouny HS, Zarins CK, Kadowaki MM et al. Hemodynamic stress and experimental aortoiliac atherosclerosis. J Vasc Surg 1994; 19:426434.

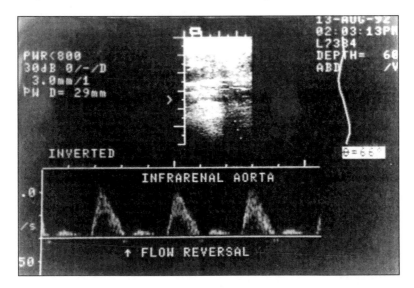

Fig. 10.3. The velocity profile of the infrarenal aorta obtained at rest demonstrates triphasic flow with reversal of flow in late systole. The reversal of flow occurs because the pressure and flow waves that are propagated forward are reflected back both from distal branches and bifurcations and from increased peripheral resistance at the level of the arterioles. Exercise decreases resistance and causes the velocity profile in the infrarenal abdominal aorta to become biphasic. Reprinted with permission from: Bassiouny HS, Zarins CK, Kadowaki MM et al. Hemodynamic stress and experimental aortoiliac atherosclerosis. J Vasc Surg 1994; 19:426-434.

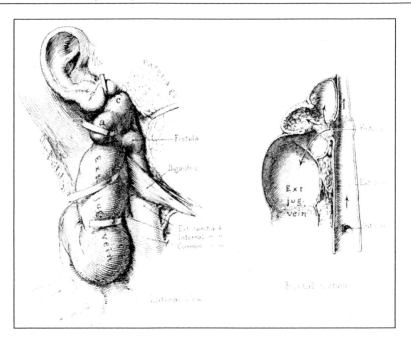

Fig. 10.4.The natural history of a traumatic arteriovenous fistula is the development of arteriomegaly and aneurysm formation in the artery proximal to the fistula. A two year old girl sustained a scissors puncture wound to the right neck. 32 years later, the patient presented with a right external carotid artery to right external jugular vein arteriovenous fistula. Operative repair revealed aneurysm formation of the right common and external carotid arteries. The external jugular vein was calcified in addition to being enlarged. Reprinted with permission from: Reid, M. Studies on abnormal arteriovenous communications, acquired and congenital. Report of a series of cases. Arch Surg 1925; 10:601-610.

finding that as the pressure wave travels distally along the aorta, the systolic pressure increases, the diastolic pressure decreases, and the pulse pressure (defined as the difference between the systolic and diastolic pressure) increases (Fig. 10.5).[8] In normal arteries both the femoral artery systolic pressure and the pulse pressure exceed the systolic and pulse pressure in the corresponding brachial artery. Under resting conditions in normal arteries, Yao found that the systolic pressure at the ankle exceeds the systolic pressure in the arm by about 10%.[9]

The increasing pulse pressure and systolic pressure,[10] the dicrotic wave of the pressure pulse,[11] and the reversed component of the flow pulse[12] each can be explained by the reflection of pressure waves from vessel branch points and bifurcations that are observed along the arterial tree, and by the high resistance peripheral arteriolar beds.[13]

Direct evidence for the role of wave reflection on the differences in pulse pressure at different locations within the arterial tree was provided by Länne and Bergentz in their examinations of the mechanical properties of normal arteries.[14] Recordings of the diameter vs. time curves in the common carotid, abdominal aorta, and common femoral arteries all revealed changes in diameter over the course of the pressure wave, but the characteristics of each recording differed depending on the artery that was examined. The diameter of the artery increased rapidly in all three vessels due to a rapid increase in arterial pressure, but the dicrotic notch (representing closure of the aortic valve) that was visible in the common

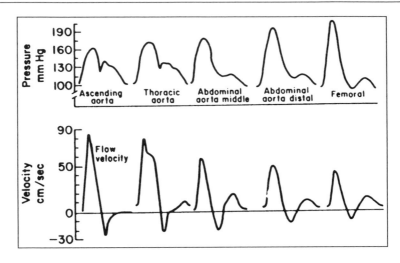

Fig. 10.5. As blood travels from the ascending aorta to the level of the femoral arteries, the mean pressure and the diastolic pressure fall, but the systolic pressure and the pulse pressure increase. The velocity wave decreases in amplitude from proximal to distal. The incident pressure and pulse waves start out identical in the ascending aorta.[13] These changes in the pressure and pulse waves can be explained by the reflection of pressure and flow waves from distal bifurcations and the high resistance peripheral arteriolar beds. The reflected pressure wave is positive, so it is additive to the subsequent forward pressure waves. The reflected pulse wave is negative, so it subtracts from the subsequent forward pulse wave. Reprinted with permission from: McDonald DA. Blood Flow in Arteries. Baltimore: Williams & Wilkins 1974:356.

carotid artery was not seen in the abdominal aorta or the common femoral artery. These regional differences were due to the propagation and reflection of pulse waves in the arterial tree.

The diastolic wave changes the distance to the sites of reflection varies.[14] The common carotid artery is in close proximity to the aortic valve. The few bifurcations and branches and low resistance of the brain provide for a lower number of wave reflections that can hide the dicrotic notch. However, this incisura representing aortic valve closure is hardly visible in the diameter vs. time curves for either the abdominal aorta or the common femoral artery, which are more peripheral in the arterial tree. The dicrotic notch in the abdominal aorta and common femoral arteries nearly is lost in the noise of the reflecting pulse waves. These reflected waves are due to a much greater peripheral resistance and a greater number of larger bifurcations and branches in the runoff from the abdominal aorta. These diastolic waves that are visible in the common carotid, abdominal aorta, and common femoral arteries represent the summation of waves that had been transmitted and subsequently reflected by distal bifurcations and high resistance arteriolar beds.

The diastolic wave also changes appearance as the speed of the reflected waves varies. Figure 3.5 (Chapter 3) demonstrates simultaneously recorded curves of mean arterial pressure vs. time and arterial diameter vs. time in the normal abdominal aorta in a 24-year-old male and a 69-year-old male.[15] In the elderly, a separate pressure wave during diastole is no longer found in the abdominal aorta. The explanation for this is that the speed at which the pressure waves travels increases with increasing age.[16,17] Reflected waves travel faster in less compliant, stiffer, older arteries.[18] Since the reflected waves travel faster, the pressure wave of late systole is accentuated and augmented in older arteries. The end result of the reflected waves travelling and returning faster in the aorta of the elderly is that the abdominal aorta and the runoff vessels of the elderly are subjected to an increased or widened pulse pressure. This increased pulse pressure delivers an increased repetitive hemodynamic stress on the arterial wall. He and Roach observed that the aneurysmal aorta is significantly stiffer than the nonaneurysmal aorta (Fig. 10.6).[19] Since pressure waves travel faster in stiffer arteries, this findings by He and Roach suggests that the pulse pressure may be further increased in arteries that develop aneurysms. In the aorta and lower extremity arteries of elderly individuals, stiffer arterial walls speed up the transmission of pulse and pressure waves, resulting in an increased pulse pressure and a greater hemodynamic stress. This repetitive stress may cause the vessel wall to fatigue and a loss of tensile strength.

Increased pulse pressure is a measurable hemodynamic stress that can help explain the location and formation of aortic aneurysms and lower extremity aneurysms. If increased pulse pressure is the critical factor, then other models of hemodynamic stress where aneurysms form, specifically arteriovenous fistula and poststenotic dilation, should be explainable in terms of pulse pressure. However, an increased pulse pressure does not necessarily exist in either arteriovenous fistulas or poststenotic dilations. In arteriovenous fistulas, the pulse pressure can be increased in the afferent artery that dilates, but the pulse pressure is decreased in the efferent vein that also dilates.[20]

An ascending aortic aneurysm beyond a stenotic aortic valve is a type of poststenotic dilation, but the pulse pressure is increased significantly in this example, primarily when the aortic valve is incompetent.[21] In thoracic aortic coarctation and other examples of poststenotic dilation, the pulse pressure may not be increased in the distal aorta where the dilation occurs. This apparent paradox can be reconciled if one recognizes that the hemodynamic condition that is common to all of these examples, including the abdominal aorta in the elderly, arteriovenous fistulas, and poststenotic dilations, is a *rapid* pressure drop.

In fluids, rapid drops in pressure produce vibrations. A widened pulse pressure is simply a rapid pressure drop between the systolic and the diastolic phases of each cardiac cycle. The rapid pressure drop across an arteriovenous fistula produces vibrations that are felt as a thrill and heard as a bruit. A rapid pressure drop across a short stenosis also produces similar vibrations. In poststenotic dilations, this rapid pressure drop also produces a bruit and a thrill.

Repeated, rapid intravascular pressure drops have been associated with other adverse cardiovascular outcomes besides the formation of aneurysms. An increased pulse pressure has been found to be predictive of the development of congestive heart failure.[22]

Shear Stress

Wall shear stress is one of the distinct forces acting on the vessel wall when pulsatile blood flows under pressure within an artery. The effects of each of these forces on the arterial wall and their relationship to each other have been well described.[24] The blood pressure within the arterial lumen distends the artery in two directions: circumferentially (θ) and longitudinally (Z) (Fig. 10.7). This is accompanied by thinning of the wall in the radial (r) direction. The deformed wall generates retractive stresses in each of these three directions. The force of blood traveling parallel to the arterial wall produces friction or wall shear stress.

Blood pressure is the driving force behind blood flow. The relationship between pulsatile blood pressure, pulsatile blood flow, and vascular resistance is analogous to Ohm's Law, where the voltage (pressure) is determined by the product of current (flow) and impedance (resistance).[11] Pressure, flow and resistance are interdependent. The mean pressure gradient from proximal to distal within the arterial tree causes the mass movement of red blood cells parallel to the arterial lumen. As liquids move over a surface, the molecules tumble over one

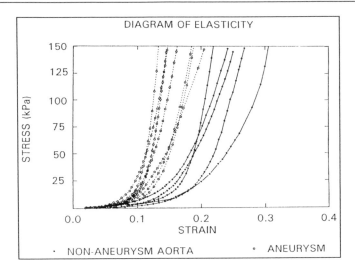

Fig. 10.6. Elastic diagram of aneurysmal (open circles) and normal, nonaneurysmal (dark circles) abdominal aortic wall. Curves for all of the aneurysmal aortic wall are shifted to the left and their slopes are greater, consistent with decreased elasticity in the aneurysmal aorta. Pressure waves travel faster in the stiffer, less elastic aortic walls and will lead to an increase in pulse pressure. Reprinted with permission from: He, CM, Roach, MR. The composition and mechanical properties of abdominal aortic aneurysms. J Vasc Surg 1994; 20:6-13.

another, and energy is needed for the liquid to deform into its new shape. Viscosity (μ) is the measurement of the resistance of the fluid to changes in shape. In blood, viscous resistance exists because energy is required to deform and move the red blood cells and proteins suspended within the plasma. The viscosity of blood is greater than water or saline. Viscosity is measured by comparing the amount of force per unit area needed to deform the fluid at a constant rate. This force per unit area is called stress.[23] The flow of blood adjacent to and parallel to the arterial wall exerts a tangential force parallel to the arterial wall surface. This force is called the wall shear stress (τ_w). Wall shear stress can be thought of as the drag, or friction of blood moving along the arterial wall.

The three major factors that determine the wall shear stress are the flow velocity, the arterial diameter, and the viscosity of the blood. Blood flow is inhibited by viscous resistance and shear stress is proportional to viscosity (μ).[23] Wall shear stress also is proportional to blood flow velocity (Q) and inversely related to the cube of the arterial radius (R). These variables permit calculation of wall shear stress for a parabolic flow profile by the Hagen-Poiseulle formula;

$$\tau_w = \frac{4\,\mu\,Q}{\pi\,R^3}$$

The velocity of red blood cells is greater at the center of the artery than the velocity of blood cells immediately adjacent to the arterial wall. If vectors are used to represent the product of red blood cell mass times velocity at various radial distances from the arterial wall, this velocity profile has a parabolic shape. While the highest velocities are at the center of the artery, the velocity of blood flowing immediately adjacent to the arterial wall approaches

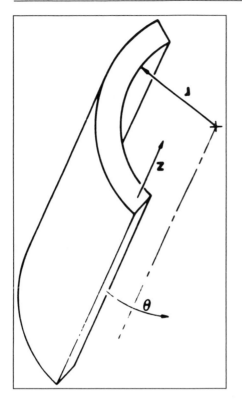

Fig. 10.7. Diagram of segment of blood vessel illustrating circumferential (0), longitudinal (Z), and radial (r) directions. Deformations and stresses are tensile in the θ and Z directions because the tissue is stretched in these directions when the vessel is pressurized. Deformation and stress in the r direction are compressive because the wall is forced to become thinner with pressurization. Reprinted with permission from: Dobrin, PB. Mechanical properties of arteries. Physiol Rev 1978; 58:397460.

zero.[26] Since shear stress is proportional to velocity, this laminar flow results in higher shear stress at the at the center of the artery compared to the blood-artery interface, where shear stress is lower.

Calculations of shear stress using the Hagen-Poiseulle formula assume a nonpulsatile flow, but instantaneous shear stress for pulsatile fluids can still be determined using this method. Shear stress can have an instantaneous value that is negative if the direction of blood flow changes. For example, in the infrarenal aorta during late systole, flow reversal can cause shear stress to be negative because of the direction of flow.

Arteriovenous fistula and aortic coarctation are two hemodynamic examples that have been cited where on an initial evaluation, increased shear stress may appear to induce aneurysmal development. In both of these models, the increased wall shear stress caused by increased blood flow velocity is followed by the development of arterial ectasia. As the arterial diameter increases, the wall shear stress subsequently returns to normal. Both arteriovenous fistula and aortic coarctation are associated with a simultaneous drop in intravascular pressure, so absolute blood pressure per se is not the force responsible for arterial dilation. In fact, low pressure systems with arteriovenous fistulas or coarctations can develop aneurysms. For example, De Vries reported an aneurysm of the pulmonary artery distal to a stenosis of the pulmonic valve.[27] Even in this low pressure system, the characteristic histologic changes typical of aneurysms: the loss and the disruption of elastin layer were observed (Fig. 10.8). These findings suggest that absolute changes in mean blood pressure appear to have less of an influence on arterial wall diameter than do changes in wall shear stress[28] and changes in pulse pressure. This is not to say that hypertension is not an aggravating factor in aneurysmal dilation, but rather that hypertension is not an antecedent for aneurysm

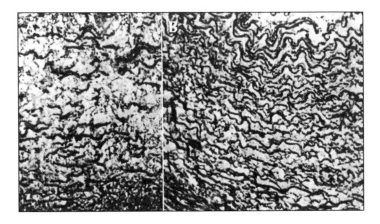

Fig. 10.8. Poststenotic dilation is accompanied by disruption of the elastin layer in the dilated segment. An eight year old boy with pulmonary valve stenosis. (A) The dilated pulmonary artery 2 cm distal to the stenosis reveals disrupted elastin fibers. (B) The normal caliber pulmonary artery 10 cm distal to the stenosis reveals normal appearing elastin fibers. (Verhoeff's elastin stain) Aneurysmal dilation in the low pressure pulmonary artery demonstrates that hypertension is not a prerequisite for aneurysm development. Reprinted with permission from: DeVries HK, Van Den Berg JW. On the origin of poststenotic dilation. Cardiologia 1958; 33:195-211.

formation.[29] Changes in mean blood pressure have a more profound effect on arterial wall thickness. The maximum change in cross-sectional area of large arteries in response to fluctuations in pressure is only 6%.[30]

Because blood flow is pulsatile and flow reverses in diastole (Fig 10.3 and 10.5), oscillations occur both in the blood pressure and in the direction of shear stresses.[30,31] Perturbations in flow occur as arteries expand and contract. Flow eddies and vortices occur at arterial bends, arterial bifurcations, and at the orifices of branches. These oscillations in pressure and flow producing rapid variations in shear stress (Fig 10.9).[32-34] Wall shear stress has been shown to vary from 5-15 dynes/cm in forward flow to +2 to -5 dynes/cm in reversed flow.[35] While absolute changes in pressure may not be essential for the development of aneurysms, this does not eliminate the importance of rapid changes in pressure induced by oscillation in flow and pressure.

Zarins et al created mid-thoracic aortic coarctations in monkeys in order to increase both blood flow velocity and wall shear stress. The artery distal to the coarctation experienced an increase in diameter such that a normal wall shear stress was achieved.[36] Alternatively, vessels exposed experimentally to decreased arterial wall shear stress, accomplished by the complete ligation a proximal artery, underwent a decrease in diameter distal to the ligature, so that normal shear stress was achieved.[28] Zarins proposed that arteries seek to achieve an optimal wall shear stress.[37]

Why should blood vessels appear to seek an optimal shear stress? In an attempt to explain these adaptations in arterial diameter to changes in blood flow, Murray proposed the principle of minimal work (PMW).[38] The principle of minimal work established a relationship between volumetric flow, flow velocity, and blood vessel radius.[39] In brief, Murray proposed that the work required to deliver blood to the tissues will be performed as efficiently as possible. In response to local changes in blood flow velocity, the arterial circulation will adapt by changing arterial diameter and circulatory volume so that the delivery of oxygen and nutrition will be maximized relative to the amount of energy expended. A balance is

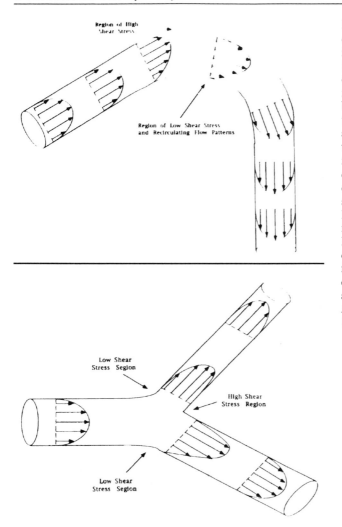

Fig. 10.9. Conservation of energy requires that blood flow velocities are greater (A) along the outside walls of curves and (B) along the inside walls at bifurcations because red blood cells at these locations must travel a greater distance. Conversely, the velocity and the mean shear stress are lowest along the inside walls of curves and along the outside walls at bifurcations. Reprinted with permission from: Panaro NJ, McIntire LV. Flow and shear stress effects on endothelial cell function.. In: Sumpio BM, ed. Hemodynamic Forces and Vascular Cell Biology. Austin: RG Landes Co., 1993:47-65.

achieved whereby the maximal delivery of oxygen and nutrition is achieved with the least amount of work. By altering the variables of volumetric flow, flow velocity, and vessel radius, the energy loss due to frictional forces, as measured by shear stress, is minimized.[39] The principle of minimal work provides a theoretic basis for the Hagen–Poiseulle formula, whereby the arterial diameter is adjusted to the cube root of the volume of flow so that the energy losses due to frictional forces are minimized relative to volumetric blood flow. The principle of minimal work is achieved at a shear stress of approximately 15 dynes/cm. Arteries tend to normalize to a wall shear stress of 15 dynes/cm in a large range of arteries in many animal species.[25,32,37]

Endothelial cells experience the effects of shear stress because of their location at the blood- arterial wall interface. Endothelial cell gene function is modulated by changes in wall shear stress, resulting in alterations in either cell morphology[40,41] or cell function.[42-44] These authors propose that changes in endothelial cell gene expression may permit the endothelium to indirectly control arterial wall diameter through the effects of these gene products.

The smooth muscle cells contained in the arterial medial layer are among the other cells in the arterial wall that could be affected by changes in endothelial cell gene expression. The medial layer of the arterial wall provides strength and compliance and directly determines arterial diameter in muscular arteries. Endothelial gene products, especially nitric oxide (NO), could exert their effects on the medial smooth muscle cells. Nitric oxide is a potent vasodilatory agent which mediates short-term vasodilatory effects through the relaxation of vascular smooth muscle cells. Increased wall shear stress caused by elevated blood flow velocity is probably the most powerful stimulus for the production of nitric oxide by the endothelium in vitro.[45] In response to changes in pressure or flow, the artery could expand or recoil by way of smooth muscle cell relaxation or contraction. Thus the smooth muscle cells and the endothelial cells could act in concert in the arterial wall through transcellular metabolism.[46,47] Alternatively, macrophages can produce interleukin-1β that exerts effects both on endothelial cell function and nitric oxide release[48] and smooth muscle cell metabolism.[49]

Arterial dilation in the setting of elevated shear, such as occurs following the creation of a coarctation of the aorta (Fig. 10.10),[50,51] is characterized histologically by a disruption of the elastin layer and a loss of smooth muscle cells (Fig. 10.11).[52] The arterial media is organized into lamellar units of smooth muscle cells interspersed between layers of elastin sheets and collagen fibers. The smooth muscle cells are responsible for the synthesis of the matrix proteins elastin and collagen, and the physical properties of elastin and collagen determine the mechanical properties of both normal and diseased arteries. These roles for elastin and collagen in exerting the retractive forces of the arterial wall have been confirmed by histologic and enzymatic degradation studies.[53] The microarchitecture of the larger aorta differs from the structure of smaller muscular arteries. The matrix protein layers in the aorta are more extensive and form continuous elastic lamellae between layers of smooth muscle cells. This organization of cells, that includes increased matrix proteins, permits this more elastic artery to better absorb greater circumferential stress. In muscular arteries, the layering of elastic fibers is less pronounced because of the relative paucity of matrix protein fibers.[25] The possibility exists that the interacting endothelial and smooth muscle cells which organize the lamellar architecture and the proportions of matrix proteins in the medial layer of the arterial wall respond abnormally or inappropriately to changes in wall shear stresses.[30] Additional cells such as macrophages may facilitate this abnormal response.[49]

Zarins observed that poststenotic dilation and aneurysm formation in the thoracic aorta of monkeys was associated with a two-fold increase aortic wall collagenase production in the region that underwent dilation.[36] Collagenase (MMP-1)[49] and type IV collagenase (MMP-2)[54] are produced predominantly by the arterial smooth muscle cells. C-Fos is a nuclear transcription factor involved in the regulation of matrix degrading enzymes.[55] Subsequent experiments by Zarins's group suggested that the upregulation of the proto-oncogene c-Fos in aortic smooth muscle cells occurred following the onset of increased wall shear stress, but that this up regulation took place prior to the onset of aortic enlargement.[56] Arterial ectasia may be the adaptation of the arterial wall to increased flow and increased wall shear stress over longer periods of time. Guzman and Zarins suggested that these increases in arterial diameter may be mediated either directly or indirectly by the effects of increased shear stress on the arterial wall and its smooth muscle cells.[56] Smooth muscle cells from aneurysmal aorta exhibited increased collagenase (MMP-2) production compared with smooth muscle cells from nonaneurysmal aorta.[57]

Aneurysms have a predilection for forming most commonly in large, lower extremity arteries at locations immediately proximal to bifurcations or large branches. The abdominal aorta, the most common location for arterial aneurysms, fits this grouping since the predominant role of the infrarenal aorta is the perfusion of the lower extremities. Since

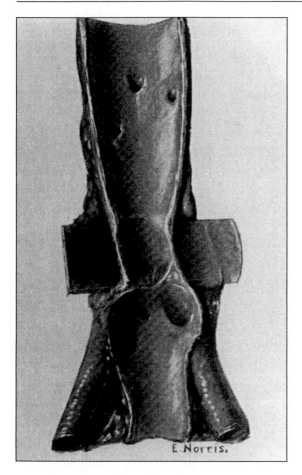

Fig. 10.10. William Halsted investigated the development of subclavian artery aneurysms in patients with cervical ribs by creating a model of aortic coarctation in dogs where he reproduced poststenotic dilation by placed a constricting metal band around the circumference of the aorta. Reprinted with permission from: Halsted WS. An experimental study of circumcised dilation of an artery immediately distal to a partially occluding band, and its bearing on the dilation of the subclavian artery observed in certain cases of cervical rib. J Exp Med 1916; 14:271286.

there are no sinuses in these arteries,[58] the flow pulse determines the local shear stress. The normal resting arterial flow pulse changes dramatically between the suprarenal and infrarenal aorta.[59] In the suprarenal aorta, the flow pulse remains positive throughout the cardiac cycle. This is because a large component of arterial flow at the level of the suprarenal aorta is to the kidneys that have a low resistance. In the infrarenal aorta and its runoff vessels, the flow pulse has triphasic flow. Initial forward flow in early and mid-systole is followed, secondly, by a reversal of the flow pulse beginning in late systole and early diastole. Thirdly, slow forward flow in late diastole completes the triphasic arterial flow cycle.[1] This swinging between forward, reversed, and forward flow in the infrarenal aorta results in oscillations both in blood flow and blood pressure, resulting in alterations in the shear stress between positive and negative values. Rapid oscillations in flow and pressure produce extremely rapid variations in pulse pressure.

Models of the abdominal aorta have shown that the flow patterns are more complex in the infrarenal aorta compared to the suprarenal aorta.[60] Blood flow oscillates within the abdominal aorta from the renal arteries to the aortic bifurcation, and requires several cardiac cycles for blood to clear from the posterior aortic wall where shear stress is lowest. Blood clears fairly rapidly from the anterior aortic wall, where shear stress is higher.[25] Using a model of the abdominal aorta[61] and magnetic resonance imaging,[62] blood flow velocity

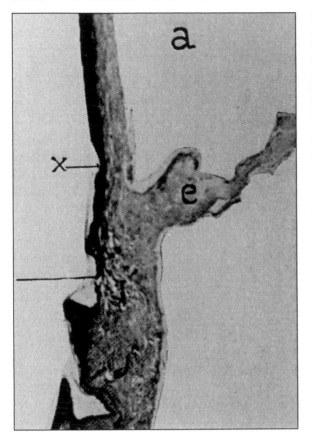

Fig. 10.11. Mont Reid first observed in Halsted's coarctation model that poststenotic dilation and aneurysm formation was associated with the loss of arterial wall elastin. (a) is the site of the metal band. (x) mark the lower limit of the aortic wall that was included in the metal band. Straight line points to luminal side of the aorta in the area of dilation showing atrophy and disruption of the elastic tissue. (e) is new wall that formed outside of the band (albumin). Weigert's elastic tissue stain. Reprinted with permission from: Reid, M. Partial occlusion of the aorta with the metallic band. Observations on blood pressures and changes in the arterial walls. J Exp Med 1916; 14:287290.

profiles were obtained around the circumference of the abdominal aorta. These blood flow velocity profiles then were used to calculate shear stress circumferentially around the aorta at sixteen points in the cardiac cycle. Under resting (nonexercise) conditions in the suprarenal aorta, average mean shear stress at the aortic wall was 1.3 dynes/cm. Under resting conditions in the infrarenal aorta, the mean shear stress oscillated at the posterior aortic wall, where the wall shear stress was negative for 75% of the pulse cycle. The mean shear stress along the posterior aortic wall was -5.0 dynes/cm. The two regions with the most negative shear stress values were the infrarenal aorta along the posterior wall and the aortic bifurcation along the lateral walls, where negative shear stress values as low as -12 dynes/cm were measured.[61] An increased mean blood flow velocity was introduced in this model to simulate exercise. The result was the elimination of the negative shear stress values. Mean shear stress in the infrarenal aorta under conditions simulating exercise ranged from 5.4 dynes/cm at the posterior wall to 10.6 dynes/cm at the lateral wall.[25,62]

The blood flow pattern in the abdominal aorta exists almost entirely in a state of high peripheral resistance, oscillating flow, a low mean velocity and therefore a low mean shear stress. The inference from these experiments by Ku and Moore is that in the abdominal aorta, the most common location for arterial aneurysms, the predominant hemodynamic state is one of oscillating shear stress and *low* mean shear stress. High resistance, oscillating flow, and low mean blood flow velocity are characteristic of blood flow in all of the lower extremity arteries where aneurysms form most commonly. These data suggest that low mean shear stress or oscillating shear stress is associated with an increased incidence of arterial aneurysms. Oyre et al performed additional experiments using magnetic resonance velocity mapping[63] and Pederson et al used laser Doppler anemometry[64] to confirm that low shear stress and oscillating shear stress were the predominant patterns in the infrarenal aorta where intimal thickening developed. Pederson et al found no correlation between high shear stress parameters and the development of infrarenal aortic disease that is believed to be the first step in aneurysm formation.[64]

These findings by Ku, Oyre, and Pederson create a paradox since low mean shear stress certainly is not common to all hemodynamic environments where arterial dilation develops or aneurysms form. For example, both the arterial dilation that forms beyond an aortic coarctation and the arteriomegaly or aneurysm that develops proximal to an arteriovenous fistula occur in the settings of *high* mean shear stress. Therefore, the low mean shear stress that is found in the infrarenal abdominal aorta at rest cannot be used to explain why aneurysms form most commonly in this location. How can this paradox of both high and low mean shear stress and the formation of aneurysms in both hemodynamic environments be resolved?

Vibration

Perhaps it is the rapid fluctuations in wall shear stress that occur at these locations, rather than the average or mean shear stress, that is important in the formation of aneurysms. Oscillating shear direction and stress is common to all of the major arteries where aneurysms form in humans. The infrarenal aorta, iliac, hypogastric, common femoral, and popliteal arteries all have oscillating shear stress at the blood-arterial wall interface by nature of the triphasic arterial flow that characterizes these arteries at rest. Because blood flow in these arteries is pulsatile and flow reverses in diastole, there are fluctuations in pressure and oscillations in the direction of the shear stresses.

Since pressure determines flow, rapid changes in pressure cause the oscillations in flow and shear. Therefore, it is not the oscillations in flow and shear, but rather the rapid fluctuations in pressure that produce vibration within the vessel and its wall. The oscillation in shear stress may be a phenomenon that is secondary to rapid changes in pressure.

These vibrations that are caused by rapid fluctuations in pressure are exaggerated in arteriovenous fistulas and coarctations, where the oscillations in pressure that produce vibrations can be palpated as a thrill and heard as a bruit. In an arteriovenous fistula, the vibrations travel proximally along the arterial and distally along the vein wall. These vibrations are stresses that induce a loss of tensile strength and cause both the artery and the vein to dilate. The pulse pressure is very high in both of these models. These rapid fluctuations in pressure that produced vibrations in the arterial wall of coarctations[65] and arteriovenous fistula[66] were determined by Robicsek to be associated not only with significant vibration, but also with significant energy loss in the region of the artery wall immediately just beyond the stenosis.[67]

Holman et al first proposed that vibration causes a structural fatigue of the arterial wall. In addition to examining poststenotic dilation in humans, Holman produced a similar phenomenon of poststenotic dilation in rubber tubes.[68] Bruns et al showed that vibration alone without any flow could induce dilation when he vibrated thin-walled, water-filled rubber tubes.[69] This vibration produced a pressure fluctuation (pulse pressure).

Boughner and Roach examined the effect of low frequency vibration on the human arterial wall. Vibration, even without any flow, resulted in significant dilation in pressurized human iliac arteries over short periods of time.[70] The vibrational frequencies that appeared to produce resonance and cause dilation in excess of the expected creep varied, depending on the age of the patient whose iliac artery underwent vibrational stimulation postmortem.

Older arteries required higher vibrational frequencies of approximately 200 Hz in order to dilate. Younger arteries required frequencies of approximately 100 Hz to dilate. Boughner and Roach proposed that vibrational energy weakened the elastin within the arterial wall. Stehbens has proposed that repetitive low frequency vibrations lead to a hemodynamic fatigue of the arterial wall.[71] The effect of this prolonged release of vibratory energy is molecular alterations within the arterial wall that lead to a loss of tensile strength.[72]

Stehbens suggested that atherosclerosis and aneurysms are just two different adaptations to the same loss of tensile strength (Fig. 10.12).[73] The oscillations in wall shear stress and pressure that lead to hemodynamic fatigue of the arterial wall can produce either compensatory intimal proliferation or arterial wall ectasia, tortuosity, and aneurysms (Fig. 10.13).[74] The apparent paradox of arterial dilation and aneurysm formation that can occur with either high or low mean wall shear stress can be reconciled by the hypothesis that hemodynamic fatigue produced by low frequency vibrations also is induced by rapid oscillations in pressure. The aneurysms that form in arteries proximal to arteriovenous fistulae and distal to arterial coarctations and the aneurysms that form in the abdominal aorta and its runoff vessels all develop in the setting of a low oscillating shear pressure and shear stress.

A higher or widened pulse pressure (or any rapid pressure drop) and the increased energy associated with this can be visualized by the example of a waterfall of increasing height: A higher pulse pressure corresponds to a higher waterfall. Just as greater energy is released from a higher waterfall, a higher pulse pressure releases more energy in a short distance to the recipient artery than does a lower pulse pressure. While the infrarenal abdominal aorta and lower extremity arteries that form aneurysms in humans exist in a setting a low mean shear stress, the oscillating shear stress, fluctuating pressure and the relatively increased pulse pressure characteristic of these vessels at rest can be sufficient to produce low frequency vibrations. The result of these vibrations is a loss of tensile strength from vibration-induced mechanical fatigue that can lead from ectasia and aneurysm formation.

The effects of vibration on resilient, elastic arteries differ from the effects of vibration on muscular arteries. Stehbens made these observations following experiments in rabbit femoral and renal arteries. Rabbit common carotid arteries are predominantly elastic, while rabbit femoral arteries are more muscular. The muscular femoral arteries appeared to be more susceptible to vibration-induced arterial dilation,[75] and the more elastic common carotid arteries appeared to have a greater tolerance to vibration. Stehbens surmised that the larger amount of elastic matrix found in larger arteries took longer to succumb to the cumulative effects of vibration.

These conclusions that vibration induces a loss in tensile strength are consistent with the principles of engineering and material failure. There is no reason to believe that biomaterials are immune to the forces that cause fatigue and a loss of tensile strength in nonbiologic materials. These examples demonstrate that mural fatigue can be induced by vibration. The basic mechanism behind bioengineering fatigue is that flow-induced tensile stresses cause macromolecules, such as elastin, to fragment. The manifestations of aneurysmal disease represent mural failure and weakness with yielding under the intraluminal pressure. Mural thickening of the wall in aneurysms may be a measure of the attempts by the vessel wall to compensate for a loss in tensile strength.[72]

Conclusion

This chapter discusses the role of blood flow, pulse pressure and other rapid changes in pressure, shear stress, and vibration in the formation of aneurysms. Shear stress alone does not adequately explain the forces that cause arterial dilation and aneurysm formation. Cause

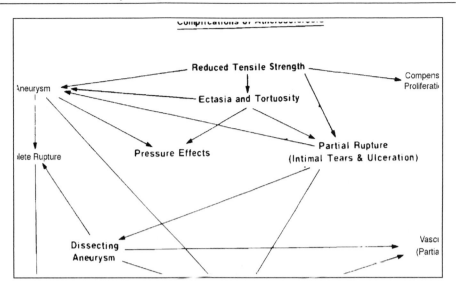

Fig. 10.12. The complications of atherosclerosis, including both aneurysmal disease and occlusive disease, can be attributed to the loss of arterial wall tensile strength. Reprinted with permission from: Stehbens WE. The Pathology of atherosclerosis and experimental atherosclerosis. In: Stehbens WE, ed. The Lipid Hypothesis of Atherogenesis. Austin: RG Landes Co., 1993:10-47.

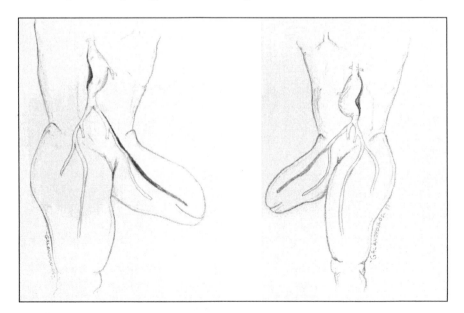

Fig. 10.13. Abdominal aortic aneurysms developed in veterans with longstanding above knee amputations at a fivefold higher rate than in nonamputees. The abdominal aorta sustained axial shifting toward a dilated iliac artery contralateral to the side of the amputation in 84% of amputees who developed aneurysms. Vascular morphology in (A) left-sided and (B) right-sided amputation. Modified with permission from: Vollmer, JF, Paas E, Pauschinger P et al. Aortic aneurysm as a late sequelae of above knee amputation. Lancet 1989; 2:834-835.

is the prerequisite without which the disease cannot occur.[76] The development of aneurysms in humans cannot be evaluated without considering the effects of rapid fluctuations in pressure and vibration as the ultimate etiology. Additional molecular, humoral, and genetic factors that lead to decrease in arterial wall tensile strength can predispose to aneurysm formation. These factors may play a role in the rate at which aneurysms grow or progress.[77]

References

1. Bassiouny HS, Zarins CK, Kadowaki MM et al. Hemodynamic stress and experimental aortoiliac atherosclerosis. J Vasc Surg 1994; 19:426434.
2. Stehbens WE. The lipid hypothesis and the role of hemodynamics in atherogenesis. Progr Cardiovasc Dis 1990; 33:119136.
3. Halsted WH. Cylindrical dilation of the common carotid artery following partial occlusion of the inanimate and ligation of the subclavian. In: The Surgical Papers of William Steward Halsted. Baltimore: The John Hopkins Press, 1928:460-468.
4. Osler W. Arterio-venous aneurysm. Lancet 1915; 1:949-955.
5. Reid, M. Studies on abnormal arteriovenous communications, acquired and congenital. Report of a series of cases. Arch Surg 1925; 10:601-610.
6. Ozcan F, Baki C, Piskin B et al. Aneurysmatic dilatation of popliteal and femoral artery due to longstanding traumatic arterio-venous fistula. VASA 1990; 19:79-80.
7. Gerbode F, Holman E, Dickenson EH et al. Arteriovenous fistulas and arterial aneurysms. Surg 1952; 32:259-274.
8. McDonald DA. Blood Flow in Arteries. Baltimore: Williams & Wilkins 1974:356.
9. Yao, JST. Hemodynamic studies in peripheral arterial disease. Br J Surg 1970; 57:761.
10. Remington JW, O'Brien LJ. Construction of aortic flow pulse from pressure pulse. Am J Physiol 1970; 218:437-447.
11. Strandness DE Jr, Sumner DS. Physics of arterial blood flow. In: Standness DE Jr, Sumner DS, eds. Hemodynamics for Surgeons. New York:Grune and Stratton, 1975:73-95.
12. Westerhoff, Sipkema P, Van Den Bos GC et al. Forward and backward waves in the arterial system. Cardiovasc Res 1972; 6:648.
13. Nichols WW, O'Rourke MF. Wave reflections. In: Nichols WW, O'Rourke MF, eds. McDonald's Blood Flow in Arteries 3rd ed. Philadelphia: Lea & Febiger, 1990; 251-269.
14. Lanne T, Bergentz SE. Imaging of arterial wall movement. In: Greenhalgh RM, ed. Vascular Imaging for Surgeons. London: WB Saunders, 1995:3-20.
15. Sonnesson B, Lanne T. The mechanical properties of the normal and aneurysmal abdominal aorta in vivo. In: Keen RR, Dobrin PB, eds. Development of Aneurysms. Austin: Eurekah.com, 2000.
16. Sonneson B, Länne T, Vernersson E et al. Sex differences in the mechanical properties of the abdominal aorta in human beings. J Vasc Surg 1994; 20:959-969.
17. Länne T, Hansen F, Mangell P et al. Differences in mechanical properties of the common carotid artery and abdominal aorta in healthy males. J Vasc Surg 1994; 20:218-225.
18. O'Rourke MF, Kelly RP, Avolio AP. In: Pine JW Jr., ed. The Arterial Pulse. Philadelphia: Lea & Febiger, 1992:47-71.
19. He, CM, Roach, MR. The composition and mechanical properties of abdominal aortic aneurysms. J Vasc Surg 1994; 20:6-13.
20. Stehbens, WE. Abnormal arteriovenous communications and fistulae. In: Stehbens WE, Lie, JT, eds. Vascular Pathology. London: Chapman Hall 1995:517-552.
21. Roach MR. Hemodynamic factors in arterial stenosis and poststenotic dilation. In: Stehbens WE, ed. Hemodynamics and the Blood Vessel Wall. Springfield: Charles C. Thomas, 1979:439-464.
22. Chae CU, Pfeffer MA, Glynn RJ et al. Increased pulse pressure and risk of heart failure in the elderly. JAMA 1999; 281:634-639.
23. Ku DN, Zhu C. The mechanical environment of the artery. In: Sumpio BE, ed. Hemodynamic Forces and Vascular Cell Biology. Austin: RG Landes Co., 1993:1-23.
24. Dobrin, PB. Mechanical properties of arteries. Physiol Rev 1978; 58:397-460.

25. Stehbens WE. Structural and architectural changes during arterial development and the role of hemodynamics. Acta Anat 1996; 157:261-274.
26. Nichols WW, O'Rourke MF. The nature of flow of a fluid. In: Nichols WW, O'Rourke MF, eds. McDonald's Blood Flow in Arteries. 3rd ed. Philadelphia: Lea & Febiger, 1990; 12-53.
27. DeVries HK, Van Den Berg JW. On the origin of poststenotic dilation. Cardiologia 1958; 33:195-211.
28. Zarins CK. Adaptation to atteroschlerosis: human versus animal. J Vasc Surg 1995; 22:86-87.
29. Stehbens WE. Pathology and pathogenesis of degenerative atherosclerotic aneurysms. In: Keen RR, Dobrin PB, eds. Development of Aneurysms. Austin: Eurekah.com, 2000.
30. Gerrard JH, Charlesworth DM. Vibration and atherosclerosis. In: Sumpio BM, ed. Hemodynamic Forces and Vascular Cell Biology. Austin: RG Landes Co., 1993:38-46.
31. Gerrard JR Charlesworth D. Atherosclerosis and disturbance in flow. Ann Vasc Surg 1988; 2:5762.
32. Glagov S, Zarins C, Giddens DP, Ku DN. Hemodynamics and atherosclerosis: Insights and perspectives gained from studies of human arteries. Arch Path Lab Med, 1988; 112:10181031.
33. Karino T, Goldsmith HL, Motomiya M et al. Flow patterns in vessels of simple and complex geometrys. Ann NY Acad Sci 1987; 516:422-441.
34. Panaro NJ, McIntire LV. Flow and shear stress effects on endothelial cell function. In: Sumpio BM, ed. Hemodynamic Forces and Vascular Cell Biology. Austin: RG Landes Co., 1993:47-65.
35. Zarins CK, Glagov S, Giddens DP et al. Hemodynamic factors and atherosclerotic changes in the aorta. In: Bergan JJ, Yao JST, eds. Aortic Surgery. Philadelphia: WB Saunders, 1989:126.
36. Zarins, C., RunyonHaas, A. Zatina, M., Lu, C. and Glagov, S. Increased collagenase activity in early aneurysmal dilatation. J Vasc Surg 1986; 3:23848.
37. Zarins CK, Zatina MA, Giddens DP et al. Shear stress regulation of artery lumen diameter in experimental atherogenesis. J Vasc Surg 1987; 5:413420.
38. Murray CD. The physiological principle of minimum work: the vascular system and the cost of blood flow. Proc Nat Acad Sci USA 1926; 12:207-214.
39. Rossitti S, Lofgren J. Vascular dimensions of the cerebral arteries follow the principle of minimum work. Stroke 1993; 24:371377.
40. Levesque MJ, Nerem RM. The elongation and orientation of cultured endothelial cells in response to shear stress. J Biochem Eng 1985; 176:341347.
41. Nerem RM, Levesque MJ, Cornhill JF. Vascular endothelial morphology as an indicator of blood flow. J Biochem Eng 1981; 104:172176.
42. Nollert MU, Panaro NJ, McIntire LV. Regulation of genetic expression in shear stress-stimulated endothelial cells. Ann NY Acad Sci 1992; 665:94-104.
43. Grabowski EF, Jaffe EA, Weksler BB. Prostacyclin production by cultured endothelial cell monolayers exposed to step increases in shear stress. J Lab Clin Med 1985; 105:3643.
44. Koller A, Sun D, Kaley G. Role of shear stress and endothelial prostaglandins in flow and viscosity induced dilation of arterioles in vitro. Circ Res 1993; 72:12761284.
45. Korenga R, Ando J, Tsuboi H et al. Laminar flow stimulates ATP and shear stress dependent nitric oxide production in cultured bovine endothelial cells. Biochem Biophy Res Commun 1994; 198:213219.
46. Lands WEM, Keen RR. Peroxide tone and its consequences. In: Reddy CC, Hamilton GA, Madyastha KM, eds. Biological Oxidation Systems. San Diego: Academic Press, 1990: 657-666.
47. Ellenby MI, Ernst CB, Carretero 0 et al. Role of nitric oxide in the effect of blood flow on neointima formation. J Vasc Surg 1996; 23:314-322.
48. Joly GA, Schini VB, Vanhoutte PM. Balloon injury and interleukin-1β induce nitric oxide syntheses activity in rat carotid arteries. Circ Res 1992; 71:331-338.
49. Keen RR, Nolan KB, Cipollone M et al. Interleukin-1β induces differential gene expression in aortic smooth muscle cells. J Vasc Surg 1994; 20:774-786.

50. Halsted WS. Partial, progressive and complete occlusion of the aorta and other large arteries in the dog by means of the metal band. J Exp Med 1909; 11:373 –391.
51. Halsted WS. An experimental study of circumcised dilation of an artery immediately distal to a partially occluding band, and its bearing on the dilation of the subclavian artery observed in certain cases of cervical rib. J Exp Med 1916; 14:271286.
52. Reid, M. Partial occlusion of the aorta with the metallic band. Observations on blood pressures and changes in the arterial walls. J Exp Med 1916; 14:287290.
53. Dobrin PB, Schwarcz Th, Baker WH. Mechanisms of arterial and aneurysmal tortuosity. Surg 1988; 104:568-571.
54. McMillan WD, Patterson BK, Keen RR et al. In situ localization and quantification of 72 kilodalton type IV collagenase in aneurysmal, occlusive, and normal aorta. J Vasc Surg 1995; 22:295-305.
55. Hsieh HJ, Li NQ, Frangos JA. Pulsatile and steady flows increase proto-oncogenes c-fos and c-myc mRNA levels in human endothelial cells. FASEB J 1992; 6:A1820.
56. Guzman RJ, Krystowiak AJ, Zarins CK. Smooth muscle cell cFos gene expression precedes shear stress induced aortic enlargement. Proceedings of Society for Vascular Surgery 1997, Boston.
57. Crowther M, Brindle NPJ, Sayers R et al. Aneurysmal smooth muscle cells exhibit increased matrix metalloproteinase2 production in vitro. Ann NY Acad Sci 1996; 800:283-285.
58. Ku DN, Zarins CK, Giddens DP, et al. Pulsatile flow and atherosclerosis in the human carotid bifurcation. Positive correlation between plaque localization and low and oscillating shear stress. Arteriosclerosis 1985; 5:292302.
59. Hale JF, McDonald DA, Womersley JR. Velocity profiles of oscillating arterial flow, with some calculations of viscous drag and the Reynold's number. J Physiol 1955; 128:629-640.
60. Holenstein R, Ku DN. Reverse flow in the major infrarenal vessels a capacitive phenomenon. Biotechnology 1988; 25:835842.
61. Ku DN, Glagov S, Moore JE et al. Flow patterns in the abdominal aorta under simulated postprandial and exercise conditions: An experimental study. J Vasc Surg 1989; 9:309316.
62. Moore JE, Ku DN, Zarins CK et al. Pulsatile flow visualization in the abdominal aorta under differing physiologic conditions: Implications for increased susceptibility to atherosclerosis. J Biomech Eng 1992; 114:391-397.
63. Oyre S, Pedersen EM, Ringgaard S, et al. In: Vivo wall shear stress measured by magnetic resonance velocity mapping in the normal human abdominal aorta. Eur J Vasc Endovasc Surg 1997; 13:263-271.
64. Pedersen EM, Agerbaek M, Kristensen IB, et al. Wall shear stress and early atherosclerotic lesions in the abdominal aorta in young adults. Eur J Vasc Endovasc Surg 1997; 13:443451.
65. Foreman, JEK, Hutchinson KJ. Arterial wall vibration distal to stenoses in isolated arteries of dogs and man. Circ Res 1970; 26:583-589.
66. Ingebrigtsen R, Lie M, Hol R et al. Dilation of the ileo-femoral artery following the opening of an experimental arterio-venous fistula I the dog. Scand J Clin Lab Invest 1973; 31:255-262.
67. Robicsek F. Poststenotic dilatation of the great vessels. Acta Medica Scandinavica 1955; 61:481486.
68. Holman E. On circumscribed dilatation of an artery immediately distal to a partially occluding band: Poststenotic dilatation. Surg 1954; 36:324.
69. Bruns DL, Connolly JE, Holman E. et al. Experimental observations on poststenotic dilatation. J Thorac Cardiovasc Surg 1959; 38:662669.
70. Boughner DR, Roach MR. Effect of low frequency vibration on the arterial wall. Circ Res 1971; 29:136144.
71. Simkins TE, Stehbens WE. Vibrations recorded from the adventitial surface of experimental aneurysms and arteriovenous fistulas. Vasc Surg 1974; 8:153-165.
72. Stehbens WE. The pathogenesis of atherosclerosis: a critical evaluation of the evidence. Cardiovasc Path 1997; 6:123-153.
73. Stehbens WE. The Pathology of atherosclerosis and experimental atherosclerosis. In: Stehbens WE, ed. The Lipid Hypothesis of Atherogenesis. Austin: RG Landes Co., 1993:10-47.

74. Vollmer, JF, Paas E, Pauschinger P et al. Aortic aneurysm as a late sequalar of above knee amputation. Lancet 1989; 2:834-835.
75. Greenhill NS, Stehbens WE. Scanning electron microscopic study of the afferent arteries of experimental femoral arteriovenous fistulae in rabbits. Pathology 1987; 19:22-27.
76. Stehbens WE. Causality in medical science with reference to coronary heart disease and atheroschlerosis. Perspect Biol Med 1992; 36:97-119.
77. Wills A, Thompson MM, Crowther M et al. Pathogenesis of abdominal aortic aneurysms Cellular and biochemical mechanisms. Eur J Vasc Endovasc Surg 1996; 12:391400.

CHAPTER 11

Infection, Atherosclerosis, and Aneurysmal Disease

Michael T. Caps, Philip B. Dobrin

The preceding chapters in this book have reviewed the epidemiology, pathophysiology, and pathologic progression of arterial aneurysms. But the critical question remains: What initiates their formation? One possibility is that aneurysms develop because of the fatiguing of connective tissues with age. This hypothesis is discussed in some detail in chapters 4 and 6. The present chapter reviews the evidence supporting the hypothesis that infection with *Chlamydia pneumoniae* (*C. pneumoniae*), cytomegalovirus (CMV), and/or herpes simplex virus, type 1(HSV-1) may play an etiological role in causing atherogenesis and, in turn, the development of aneurysms.

Atherosclerosis and Aneurysm

For many years, it has been accepted that atherosclerotic degeneration of the arterial wall causes the formation of aneurysms. The histologic evidence for this view is discussed in detail in the chapter in this book by Stehbens. What is not certain is why, in the majority of cases, atherosclerosis causes occlusive symptoms and not the formation of an aneurysm. Zarins and Glagov proposed that the local response of the media to injury determines the outcome of atherosclerotic injury.[1] These investigators concluded that if the media heals fully, then an obstructing lesion will remain. By contrast, if adequate healing fails to occur and the media becomes atrophic, then aneurysmal degeneration of the wall will occur. Bengtsson's epidemiological studies in patients are consistent with this interpretation.[2] These investigators found that 13.5% of patients who underwent carotid surgery for occlusive carotid disease also had aortic enlargement or an abdominal aortic aneurysm. However, the factors that define whether an atherosclerotic lesion will become occlusive or will cause medial degeneration remain unclear. And, of course, thoracic aortic aneurysms are often devoid of atherosclerosis.

A large body of evidence suggests an etiologic association between atherosclerosis and abdominal aortic aneurysm (AAA). In large, population-based epidemiologic studies, many of the classical risk factors for atherosclerosis also are strongly associated with AAA. In the Cardiovascular Health Study (CHS), for example, cigarette smoking, elevated LDL and reduced HDL cholesterol, and diabetes were associated with an increased prevalence of AAA.[3] AAA was more common among those patients with clinical coronary artery disease, and among those with sub-clinical measures of atherosclerosis such as a reduced ankle-arm pressure indices and increased carotid artery intima-media thickness. In the Rotterdam Study,[4] similar associations were observed.

Development of Aneurysms, edited by Richard R. Keen and Philip B. Dobrin.
©2000 Eurekah.com.

Atherosclerotic plaques are virtually always seen in the walls of AAAs.[5] Zarins and co-workers demonstrated that graded destruction of the elastic lamellae in pigs produces aneurysms.[6] Such destruction could occur with atherosclerotic degeneration of the wall. In a related experiment, these investigators[7] fed cynomolgus and rhesus monkeys an atherogenic diet. After one and a half to two years, they placed the animal on a diet designed to cause regression of atherosclerosis. With regression, 13% of the cynomolgus monkeys and one percent of the rhesus monkeys developed aneurysms. This suggests that regression of the plaque including the matrix proteins present in the plaque, or perhaps atherosclerotic destruction of the normal load-bearing connective tissue elements in the wall, led to the formation of aneurysms.

Although atherosclerosis may contribute to the aneurysmal process, it may not be sufficient to cause the destruction of the aortic media seen in patients with AAA. As noted above, atherosclerosis leads more frequently to occlusive disease than to aneurysmal lesions. When patients with AAA are compared with patients having aortic occlusive disease (AOD), there are distinct differences. The aortic media in AAA patients is thinner than the media in patients with AOD owing to a loss of elastin and a reduction in the number of smooth muscle cells between the elastic lamellae.[5] Patients with AOD have a significantly higher prevalence of lower extremity, carotid, and coronary occlusive disease than AAA patients, diabetes is more common in AOD than in AAA, and cigarette smoking appears to have a stronger association with AAA than with AOD.[8] The discrepancy in the association with cigarette smoking between AOD and AAA is particularly interesting because cigarette smoking is known to cause an elevation of circulating proteases such as elastase.[9,10]

Atherosclerosis and aneurysmal disease are associated with inflammation. Several markers of the systemic inflammatory response, including the white blood cell count, fibrinogen, and C-reactive protein levels are elevated in patients with clinical coronary, cerebral, and peripheral vascular disease compared to that observed in control patients.[11-15] When aortic histology is compared between patients with AAA and AOD, there are notable differences. Patients with both AOD and AAA have inflammatory cell infiltrates within the aortic wall, but macrophages are distributed throughout the aortic wall in AAA whereas they are found chiefly in the plaque in patients with AOD.[16,17] Macrophages are of particular interest because they are capable of producing proteases that can degrade the matrix connective tissues.

The strong familial aggregation of AAA is another factor arguing in favor of a causal pathway for AAA that is separate from AOD. In 1986, Johansen and Koepsell found that first degree relatives of patients with AAA were 11.6 times more likely to be afflicted with AAA than age- and sex-matched controls.[18] The inheritance pattern is likely polygenetic involving both X-linked and autosomal dominant components, and a putative gene has not yet been identified.[19] This is discussed in Chapter 13 by Kuivaniemi and Tromp.

In summary, the etiology of AAA is complex and may involve atherosclerosis and other nonatherosclerotic factors leading to destruction of the load-bearing matrix proteins of the aortic media. As described in the preceding chapters, inflammation may play an important role in this process.

Infection and Atherosclerosis

At the turn of the century, it was suggested that an infectious process may be the cause of a variety of "constitutional" arterial diseases. Infection as a potential cause of atherosclerosis is compatible with the "response to injury" hypothesis:[16] Endothelial cell injury leads to monocyte adherence and migration, platelet attachment, smooth muscle cell proliferation, and cholesterol sequestration. It has been suggested that infection with *C. pneumoniae* and the herpes viruses, including CMV and HSV-1, is capable of initiating atherosclerosis. The evidence in support of a causal relationship between infection with these organisms and atherosclerosis comes primarily from seroepidemiologic, pathologic, and animal experimental studies. These studies are briefly summarized here.

Chlamydia pneumoniae

In 1986, Grayston et al described a new species of Chlamydia, *Chlamydia pneumoniae*.[21] This species has been shown to be a common cause of sinusitis, pharyngitis, bronchitis, and pneumoniae in both adults and children.[22] The organism is transmitted from person-to-person via aerosolized droplets and there is no known animal host. The prevalence of antibody directed against *C. pneumoniae* in the general population approaches 80%-85% in those ≥ 65 years of age. In 1988, an association was demonstrated between antibody to *C. pneumoniae* and coronary heart disease,[23] and subsequent studies have shown similar relationships between antibody directed against this organism and atherosclerotic occlusive disease in coronary and carotid arteries.[22,24-28] Early, preclinical atherosclerosis, as manifested by ultrasonically-detected thickening of the carotid artery, is more common among antibody-positive persons than antibody-negative controls.[28] In a recent large case-control study from the United Kingdom,[29] use of certain antibiotics within a prior 3-year time period was found to be associated with a reduced risk of a first-time anterior myocardial infarction. This protective effect was restricted to tetracyclines (OR = 0.7, 95% CI 0.55 to 0.90) and pneumoniae activity. Use of erythromycin, beta-lactams, and sulfonamides was not associated with a significant reduction in coronary risk and these antibiotics are considered ineffective against *C. pneumoniae*. Recently, viable *Chlamydia pneumonia* has been demonstrated in atherosclerotic plaques of carotid arteries using reverse transcriptase polymerase chain reaction, although it remains to be clarified as to whether the microorganism causes or colonizes existing atherosclerotic plaques.[29a]

C. pneumoniae organisms are capable of infecting human endothelial cells as well as smooth muscle cells and macrophages in vitro; in fact, they replicate in smooth muscle cells and macrophages in vivo.[30] *C. pneumoniae* induces the formation of foam cells by human monocyte-derived macrophages.[31] The organisms have been demonstrated in a variety of atheromatous tissues by immunocytochemistry (ICC), polymerase chain reaction (PCR), and electron microscopy.[32] Chlamydial heat shock protein (HSP) 60 frequently colocalizes with human HSP 60 in atherosclerotic lesions, and both induce TNF-alpha and matrix-degrading metalloproteinases by macrophages.[33] In contrast, *C. pneumoniae* antigens are generally not detectable in normal arterial walls or in nonatherosclerotic segments of arteries in patients with atherosclerotic disease.

Murine and rabbit models of *C. pneumoniae* infection have recently been developed which closely resemble the self-resolving interstitial pneumonia seen in humans.[34,35] In the murine model, infection in the setting of an atherogenic diet or genetic susceptibility leads to early atherosclerotic lesions containing *C. pneumoniae* organisms.[35] Muhlestein and co-workers[36] administered intranasal inoculations of *C. pneumoniae* versus saline to New Zealand White Rabbits fed an atherogenic diet. They found that intimal thickness in the infected animals was more than three-fold greater than that observed in the control animals. Furthermore, infected animals treated with the nonchlamydial agent, azithromycin, showed no difference in intimal thickness as compared to control animals.

Another important line of evidence supporting an association between *C. pneumoniae* infection and atherosclerosis comes from a recent randomized, prospective clinical trial reported by Gurfinkel and colleagues[37] in which 202 patients with recent myocardial infarction were randomized to receive an antibiotic, roxithromycin, or placebo. Patients who received the antibiotic experienced a significant reduction in the risk of recurrent ischemia, recurrent myocardial infarction, and ischemic death compared to the placebo group. Larger trials are currently underway in Europe and the United States to examine the efficacy of antimicrobial therapy in this setting.

Cytomegalo Virus

Herpes viruses, including CMV, were first hypothesized to play an etiologic role in human atherosclerosis because atherosclerosis can be induced in chickens by the avian herpes virus, Marek's Disease Virus (MDV).[38,39] Benditt et al[38] demonstrated the presence of nucleic acid sequences of MDV in the arteries of infected chickens. The atherosclerotic lesion that developed was remarkably similar to human atherosclerosis. Cholesterol-feeding accelerated the disease; however, such lesions did not occur in uninfected chickens, even with high blood cholesterol levels. Moreover, prior immunization with an antigenically-related herpesvirus prevented virus-induced atherosclerosis.[40] Another similar model has been developed in the Japanese quail.[41] The herpes family of viruses is ubiquitous in humans, and may be identified even in normal appearing cells.[40] Infection is widespread in human populations, often beginning in childhood.

Persons and co-workers[42] studied the effect of rat CMV (RCMV) on atherogenesis in rats. The left common carotid artery was subjected to balloon injury. Fourteen to 17 days later the rats were infected intravenously with RCMV or a control solution. Active RCMV infection was demonstrated in the neointima of the injured arteries. Medial cells and neointimal cells covered by endothelial cells were not infected. No infection was seen in the uninjured right carotid arteries. The majority of infected cells were neointimal smooth muscle cells containing actin; medial smooth muscle cells were much less susceptible. CMV infection in the rat produces changes in the endothelium detectable by electron microscopy. The changes are similar to those induced by hypercholesterolemia.[43,44] Many of the histological changes seen in rats resemble those seen in human atherosclerosis, i.e., swollen endothelial cells, bleb and microvilli formation, widening of gap junctions, adhesion of leukocytes and macrophages in the sub-endothelium, and lipid accumulation ("foam cells"). A characteristic phenomenon was loosening of the endothelial cells from the underlying basement membrane. CMV infection of smooth muscle leads to proliferation similar to that seen in benign neoplasms.[45]

CMV has been isolated from many human tissues, including lymphocytes which have prolonged contact with the endothelial cells lining the walls of arteries. CMV antigens and nucleic acid sequences have been detected in atherosclerotic tissues and in the walls of nonatherosclerotic arterial segments from several sites in patients with symptomatic arterial occlusive disease.[38,46-48] Several studies have demonstrated genomic sequences in human coronary arteries, aorta, and femoral samples that resembled that of herpes viruses including CMV. These were often seen in regions of arteries exhibiting early atherosclerotic changes, i.e., intimal thickening with smooth muscle cell proliferation and lymphocytic infiltration; however, no intact virus could be identified.[47,49] Because the nucleic acids are often detectable in the absence of antigen, the arterial wall may be the site of latent CMV infection.[46,50] Reactivation of the latent CMV organisms may occur in the coronary arteries of CMV-infected immunosuppressed transplant patients who frequently demonstrate a rapidly progressing form of atherosclerosis with CMV organisms detectable within their atheroma.[51]

A number of studies have demonstrated the chemical changes induced by CMV infection. These include increased plasma levels of Von Willebrand factor, factor VIII, and protein, and decreased activated partial thromboplastin time. CMV titers are also associated with increased antithrombin III and fibrinogen levels. CMV appears to have a net procoagulant effect.[52]

Several seroepidemiologic studies have suggested an association between cytomegalovirus (CMV) infection and atherosclerosis in humans.[39,43,44] Adam and co-workers[53] screened 157 male subjects undergoing vascular surgery for atherosclerosis versus a matched control group for antibodies to CMV. A significantly greater proportion of the surgical group had high CMV titers compared to the control group (57% versus 26%). In a nested case-control study of patients from the population-based Atherosclerosis Risk In Communities (ARIC) cohort, Nieto and colleagues detected an odds ratio for CMV seropositivity of 5.3 (95% CI = 1.5,18) among patients with increased carotid wall thickness.[54] In studies of patients undergoing coronary artery balloon angioplasty, Blum and co-workers[55] and Zhou and coworkers[56] found that those with high nonCMV antibody tiers were more likely to develop restenosis than patients with low or no antibody titers.

Herpes Simplex Virus, Type 1

Initial interest in the potential association between HSV-1 infection and atherosclerosis was stimulated by studies involving the Marek's Disease Virus (MDV) virus in chickens. In addition to CMV, human herpes simplex virus nucleic acid sequences have been identified in atherosclerotic tissue removed from aortas in patients undergoing aortofemoral bypass.[38,57] Yamashiroya and co-workers,[49] using DNA probes and monoclonal antibodies, found HSV and CMV DNA and antigens in the coronary arteries of 8 of 20 trauma victims. In 7 of the 8 cases, the DNA or antigen was found in areas of focal intimal thickening due to smooth muscle cell proliferation and in areas of lymphocytic infiltration.

HSV-1 infection of vascular endothelium has many effects including the stimulation of granulocyte attachment with subsequent detachment of the endothelial cell from its substratum.[57,58] In vitro studies have shown that HSV-1 infection may have a pro-coagulant effect. Human endothelial cells infected by HSV-1 exhibit excessive thrombin formation, produce virtually no prostacyclin, and display an increased adherence to platelets and granulocytes.[59-61] Furthermore, infection of human umbilical vein endothelium with HSV-1 may induce a procoagulant state by reducing thrombomodulin activity and stimulating the synthesis of tissue factor.[62,63]

The seroepidemiologic evidence of association with atherosclerosis is less compelling for HSV-1 than for *C. pneumoniae* or CMV. In the ARIC study for example, there was only a weak relationship between HSV-1 antibody and carotid artery wall thickening (odds ratio = 1.41, P=.07) and no association between HSV-2 infection and wall thickness.[39]

While these findings are intriguing, it is not known whether there is a direct causal relationship between arterial infection and atherosclerosis. An alternative hypothesis is that arteries with atherosclerosis are more susceptible to infection with these organisms than are normal arteries. Additional clinical trials of antibiotic or antiviral therapy to prevent or retard atherosclerosis may ultimately be the only way to resolve this uncertainty.

Infection and Aneurysm

It is reasonable to hypothesize that infection with *C. pneumoniae*, CMV, and/or HSV-1 could result in AAA formation, possibly due to stimulation of atherogenesis, by direct destruction of the aortic media, or possibly due to the inflammatory changes associated with atherosclerosis. Arterial infection has long been appreciated as an etiologic factor in a small subset of aortic aneurysms. Patients with advanced syphilis have a propensity to develop aneurysms of the thoracic aorta. The aneurysms are characterized by necrosis of the arterial media. They have a rapid growth rate and often rupture. Fortunately, they have become extremely rare with the advent of effective antibiotic treatment of syphilis. The so-called "mycotic" aortic aneurysm typically results either from bacteremic infection of a pre-existing sterile aneurysm or via hematogenous infection of a nonaneurysmal but atherosclerotic artery.[64] The responsible pathogens typically include *Salmonella*, *Streptococcus*, and

Staphylococcus species. Mycotic aortic aneurysms comprise only 1-2% of all AAAs requiring surgical repair.

The inflammatory changes seen in the walls of AAAs are striking. This is most evident in thick-walled, inflammatory aneurysms,[65,66] but inflammation often is seen even in ordinary degenerative aneurysms. Russell bodies and immunoglobulin are found in the aneurysmal wall,[67] as are macrophages and B-cells.[16] There is abundant elastolytic and collagenolytic activity[68] in the form of metalloproteinase-9.[69,70] Plasmin and activation of metalloproteinase by plasminogen activator are also present in the aneurysmal wall,[71,72] as are IL-1β and TNF-α.[72]

Direct evidence of an association between infection with *C. pneumoniae* and AAA comes primarily from pathologic analysis of human aortic aneurysms. In three pathologic studies, *C. pneumoniae* DNA was detected in 26 of 51,[73] in 14 of 32,[74] and in 6 of 6[75] aortas from patients who underwent AAA repair. Peterson and colleagues[76] used a nested polymerase chain reaction (PCR) method to investigate the presence of *C. pneumoniae* DNA in the walls of 40 patients operated on for infrarenal AAA. These investigators compared this group with 40 deceased controls without AAA. The detection of *C. pneumoniae*-specific DNA was significantly higher in the AAA group than among controls (35% versus 5%, P=.001). By contrast, Lindholt and co-workers failed to detect *C. pneumoniae* DNA in any of the 20 aneurysms that were assessed. They concluded that their results support the hypothesis that AAA and atherosclerosis are two separate disease entities. It should be mentioned that the true prevalence of *C. pneumoniae*-infected aneurysms in all of these studies may have been higher because of the possibility of false-negatives which have been encountered using both the immunocytochemistry (ICC) and PCR techniques.

Recently, a 40 kDa protein has been purified from the walls of human abdominal aortic aneurysms termed aneurysm-associated antigenic protein (AAAP-40).[77] This protein is immunoreactive with immunoglobulin from the walls of AAAs, and it resembles a similar protein in the pig where it is limited to the aorta.[78] If these autoantigenic proteins are homologous, it might explain why aneurysms are most frequently located in the aorta and its branches. AAAP-40 shares motifs with the herpes viruses, including CMV and HSV, suggesting that an infectious process may be involved in the degenerative aspects of atherosclerosis, and possibly the formation of aneurysms, possibly by molecular mimicry. CMV DNA was recently detected in a large fraction of aneurysmal aortas, and were more common in aneurysms classified as "inflammatory" based on the histologic findings of extensive adventitial fibrosis and mononuclear cell infiltrate (86%) than in degenerative "atherosclerotic" aneurysms (65%).[79]

Few seroepidemiologic studies have examined the possibility of an association between AAA and antibody to *C. pneumoniae* and the herpes viruses. Preliminary analysis of a large (n=590) case-control study conducted in a population-based cohort of adults aged 65 and older has demonstrated an odds ratio of AAA = 2.4 associated with high anti-*C. pneumoniae* IgG antibody titer, an association that persisted after controlling for potential confounding factors such as cigarette smoking (unpublished data). Additional seroepidemiologic studies will be required to confirm this relationship and to explore possible associations between nonherpesvirus antibody and the risk of AAA.

Conclusions

The initiating events in the development of atherosclerosis and aneurysmal disease are not definitively known. There is a well-established association between infection with *C. pneumoniae* and the herpes viruses, including CMV and HSV-1, and atherosclerosis, but it is unclear whether this relationship is causal. There is also a less well-characterized association between infection with these pathogens and the presence of AAA. This relationship may be mediated by the association between infection and atherosclerosis or by other nonatherogenic mechanisms. Additional epidemiologic, pathologic and animal research is required to elucidate these mechanisms. Important evidence also will be provided by the performance of randomized prospective clinical trials to examine the efficacy of antimicrobial and antiviral therapy in reducing the complications of atherosclerotic and aneurysmal arterial disease.

References

1. Zarins CK, Glagov S. Aneurysms and obstructive plaques. Differing local responses to atherosclerosis. In: Aneurysms, Diagnosis and Treatment. Bergan JJ and Yao JST eds. Grune and Stratton 1982, pp. 61-82.
2. Bengsston H, Ekberg O, Aspelin P et al. Abdominal aortic dilatation in patients operated on for carotid stenosis. Acta Chir Scand 1988; 154:441-445.
3. Alcorn HG, Wolfson SK, Sutton Tyrrell K et al. Risk factors for abdominal aortic aneurysms in older adults enrolled in the cardiovascular health study. Art Throm Vasc Biol 1996; 16:963-970.
4. Pleumeekers HJ, Hoes AW, van der Does E et al. Aneurysms of the abdominal aorta in older adults. The Rotterdam Study. Am J Epidemiol 1995; 142:1291-1299.
5. MacSweeney ST, Powell JT, Greenhalgh RM. Pathogenesis of abdominal aortic aneurysm. Br J Surg 1994; 81:935-941.
6. Zatina MA, Zarins CK, Gewertz BL et al. Role of medial lamellar architecture in the pathogenesis of aortic aneurysms. J Vasc Surg 1984; 1:442-448.
7. Zarins CK, Glagov S, Vesselinovitch D et al. Aneurysm formation in experimental atherosclerosis: Relationship to plaque evolution. J Vasc Surg 1990; 12:246-256.
8. Patel MI, Hardman DT, Fisher CM et al. Current views on the pathogenesis of abdominal aortic aneurysms. J Am Coll Surg 1995; 181:371-382.
9. Senior RM, Griffin GL, Mecham RP. Chemotactic activity of elastin derived peptides. J Clin Invest 1980; 66:859-862.
10. Senior RM, Connoly HL, Curry JD et al. Elastin degradation by human alveolar macrophages: A prominent role of metalloprotein activity. Am Rev Respir Dis 1989; 139:1251-1256.
11. Violi F, Criqui M, Longoni A et al. Relation between risk factors and cardiovascular complications in patients with peripheral vascular disease. Results from the A.D.E.P. study. Atherosclerosis 1996; 120:25-35.
12. de Maat MP, Pietersma A, Kofflard M et al. Association of plasma fibrinogen levels with coronary artery disease, smoking and inflammatory markers. Atherosclerosis 1996; 121:185-191.
13. Ridker PM, Cushman M, Stampfer MJ et al. Plasma concentration of C-reactive protein and risk of developing peripheral vascular disease. Circulation 1998; 97:425-428.
14. Ridker PM, Rifai N, Pfeffer MA et al. Inflammation, pravastatin, and the risk of coronary events after myocardial infarction in patients with average cholesterol levels. Cholesterol and Recurrent Events (CARE) Investigators. Circulation 1998; 98:839-844.
15. Heinrich J, Schulte, Schonfeld R et al. Association of variables of coagulation, fibrinolysis and acute-phase with atherosclerosis in coronary and peripheral arteries and those arteries supplying the brain. Thromb Haemost 1995; 73:374-379.
16. Koch AE, Haines GK, Rizzo RJ et al. Human abdominal aortic aneurysms. Immuno-phenotypic analysis suggesting an immune-mediated response. Am J Pathol 1990; 137:119-1213.
17. Koch AE, Kunkel SL, Pearce WH et al. Enhanced production of the chemotactic cytokines interleukin-8 and monocyte chemoattractant protein-1 in human abdominal aortic aneurysms. Am J Pathol 1993; 142:1423-1431.
18. Johansen K. Koepsell T. Familial tendency for abdominal aortic aneurysms. JAMA 1986; 256:1934-1936.
19. Tilson MD, Seashore MR. Fifty families with abdominal aortic aneurysms in two or more first-order relatives. Am J Surg 1984; 147:551-553.
20. Ross R. The pathogenesis of atherosclerosis-an update. N Engl J Med 1986; 314:488-500.

21. Grayston JT, Kuo CC, Wang SP et al. A new Chlamydia psittaci strain, TWAR, isolated in acute respiratory tract infections. N Engl J Med 1986; 315:161-168.

22. Grayston JT, Kuo CC, Campbell LA et al. *Chlamydia pneumoniae*, strain TWAR and atherosclerosis. Eur Heart J 1993; 14 Suppl K:66-71.

23. Saikku P, Leinonen M, Mattila K et al. Serological evidence of an association of a novel Chlamydia, TWAR, with chronic coronary heart disease and acute myocardial infarction. Lancet 1988; 2:983-986.

24. Linnanmaki E, Leinonen M, Mattila K et al. *Chlamydia pneumoniae*-specific circulating immune complexes in patients with chronic coronary heart disease [see comments]. Circulation 1993; 87:1130-1134.

25. Saikku P, Leinonen M, Tenkanen L et al. Chronic *Chlamydia pneumoniue* infection as a risk factor for coronary heart disease in the Helsinki Heart Study. Ann Intern Med 1992; 116:273-278.

26. Thom DH, Grayston JT, Siscovick DS et al. Association of prior infection with *Chlamydia pneumoniae* and angiographically demonstrated coronary artery disease. JAMA 1992; 268:68-72.

27. Thom DH, Wang SP, Grayston JT et al. *Chlamydia pneumoniae* strain TWAR antibody and angiographically demonstrated coronary artery disease. Arterioscler Thromb 1991; 11:547-551.

28. Melnick SL, Shahar E, Folsom AR et al. Past infection by *Chlamydia pneumoniae* strain TWAR and asymptomatic carotid atherosclerosis. Atherosclerosis Risk in Communities (ARIC) Study Investigators. Am J Med 1993; 95:499-504.

29. Meier CR, Derby LE, Jick SS et al. Antibiotics and risk of subsequent first-time acute myocardial infarction. JAMA 1999; 281:427-431.

29a. Esposito G, Blasi F, Allegra L et al. Demonstration of viable *Chlamydia pneumoniae* in atherosclerotic plaques of carotid arteries by reverse transcriptase polymerase chain reaction. Ann Vasc Surg 1999; 13:421-425.

30. Godzik KL, O'Brien ER, Wang SK et al. In Vitro susceptibility of human vascular wall cells to infection with *Chlamydia pneumoniae*. J Clin Microbiol 1995; 33:2411-2414.

31. Kalayoglu MV, Byrne GI. Induction of macrophage foam cell formation by Chlamydia pneumoniae. J Infect Dis 1998; 177:725-729.

32. Grayston JT, Kuo CC, Coulson A et al. *Chlamydia pneumoniae* (TWAR) in atherosclerosis of the carotid artery. Circulation 1995; 92:3397-3400.

33. Kol A, Sukhova GK, Lichtman AH et al. Chlamydial heat shock protein-60 localizes in human atheroma and regulates macrophage tumor necrosis factor-alpha and matrix metalloproteinase expression. Circulation 1998; 98:300-307

34. Moazed TC, Kuo CC, Patton DL et al. Experimental rabbit models of *Chlamydia pneumoniae* infection. Am J Pathol 1996; 148:667-676.

35. Moazed TC, Kuo C, Grayston JT et al. Murine models of *Chlamydia pneumoniae* infection and atherosclerosis. J Infect Dis 1997; 175:883-890.

36. Muhlestein JB, Anderson JL, Hammond EH et al. Infection with *Chlamydia pneumoniae* accelerates the development of atherosclerosis and treatment with azithromycin prevents it in a rabbit model. Circulation 1998; 97:633-636.

37. Gurfinkel E, Gozovich G, Daroca A et al. Randomized trial of roxithromycin in non-Q-wave coronary syndromes: ROXIS Pilot Study, ROXIS Study Group. Lancet 1997; 350:404-407.

38. Benditt EP, Barrett T, McDougall JK. Viruses in the etiology of atherosclerosis. Proc Natl Acad Sci USA 1983; 80:6386-6389.

39. Sorlie PD, Adam E, Melnick SL et al. Cytomegalovirus/herpesvirus and carotid atherosclerosis: The ARIC Study. J Med Virol 1994; 42:33-37.

40. Fabricant CG, Fabricant J, Minick CR et al. Herpesvirus-induced atherosclerosis in chickens. Fed Proc 1983; 42:2476-2479.

41. Shih JC, Pullman EP, Kao KJ. Genetic selection, general characterization, and histology of atherosclerosis-susceptible and -resistant Japanese quail. Atherosclerosis 1983; 49:41-53.

42. Persoons MC, Daemen MJ, Bruning JH et al. Active cytomegalovirus infection of arterial smooth muscle cells in immunocompromised rats. A clue to herpesvirus-associated atherogenesis? Cir Res 1994; 75:214-220.
43. Span AH, Frederik PM, Grauls G et al. CMV induced vascular injury: An electron-microscopic study in the rat. In vivo 1993; 7:567-573.
44. Span AH, Grauls G, Bosman F et al. Cytomegalovirus infection induces vascular injury in the rat. Atherosclerosis 1992; 93:41-52.
45. Benditt EP, Benditt JM. Evidence for a monoclonal origin of human atherosclerotic plaques. Proc Natl Acad Sci U S A 1973; 70:1753-1756.
46. Hendrix MG, Salimans MM, Van Boven CP et al. High prevalence of latently present cytomegalovirus in arterial walls of patients suffering from grade III atherosclerosis. Am J Pathol 1990; 136:23-28.
47. Hendrix MG, Dormans PH, Kitslaar P et al. The presence of cytomegalovirus nucleic acids in arterial walls of atherosclerotic and nonatherosclerotic patients. Am J Pathol 1989; 134:1151-1157.
48. Melnick JL, Hu C, Curek J et al. Cytomegalovirus DNA in arterial walls of patients with atherosclerosis. J Med Virol 1994; 24:170-174.
49. Yamashiroya HM, Ghosh L, Yang R et al. Herpesviridae in the coronary arteries and aorta of young trauma victims. Am J Pathol 1998; 130:71-79.
50. Melnick JL, Adam E, DeBakey ME. Cytomegalovirus and atherosclerosis. Eur Heart J 1993; 14 Suppl K:30-38.
51. Loebe M, Schuler S, Spiegelsberger S et al. [Cytomegalovirus infection and coronary sclerosis after heart transplantation]. Dtsch Med Wochenschr 1990; 115:1266-1269.
52. Nieto FJ, Sorlie P, Comstock GW et al. Cytomegalovirus infection, lipoprotein(a), and hypercoagulability: An atherogenic link? Arterioscler Thromb Vasc Biol 1997; 17:1780-1785.
53. Adam E, Melnick JL, Probtsfield JL et al. High levels of cytomegalovirus antibody in patients requiring vascular surgery for atherosclerosis. Lancet 1987; 2:291-293.
54. Nieto FJ, Adam E, Sorlie P et al. Cohort study of cytomegalovirus infection as a risk factor for carotid intimal-medial thickening, a measure of subclinical atherosclerosis. Circulation 1996; 94:922-927.
55. Blum A, Giladi M, Weinberg M et al. High noncytomegalovirus (CMV) IgG antibody titer is associated with coronary artery disease and may predict postcoronary balloon angioplasty restenosis. Am J Cardiol 1998; 81:866-868.
56. Zhou YF, Leon MB, Waclawiw MA et al. Association between prior cytomegalovirus infection and the risk of restenosis after coronary atherectomy. N Engl J Med 1996; 335:624-630.
57. Etingin OR, Silverstein RL, Hajjar DP. Identification of a monocyte receptor on herpesvirus-infected endothelial cells. Proc Natl Acad Sci U S A 1991; 88:7200-7203.
58. Visser MR, Vercellotti GM, McCarthy JB et al. Herpes simplex virus inhibits endothelial cell attachment and migration to extracellular matrix proteins. Am J Pathol 1989; 134:223-230.
59. Visser MR, Jacob HS, Goodman JL et al. Granulocyte-mediated injury to herpes simplex virus-infected human endothelium. Lab Invest 1989; 60:296-304.
60. Visser MR, Tracy PB, Vercellotti GM et al. Enhanced thrombin generation and platelet binding on herpes simplex virus-infected endothelium. Proc Natl Acad Sci U S A 1988; 85:8227-8230.
61. Hajjar DP, Pomerantz KB, Falcone DJ et al. Herpes simplex virus infection in human arterial cells. Implications in arteriosclerosis. J Clin Invest 1987; 80:1317-1322.
62. Key NS, Bach RR, Vercellotti GM et al. Herpes simplex virus type I does not require productive infection to induce tissue factor in human umbilical vein endothelial cells. Lab Invest 1993; 68:645-651.
63. Key NS, Vercellotti GM, Winkelmann JC et al. Infection of vascular endothelial cells with herpes simplex virus enhances tissue factor activity and reduces thrombomodulin expression. Proc Natl Acad Sci U S A 1990; 87:7095-7099.

64. Ellenby MI, Ernst CB. Surgical treatment of infected abdominal aortic aneurysms, in Ernst CB, Stanley JC (eds): Current therapy in vascular surgery. Mosby, 1995; 232-235.
65. Tanaka S, Komori K, Okadome K et al. Detection of active cytomegalovirus infection in inflammatory aortic aneurysms with RNA polymerase chain reaction. J Vasc Surg 1994; 20:235-243.
66. Sterpetti AV, Hunter WJ, Feldhaus RJ et al. Inflammatory aneurysms of the abdominal aorta: Incidence, pathologic, and etiologic considerations. J Vasc Surg 1989; 9:643-9.
67. Brophy CM, Reilly JM, Smith GJ et al. The role of inflammation in nonspecific abdominal aortic aneurysm disease. Ann Vasc Surg 1991; 5:229-233.
68. Newman KM, Jean Claude J, Li H et al. Cellular localization of matrix metalloproteinases in the abdominal aortic aneurysm wall. J Vasc Surg 1994; 20:814-820.
69. Herron GS, Unemori E, Wong M et al. Connective tissue proteinases and inhibitors in abdominal aortic aneurysms. Involvement of the vasa vasorum in the pathogenesis of aortic aneurysms. Arterioscler Thromb 1991; 11:1667-1677.
70. Jean Claude J, Newman KM, Li H et al. Possible key role for plasmin in the pathogenesis of abdominal aortic aneurysms. Surgery 1994; 116:472-478.
71. Reilly JM, Sicard GA, Lucore CL. Abnormal expression of plasminogen activators in aortic aneurysmal and occlusive disease. J Vasc Surg 1994; 19:865-872.
72. Newman KM, Jean Claude J, Li H et al. Cytokines that activate proteolysis are increased in abdominal aortic aneurysms. Circulation 1994; 90:20;814-820.
73. Blasi F, Denti F, Erba M et al. Detection of *Chlamydia pneumoniae* but not Helicobacter pylori in atherosclerotic plaques of aortic aneurysms. J Clin Microb 1996; 34:2766-2769.
74. Ong G, Thomas BJ, Mansfield AO et al. Detection and widespread distribution of *Chlamydia pneumoniae* in the vascular system and its possible implications. J Clin Path 1996; 49:102-106.
75. Juvonen J, Juvonen T, Laurila A et al. Demonstration of *Chlamydia pneumoniae* in the walls of the abdominal aortic aneurysms. J Vasc Surg 1997; 25:499-505.
76. Petersen E, Boman J, Persson K et al. *Chlamydia pneumoniae* in human abdominal aortic aneurysms. Eur J Vasc Endovasc Surg 1998; 15:138-142.
77. Ozsvath KJ, Hirose H, Xia S et al. Expression of two novel recombinant proteins from aortic adventitia (kappafibs) sharing amino acid sequences with cytomegalovirus. J Surg Res 1997; 69:277-282.
78. Kobayashi R, Tashima Y, Masuda H et al. Isolation and characterization of a new 36kDa microfibril-associated glycoprotein from porcine aorta. J Biol Chem 1989; 264:17437-17444.
79. Yonemitsu Y, Nakagawa K, Tanaka S et al. In situ detection of frequent and active infection of human cytomegalovirus in inflammatory abdominal aortic aneurysms: Possible pathogenic role in sustained chronic inflammatory reaction. Lab Invest 1996; 74:723-736.

CHAPTER 12

A Perspective on the Etiology of Abdominal Aortic Aneurysms

David K.W. Chew, James Knoetgen III, M. David Tilson III

Introduction

An aneurysm is a permanent focal dilatation of a blood vessel to a size that is 50% or greater than the normal diameter of the vessel.[1] The most commonly affected artery is the abdominal aorta and the incidence of this disease in Western countries has increased over the second half of this century.[2] Aneurysms rank as the 13th leading cause of death in the USA and are responsible for almost 1% of all deaths.[3]

Although the etiology of this disease is still unknown, major advances have been achieved in understanding its pathophysiology over the last 15 years. We will focus here on the nonspecific abdominal aortic aneurysm (AAA), which is the most common type encountered in clinical practice.

Normal Structural Support of the Aorta

The aorta must withstand surges of blood pressure that are generated with each contraction of the heart. In addition, the elastic recoil of this large artery during diastole propels blood throughout the rest of the systemic circulation, serving the function of a "secondary heart." Histologically, the wall of the aorta is made up of the tunica intima, tunica media and tunica adventitia. Structural support is provided by elastin and collagen fibers, with their associated microfibrillar proteins. The distribution of these important elements differs with each layer.

In the nondiseased aorta, the intima comprises a smooth carpet of endothelial cells, a basement membrane, and a thin subendothelial layer of connective tissue and smooth muscle cells. The tunica media is made up of an internal elastic lamina and there are 27 elastin lamellae alternating with collagen and smooth muscle cells.[4] This layer is relatively thick in the aorta, forming about half of the substance of the wall. The inner layer of the adventitia consists of compacted, alternating layers of elastin and collagen fibers. Towards the outer aspects of the adventitia, elastin diminishes in quantity and the strength of this layer is provided by collagen fibers. Scanning electron microscopy has shown that the elastin fibers form sheets that are oriented longitudinally in the direction of blood flow. These fibers are interlinked with finer elastin fibers that run circumferentially across the elastin layer. These elastic lamellae alternate with sheets of collagen fibers that are oriented in an oblique, circumferential spiral wrap.[5]

Development of Aneurysms, edited by Richard R. Keen and Philip B. Dobrin.

Elastin is notable for its stretchability—it can be stretched to one and a half times its original length by a force of 20-30 kg/cm and then return to its original dimensions when the tension is released.[6] It is a very stable protein with a biological half-life of 70 years.[7] Some authorities believe that after maturity, no further significant synthesis occurs in adult life.[8] Collagen, on the other hand, is much stiffer with a Young's modulus of more than four orders of magnitude greater than that of elastin.[9] The tensile strength of collagen is also over 20 times that of elastin.[7,9] Because of the inherent stiffness of collagen, it plays the major role in the structural support of the aortic wall at its limit diameters.[10] Types I and III collagen are the main subtypes in aorta, and collagen turnover continues throughout life.

How do elastin and collagen interact to perform the load-bearing function? At physiological arterial pressures, most of the load is borne by the elastin lamellae with only about 1% of the collagen component being "uncoiled." This has also been demonstrated functionally by comparing the stiffness of the intact vessel with the stiffness of collagen. With progressive increases in pressure, elastin becomes stretched and more collagen is then recruited to bear the load.[11] If the elastin layer were to fail completely, the vessel would dilate to the limit diameter imposed by the stiff collagen fibers. Therefore, in order for an artery to achieve aneurysmal proportions, the collagen fibers must fail as well.[12,13] A further discussion of the mechanical properties and roles of elastin and collagen may be found in Chapter 4, and in references 8-18.

The Enzymatic Basis for Matrix Destruction

A readily identifiable feature of AAA is the loss of elastin in the media and adventitia of the aortic wall. The effectors of destruction are a group of enzymes capable of degrading the major connective tissue components: collagen, elastin, fibronectin, laminin and the proteoglycans.[17,18] Many of these proteinases are released by macrophages and mesenchymal cells found in the AAA wall, including the family of the matrix metalloproteinases or matrixins (MMP).[19-22] Table 12.1 lists some of the better characterized metalloproteinases.

The MMPs are inhibited by the family of tissue inhibitors of metalloproteinases (TIMPs), including TIMP-1 and TIMP-2. Recent evidence suggests that the MMPs are secreted in bound complexes with TIMP.[23] An imbalance between the activated MMPs and their natural inhibitors may play a role in the destruction of the aortic wall. Preliminary studies suggested that AAA patients had a diminished level of TIMP-1 protein in the aortic walls,[24] which may be the case, but subsequent findings have not detected a decrease in TIMP mRNA levels.[25]

The Major Elastase(s) Present in AAA

Although early data on the identity of the major elastase in AAA was conflicting,[26] it now appears that the major enzyme is a metalloproteinase (rather than a serine protease). This was originally proposed by Brown et al[27] and Campa et al.[15] We have confirmed this hypothesis and using a novel approach for purification of MMPs by affinity to r-TIMP, we have isolated three caseinases with MWs of 80 kDa, 50 kDa and 32 kDa.[19] Using a monoclonal antibody to MMP-9 in immunoblots, the 80 kDa and 50 kDa activities were attributable to MMP-9. MMP-9 is elastolytic[28] and has a MW of about 80 kDa. An active form of recombinant MMP-9 has also been described at a MW of 40-50 kDa.

Identification of the Principal Collagenase in AAA

As collagen failure is a crucial event in the formation of an aneurysm, significant efforts have been expended in the search for collagenolytic activity in the AAA wall. The activity was first reported by Busuttil et al in 1980, but the enzyme was not successfully solubilized.[29] Other investigators were also unable to extract the collagenase (MMP-1).[30-32] MMP-1 binds

Table 12.1. Names and molecular weights of several well-characterized MMPs[77]

	Other Names	Molecular Mass (kDa)
MMP-1	Interstitial Collagenase Vertebrate Collagenase	42
MMP-2	Human 72-kDa Gelatinase Type IV Collagenase	66
MMP-3	Stromelysin Proteoglycanase	48
MMP-7	PUMP-1	19
MMP-8	Neutrophil Collagenase	65
MMP-9	Human 92 kDa Gelatinase Type IV Collagenase	84
MMP 10	Human Stromelysin 2	47

to its natural inhibitor TIMP, thus masking its detection in some assays.[26] Zarins et al reported increased extractable collagenase activity in an animal model.[33] Vine and Powell also reported the detection of collagenase by ^{14}C collagen assay and also by a specific antibody in extracts of AAA tissue.[34]

Adopting another approach in our laboratory, using immunoblots of soluble AAA protein extracts, strong immunoreactivities against antibody to MMP-1 were detected at MWs between 70-106 kDa and 52-57 kDa respectively.[20] The latter corresponds to the secreted isoforms of MMP-1 while the former probably represents MMP-1 complexed with other MMPs and TIMP. Control aortas had low, but detectable, amounts of MMP-1 immunoreactive material. Immunohistochemical studies have also localized the MMP-1 to cells and matrix in the adventitia of the AAA wall. MMP-1 is produced by both macrophages and cells of mesenchymal origin. Therefore, authentic interstitial collagenase (MMP-1) appears to be present in AAA tissue.

Other Proteinases

MMP-3 (Stromelysin-1), an activator of other proteinases such as MMP-1 and MMP-9, has also been shown to be present in the AAA wall. Immunohistochemical techniques have localized its source to macrophage-like mononuclear cells.[19,21,22]

Plasmin, a serine protease, is a major activator of members of the MMP family. Its presence has been demonstrated in AAA extracts and in the matrix of the AAA wall.[35,36] Plasminogen binds to extracellular matrix, where it is accessible to cell-surface receptors (e.g., urokinase-type plasminogen activator uPA) on infiltrating cells. This may be important for plasminogen activation in tissue. Reilly et al reported that tissue plasminogen activator (tPA is abundant in AAA tissue by comparison with normal or occlusive disease aorta and that tPA is detectable in resident macrophages in the adventitia of AAA).[37] Tromholt et al have shown in an in vivo study, utilizing indium-labeled monoclonal antibody against human tPA that tPA accumulates in the wall of aneurysms.[38] Louwrens et al showed that both uPa and tPa are also expressed by smooth muscle cells from AAA walls.[36] These increases in plasminogen activators without a concomitant rise in plasminogen activator inhibitors may swing the delicate balance towards proteolysis of the matrix.

Inflammation—The Machinery of Destruction

One of the predominant pathological features of AAA disease is the presence of a chronic inflammatory cell infiltrate distributed in the outer media and adventitia of the aortic wall. The most extreme example of this is seen in the "inflammatory AAA (IAAA)," a condition usually diagnosed by the presence of a marked fibrotic reaction around the aneurysm and surrounding retroperitoneum. The cellular infiltrate is mainly mononuclear, comprised of plasma cells, lymphocytes and macrophages. Even Russell bodies, which are aggregates of immunoglobulin, may also be seen in the nonspecific AAA.[39] These cells do not consistently codistribute with the areas of elastin degradation. Using immunohistochemical techniques, Koch et al characterized the infiltrate as comprising T and B lymphocytes with an increased T helper:suppressor ratio.[40] Monocytes and macrophages were found throughout the diseased tissues, often surrounded by lymphoid aggregates.[40] The degree of cellular infiltration was greatest in the IAAA group and the surrounding lymph nodes showed follicular hyperplasia.[41]

Animal models of experimental aortic aneurysms have confirmed the role that inflammation plays in AAA formation.[42,43] After a brief latent period following an initiating injury to the wall of the aorta, subsequent aneurysm development correlates with the degree of the inflammatory infiltrate and the multiple endogenous proteinases that are activated.

An Immune-Mediated Response?

The triggering events that ignite inflammation in human AAA disease are still unknown. However, there is evidence to suggest an immune-mediated mechanism, possibly directed against autoantigens found in the aortic wall. Firstly, the polyclonal T and B lymphocytes present in the aneurysmal wall are associated with a large amount of immunoglobulins compared to normal and atherosclerotic control extracts.[39] Secondly, the IgGs purified from AAA extracts have been found to be reactive with a component of the matrix at the periphery of adventitial elastic fibers (consistent with the distribution of elastin-associated microfibrillar glycoproteins).[44] Using immunoblotting techniques after gel electrophoresis of the soluble extracts, the putative antigen has a MW of about 80 kDa. Whether autoimmunity is the primary event initiating the inflammatory cascade or secondary to destruction of the aortic matrix with exposure of "hidden" antigens is still under investigation. Fibronectin as well as collagen and elastin degradation products are all known to be chemotactic to monocytes and macrophages.[45-47]

Role of Cytokines in AAA Disease

Once initiated, perpetuation and amplification of the inflammatory cascade occurs with the aid of cytokines released by the inflammatory cells. Studies have shown elevations of interleukin-1β (IL-1β), IL-6, IL-8, monocyte chemoattractant protein-1 (MCP-1), tumor necrosis factor-α (TNF-α), and interferon-γ (IFN-γ) in AAA tissue.[48-50]

TNF-α is produced primarily by activated macrophages and monocytes, although lymphocytes and vascular smooth muscle cells can synthesize it. Its effects include the induction of MMP expression, IL-1β, and prostaglandin production in vascular smooth muscle.

IL-1β or lymphocyte activating factor is produced by macrophages, B cells, endothelial cells and fibroblasts. It has a wide variety of biological activities including the induction of prostaglandin synthesis in endothelial cells and smooth muscle cells. IL-1β can stimulate MMP production, such as collagenase in smooth muscle and fibroblastic cells. TIMP expression is concomitantly reduced. IL-1β also acts on macrophages and monocytes to induce its own synthesis and the production of TNF-α. IL-1β can induce vascular smooth muscle cells to secrete IL-6, resulting in lymphocyte proliferation.

Chemotactic factors, by recruiting inflammatory cells into the aortic wall, are important in perpetuating the inflammatory process. IL-8 has chemotactic and activating activity for neutrophils, lymphocytes and endothelial cells. It is also a potent angiogenic factor. This cytokine is produced mainly by macrophages located in all layers of the aortic wall. Some endothelial cells also stain positive for IL-8 in immunohistochemical studies. Cultured human aortic smooth muscle cells also produce IL-8 upon exposure to IL-1β, TNF-α, and lipopolysaccharide. Davis et al have shown that intercellular adhesion molecule (ICAM-1) is expressed in the endothelium of AAAs to a greater extent than normal aortas.[51] This facilitates the migration of lymphocytes into the aneurysm wall in response to the IL-8 signal. MCP-1 is both chemotactic and activating for monocytes. The predominant cells expressing this cytokine are macrophages and smooth muscle cells. Other possible chemotactic factors include collagen, elastin, fibronectin breakdown products and cytokine transforming growth factor-β (TGF-β).

IL-6 is mainly produced by monocytes/macrophages, as well as fibroblasts, endothelial and smooth muscle cells. It activates T and B lymphocytes and may be responsible for the accumulation of immunoglobulins in the aneurysm wall.

Lastly, IFN-γ is found in higher concentrations in AAA compared to normal or aortas with occlusive disease. It is produced by lymphocytes and has various effects on vascular cells. IFN-γ inhibits endothelial and smooth muscle cell proliferation, induces MHC class II antigen expression on these cells, and has been implicated in aneurysm formation in Kawasaki's disease.[52,53] It also stimulates macrophages, T and B lymphocytes, and thus plays a role in cellular activation.

The Search for Aortic Autoantigens

The autoimmune theory of abdominal aortic aneurysms began with the identification of Russell bodies in histologic sections of aneurysm tissue.[39] This observation led to Western blots of soluble aortic extracts probed with labeled Protein A, which demonstrated a greater quantity of immunoglobulins in aneurysm tissue than in controls. A subsequent study demonstrated that IgG extracted from the wall of AAA specimens codistributed immunohistochemically with elastin-associated microfibrils in the adventitia of the aortic wall.[44] Gregory et al[44] also demonstrated a protein with a molecular weight of 80 kDa, which was immunoreactive with the IgGs from AAA homogenates in 11 (79%) of 14 patients with AAA disease and only I (11%) of nine control subjects (p=0.002).

Purification of this protein revealed a motif previously described in a 36 kDa bovine aortic elastin microfibril-associated-glycoprotein (MAGP-36).[54] MAGP-36 has a tissue distribution that is limited to the aorta, as opposed to the other MAGPs which are widely distributed with elastin. Since MAGP-36 exists as a dimer in vivo, it was hypothesized that this newly identified 80 kDa protein was the human homologue in dimeric form. Repeating the aortic tissue extraction under reducing conditions (to disrupt disulfide bonds) yielded a 40 kDa protein which was termed aortic aneurysm-associated protein-40 (AAAP-40).[55] AAAP-40, more appropriately termed MAGP-3 since it is the third human microfibrillar protein identified, is immunoreactive with IgG isolated from the aortic wall and serum of patients with AAA.[56]

The protein was partially sequenced and several regions of interest were recognized. A fibrinogen-like domain was identified with homologous sequences in the alpha, beta and gamma chains of fibrinogen. Another motif with 12 residues in common with human vitronectin was recognized, as well as a putative calcium-binding domain with homologous sequences seen in calcium-binding myeloid-related protein, bovine aggrecan and the calcium-binding domain of human fibrinogen-beta.[55] The partial sequence of AAAP-40 is aligned with corresponding amino acid sequences from relevant proteins in Table 12.2. Glycan differentiation analysis has recently demonstrated that AAAP-40 is a glycoprotein.[56]

The unique localization of MAGP-36 and possibly AAAP-40 to the aorta may account for an autoimmune reaction limited to the wall of the aorta. One may therefore hypothesize that a genetic predisposition to AAA disease may be expressed as: 1) an amino acid mutation rendering AAAP-40 antigenic, thus initiating an autoimmune response in the aorta; or 2) specific repertoires of MHC class II alleles that see AAAP-40 as nonself under certain conditions. It has not been ruled out, however, that the possibility that autoimmunity to aortic antigens is a secondary phenomenon accelerating matrix destruction after initiation by other factors.

The Genetics of AAA Disease

Clifton's seminal paper in 1977 of three brothers with similar blood types surgically treated for ruptured abdominal aortic aneurysms was the first suggestion in the literature that this disease is genetically influenced.[57] Several investigators reported additional series and established familial abdominal aortic aneurysm (F-AAA) as a distinct clinical entity. Tilson and Seashore[58] reported 50 families with two or more first-degree relatives afflicted with abdominal aortic aneurysms. Further investigations by Tilson and Seashore[59] proposed possible X-linked and autosomal dominant variants but did not rule out a multifactorial mechanism.

A 9-year prospective study by Darling et al[60] noted a 15.1% incidence of F-AAA among 542 patients treated for AAA with a control group incidence of 1.8%. In a multicenter, prospective study of AAA data, Johnston and Scobie[61] reported a 6.1% incidence of F-AAA. Nörrgard et al[62] found that 18% of a Swedish population with AAA had a family history of AAA. Multiple modes of inheritance have been suggested, including a single gene with dominant inheritance suggested by a segregation analysis performed by Verloes et al.[63]

The search for the putative "aneurysmogenic" gene has proceeded in several directions. The first begins with an understanding of Ehlers-Danlos syndrome type IV, a rare genetic disorder which can result in aortic rupture. It is a syndrome frequently associated with arterial aneurysms and rupture of hollow organs as a result of a mutation in the gene for type III procollagen (COL3AI). A mutation that converted the codon for glycine 619 to the codon for arginine was noted to cosegregate with aortic aneurysms in one family.[64] The resulting procollagen had a decreased temperature for thermal unfolding of the protein (e.g., "melting"). A subsequent investigation, however, evaluated 54 patients from 50 families and found nucleotide changes in only two patients resulting in alteration of protein structure.[65] Thus, this gene has been excluded as the cause of more than a small subset of nonspecific AAAs.

Another study demonstrated a deficiency allele for alpha-1 antitrypsin (MZ) in 5 (11%) of 47 patients.[66] Again, this genetic influence could not explain more than a small subset of AAA cases. Tissue inhibitors of metalloproteinases (TIMPs) are inhibitors of several enzymes destructive to connective tissue. A single base pair substitution in the cDNA for TIMP was identified in two of six patients, but this occurred in the third position of the codon conserving the amino acid.[67] Accordingly, the genetic mechanism responsible for the predisposition to AAA disease in most cases remains to be discovered.

Molecular Mimicry—The Shared Epitope Hypothesis

Genes within the class II region of the major histocompatibility complex (MHC) encode for cell surface proteins mainly found on B cells and macrophages which enable them to bind antigenic peptides. This region extends to nearly 1,000,000 base pairs on chromosome 6 and includes at least 14 different genes, generally found in one of three major subregions:

Table 12.2. Partial amino acid sequence of AAAP-40 compared with homologous regions of MFAP-4, MAGP-36, and Hum Fib-a and b

		!	(Y)F P(F)V	DLMVM	ANQPM	AAAP-40
122	TLKQK		YELRV	DLEDF	ENNTA	MFAP-4
	TL LK		YELRV	DLEDF	EXNTA	MAGP-E6
			LRV	ELED.	A.N.A	HUM Fib-a
	GE!YY		DFFQY	TXGMA	KEYDGFQ	AAAP-40
142	YAKYA		DFSIS	PNAVS	AEEDG	MFAP-4
	FAKYA		DFSIS	PNAVS	AEEDG	MAGP-36
	YTXGM		AK(IY)A	GNALM	DGASGLM	AAAP-40
162	YTLFV		AGFED	GGAGD	SLSYH	MFAP-4
	YTLYV		SGFED	GGAGD	SLTYH	MAGP-36
	Y.I.V		.K TA	GNAL		Hum Fib-b

This table shows the sequence of AAAP-40, as determined experimentally. The sequence of AAAP-40 is aligned along a continuous sequence of MAGP-36, beginning at residue 140. Homologous regions of MAGP-36 (bovine) and fibrinogen alpha (residues 120-132) and beta residues 338-353 (human) are also shown. "( )" is used to designate an ambiguous residue; "." denotes a nonconserved residue; and "!" denotes a tryptic cleavage site.

DR, DQ and DP. The genes of the DR and DQ subregions are very closely linked and are frequently inherited together.

The shared epitope theory is based on the hypothesis that class II agents are responsible for the initiation of autoimmune disease. Pathogenic antigens are recognized by antigen-presenting cells (whose surface contains class II molecules) thus activating T cells and an immune response. A molecular mimic is a normal host protein which so closely resembles the epitope of a microbial pathogen that infection with the organism may incite an autoimmune process against tissues possessing the normal host protein.

Molecular mimicry has been implicated in several diseases. The pathogenesis of multiple sclerosis (MS) has been explained by autoimmunity against several myelin proteins such as myelin basic protein (MBP), proteolipid protein, myelin-associated glycoprotein, and myelin-oligodendrite glycoprotein. This autoreaction, however, may be a secondary response to myelin breakdown and may not be the disease's inciting event. Susceptibility to the disease has been linked to the MHC class II loci DRA, DRB1*1501, DRB5*0101 and the DQ6 subtypes DQAl*0102andDQBI*0602.[68] Seven viral proteins have been identified which activate human T cell clones specific for MBP, one of which is a peptide from Influenza Type A.[69] This may explain why natives of the Faeroe islands previously unexposed to Western viruses suffered an outbreak of multiple sclerosis after British soldiers inhabited the island during World War II.

The shared epitope hypothesis has also been applied to the etiology of rheumatoid arthritis (RA). A susceptibility to RA based on molecular genetics was first proposed by Stastny who demonstrated an elevated frequency of MHC DR4 in RA patients.[70] DR4, however, also predisposes certain populations to risk for other autoimmune diseases not associated with RA. Also, RA patients who lack DR4 haplotypes but have DRI have been weakly associated with susceptibility to RA.[71] Examination of the DR1 sequence reveals an identical match with the Dw14 subtype of DR4 in the third hypervariable region. Gregersen et al have isolated a variant DR1 allele from a patient with RA, suggesting that the third hypervariable region epitopes are required for disease susceptibility.[71]

To assess the possible etiologic significance of viruses in AAA disease, Tanaka et al performed polymerase chain reaction to detect herpes simples virus (HSV types 1 and 2) and cytomegalovirus (CMV).[72] AAA specimens demonstrated HSV in 10 (27%) of 37 aortas versus 1 of 16 controls. CMV was detected in 24 (65%) of 37 aortas versus 5 (16%) of 31 controls. The presence of these viruses in the aortic wall is interesting when the shared epitope hypothesis is considered in connection with the following recombinant AAAP-40 experimental data (Table 12.3).

The production of recombinant AAAP-40 (r-AAAP-40) could potentially have important diagnostic and therapeutic consequences. An abundant source of the protein would permit an ELISA to titer serum anti-AAAP-40 IgG to detect susceptibility to AAA disease before the development of an aneurysm. The availability of r-AAAP-40 would also be useful in attempts to induce tolerance to the antigen to prevent the onset of the disease in predisposed persons.

Tilson's lab constructed a cDNA library from human aortic RNA and screened the library with polyclonal antibodies to vitronectin and fibronectin.[73] AAAP-40 itself has yet to be cloned, but the first five clones have yielded hypothetical proteins with motifs that resemble AAAP-40 and also occur in other proteins. Of interest to the shared epitope hypothesis are common motifs with cytomegalovirus in regions 2 and 5 of clones 1 and 5. The amino acid sequence of the most recently reported r-AAAP also contained a sequence "WGFTLHPCAC" which is similar to "WGFT ... AC" in herpesvirus.[56]

If a complete r-AAAP-40 clone demonstrates homologous regions with CMV and HSV that are antigenically significant, then an infection with these viruses in a genetically susceptible person hypothetically could induce an autoimmune reaction of the aortic wall creating an aneurysm.

MHC Class II DR Locus as a Candidate for a AAA Susceptibility Gene

Several autoimmune diseases have been linked to class II MHC DR regions. Rheumatoid arthritis is associated with class II MHC DR4 and Weynand et al reported that 98 of 102 (96%) patients express one of the major North American disease-linked polymorphisms (*04,*0101,or*1402).[74] More severe forms of rheumatoid disease were seen inpatients with double doses of the implicated alleles.

Since the class II MHC DR locus has affiliations with multiple autoimmune diseases and AAAs demonstrate many features of autoimmunity, attempts have been made to link AAA susceptibility gene(s) to a DR locus. A pilot study performed by Tilson's group has tissue typed 26 AAA patients including 5 African-Americans.[73] This is particularly notable since AAAs occur in African-Americans with much less frequency than in Caucasians. The hypothesis is that Americans of color with AAA disease have a double dose of the susceptibility allele.

The results of allele frequency analysis in the five African-American patients compared with allele frequency in North American Blacks without AAA disease is presented in Table 12.4. The expected frequencies were derived from the Eleventh International Histocompatibility Workshop.[75] The ten haplotypes revealed three DR alleles: 2, 12 and 13, which occurred significantly more often than expected by chance alone (p=.0003). A phenylalanine (phe) at positions 31 and 47 of the second hypervariable region was seen in all of the amino acid sequences produced by the most common alleles of these three DRB I types. DR3 and DR11 also have phe at positions 31 and 47. A phenylalanine at positions 31 and 47 also appeared in 16 (75%) of the 21 nonblack patients and 4 of these have a double

Table 12.3. *Partial amino acid sequence of AAAP-40 compared with homologous regions of VN, MFAP-4, MAGP-36, CaBP-M, Agg, and Fib-b*

```
Q E L E K                         S F E D G V L D P D Y P     AAAP-40
                                  R F E D G V L D P D Y P     VN
F C L Q Q P L D C D D I Y A Q G Y Q S D G VYL I Y P S        MFAP-4
S E L Q L P L D E D D I Y A Q G Y Q A D G VYL I   P S        MAGP-36
T E L . . . L . E . D V Y . . . Y . . D                      CaBP-M
        P . D E . D V Y                                      Agg
S E L E K H Q L . . D . T                                    Fib-b
```

Alignment of experimentally determined sequences of AAAP-40 on sequences from human vitronectin (VN, residues 230-240) and MFAP-4 (beginning at residue #34). Alignments with other calcium-binding proteins are shown, with the most highly conserved residues highlighted in bold type: Calcium-binding myeloid-related protein[78]= CBP-M; aggrecan (bovine) = Agg;[79] human fibrinogen beta (residues 144-157 from calcium-binding domain) = Fib-b.

Table 12.4. *Allele frequencies in five Americans of color with AAA compared with allele frequencies in North American blacks*

Allele frequencies in five Americans of Color with AAA (AOC-AAA, ten haplotypes) with the expected frequencies for these alleles in North American Blacks (NAB) as reported by the International Workshop in 1991 (Table 12, 132 haplotypes). The cumulative probability of this result is p = 3×10^{-4}. Probabilities were calculated by Fisher's Exact Test for 2x2 Contingency Tables.

Allele	Frequency in AOC-AAA	Frequency in NAB	P
DR131*02	40%	12%	.037
DRB 1 12	30%	5%	.023
DRB 1 13	30%	15%	.370

dose of one of the putative "aneurysmogenic" alleles. If a phenylalanine at position 37 rather than 47 is accepted as a putative aneurysmogenic allele, then 21 out of 21 nonblack patients have the allele and 50% have a double dose.

Giant cell arteritis (GCA) shares several clinical and pathologic similarities with inflammatory AAAs. Both diseases also tend to cluster within families and certain ethnic groups. Weyand et al demonstrated that 60% of 42 patients with GCA expressed the B1*0401 or B I * 0404/8 variant of the HLA-DR4 haplotype.[76] A subsequent investigation by Rasmussen et al demonstrated increased frequencies of HLA DRB1 alleles B1*15 and B1*0404 among patients with inflammatory AAAs compared with controls (47 vs 27%, and 14 vs 3%, respectively, p < 0.05).

The data from these investigations suggest a testable hypothesis, which may be studied by HLA typing of additional patients prospectively to investigate the possible susceptibility alleles in AAA disease.

References

1. Johnston KW, Rutherford RB, Tilson MD et al. Suggested standards for reporting on arterial aneurysms. J Vasc Surg 1991; 13:452-458.
2. Melton LJ, Bickerstaff LK, Hollier LH et al. Changing incidence of abdominal aortic aneurysms. A population based study. Am J Epiderniol 1984; 120:379-386.
3. Silverberg E, Lubera JA. A review of the American Cancer Society estimates of cancer cases and deaths. Can. Cancer J Clin 1983; 33:2-8.
4. Wolinsky H, Glagov S. A lamellar unit of aortic medial structure and function in mammals. Circ Res 1967; 20:99-111.
5. White JV, Haas K, Phillips S et al. Adventitial elastolysis is a primary event in aneurysm formation. J Vasc Surg 1993; 17:371-379.
6. Fawcett DW. Bloom Textbook of Histology: Connective Tissue. New York: Chapman and Hall; 1994:143.
7. Dobrin PB. Vascular mechanics. In: Shepard JT, Abboud FM, eds. The Cardiovascular System. Handbook of Physiology. WashingtonDC: American Physiological Society; 1983:65-102.
8. Rucker RB, Tinker D.. Structure and metabolism of arterial elastin. Int Rev Exp Pathol 1977; 17:1-47.
9. Dobrin PB. Mechanical properties of arteries. Physiol Rev 1978; 58:397-460.
10. Dobrin PB. Pathophysiology and pathogenesis of atherosclerotic aneurysms. Current concepts. Surg Clin N Amer 1989; 69:687-703.
11. Patel MI, Hardman DTA, Fisher CM et al. Current views on the pathogenesis of abdominal aortic aneurysms. J Am Coll Surg 1995; 181:371-382.
12. Tilson MD, Elefteriades J, Brophy C. Tensile strength and collagen in abdominal aortic aneurysm disease. In: Greenhalgh RM, Mannick JA and Powell JT, eds. The cause and management of aneurysms. London: WB Saunders; 1990:97-104.
13. Dobrin PB, Mrkvicka R. Failure of elastin and collagen as possible critical connective tissue alterations underlying aneurysmal dilatation. Cardiovasc Surg 1994; 2:484-488.
14. Sumner DS, Hokanson DE, Strandness DE. Stress-strain characteristics and collagen-elastin content of abdominal aortic aneurysms. Surg Gynecol Obstet 1970; 130:459-466.
15. Campa JS, Greenhalgh RM, Powell JT. Elastin degradation in abdominal aortic aneurysms. Atherosclerosis 1987; 65:13-21.
16. Baxter BT, McGee GS, Shively VP et al. Elastin content, cross links and mRNA in normal and aneurysmal human aorta. J Vasc Surg 1992; 16:192-200
17. Busuttil RW, Rinderbriecht H, Flesher A et al. Elastase activity: the role of elastase in aortic aneurysm formation. J Surg Res 1982; 32:214-217
18. Cannon DJ and Read RC. Blood elastolytic activity in patients with aortic aneurysm. Ann Thorac Surg 1982; 34:10-15.
19. Newman KM, Ogata Y, Malon AM et al. Identification of matrix metalloproteinases 3 (Stromelysin-1) and 9 (Gelatinase B) in abdominal aortic aneurysm. Arteriosclerosis Throm 1994; 14:1315-1320.
20. Irizarry E, Newman KM, Gandhi RH et al. Demonstration of interstitial collagenase in abdominal aortic aneurysm disease. J Surg Res 1993; 54:5 71-574.
21. Newman KM, Malon AM, Shin RD et al. Matrix metalloproteinases in abdominal aortic aneurysm: characterization, purification and their possible sources. Conn Tiss Res 1994; 30:265-276.
22. Newman KM, Jean-Claude J, Hong Li et al. Cellular localization of matrix metalloproteinases in the abdominal aortic aneurysm wall. J Vasc Surg 1994; 20:814-820.
23. Goldberg GI, Strongin A, Collier IE et al. Interaction of 92 kDa type IV collagenase with the tissue inhibitor of metalloproteinases prevents dimerization, complex formation with interstitial collagenase, and activation of the proenzyme with stromelysin. J Biol Chem 1992; 267:4583-4591.

24. Brophy CM, Marks WH, Reilly JM et al. Decreased tissue inhibitor of metalloproteinases (TIMP) in abdominal aortic aneurysm tissue: A preliminary report. J Surg Res 1991; 50:653-657.
25. Keen RR, Nolan KD, Cipollone M et al. Interleukin-1(induces differential gene expression in aortic smooth muscle cells. J Vasc Surg 1994; 20:774-786.
26. Tilson MD, Newman KM. Proteolytic mechanisms in the pathogenesis of aortic aneurysms. In: Yao JST, Pearce WH, eds. Aneurysms New Findings and Treatments. Connecticut: Appleton and Lange; 1994:3-9.
27. Brown SL, Backstrom B, Busuttil RW. A new serum proteolytic enzyme in aneurysm pathogenesis. J Vasc Surg 1982; 2:393-399.
28. Senior RM, Griffin GL, Fliszar CJ et al. Human 92- and 72-kDa Type IV collagenases are elastases. J Biol Chem 1991; 266:7870-7875.
29. Busuttil RW, Abou-Zamzarn AM, Machleder HI. Collagenase activity of the human aorta: A comparison of patients with and without abdominal aortic aneurysms. Arch Surg 1980; l 15:1373-1378.
30. Webster MW, McAuley CE, Steed DL et al. Collagen stability and collagenolytic activity in the normal and aneurysmal human abdominal aorta. Am J Surg 1991; 161:635.
31. Menashi S, Campa JS, Greenhalgh RM et al. Collagen in abdominal aortic aneurysm: typing, content and degradation. J Vasc Surg 1987; 6:578.
32. Herron GS, Unemori E, Wong M et al. Connective tissue proteinases and inhibitors in abdominal aortic aneurysms. Arteriosclerosis Throm 1991; 11:1667.
33. Zarins CK, Runyon-Hass A, Zatina MA et al. Increased collagenase activity in early aneurysmal dilatation. J Vasc Surg 1986; 3:238.
34. Vine N, Powell JT. Metalloproteinases in degenerative aortic disease. Clin Sci 1991; 81:233.
35. Jean-Claude J, Newman KM, Hong Li et al. Possible key role for plasmin in the pathogenesis of abdominal aortic aneurysms. Surgery 1994; 116:472-478.
36. Louwrens HD, Kwaan HC, Pearce WH et al. Plasminogen activator and plasminogen activator inhibitor expression by normal and aneurysmal human aortic smooth muscle cells in culture. Eur J Vas Endovas Surg 1995; 10:289-293.
37. Reilly JM, Sicard GA, Lucore CL. Abnormal expression of plasminogen activators in aortic aneurysmal and occlusive disease. J Vasc Surg 1994; 119:865-872.
38. Tromholt N, Jorgensen SJ, Hesse B et al. In vivo demonstration of focal fibrinolytic activity in abdominal aortic aneurysms. Eur J Vas Surg 1993; 7:675-679.
39. Brophy CM, Reilly JM, Walker Smith GJ et al. The role of inflammation in nonspecific abdominal aortic aneurysm disease. Ann Vasc Surg 1991; 5:229-233.
40. Koch AE, Haines GK, Rizzo RJ et al. Human abdominal aortic aneurysms. Immunophenotypic analysis suggesting an immune-mediated response. Am J Path 1990; 137:1199-1213.
41. Lieberman J, Scheib JS, Googe PB et al. Inflammatory abdominal aortic aneurysm and the associated T cell reaction: A case study. J Vasc Surg 1992; 15:569-572.
42. Anidjar S, Dobrin PB, Eichorst M et al. Correlation of the inflammatory infiltrate with the enlargement of experimental aortic aneurysms. J Vasc Surg 1992; 16:139-147.
43. Halpern VJ, Nackman GB, Gandhi RJ et al. The elastase infusion model of experimental aortic aneurysms: Synchrony of induction of endogenous proteinases with matrix destruction and inflammatory cell response. J Vasc Surg 1994; 20:51-60.
44. Gregory AK, Yin NX, Capella J et al. Features of autoimmunity in the abdominal aortic aneurysm. Arch Surg 1996; 13:85-88.
45. Postlethwaite AE, Kang AH. Collagen and collagen peptide-induced chemotaxis of human blood monocytes. J Exp Med 1976; 143:1299-1307.
46. Norris DA, Clark RAF, Swigart LM et al. Fibronectin fragments are chemotactic for human peripheral blood monocytes. J Immunol 1982; 129:1612-1618.
47. Senior RM, Griffin GL, Mecharn RP. Chemotactic activity of elastin-derived peptides. J Clin Invest 1980; 66:859-862.
48. Newman KM, Jean-Claude J, Hong Li et al. Cytokines that activate proteolysis are increased in abdominal aortic aneurysms. Circulation 1994; 90:11-224-227.

49. Koch AE, Kunkel SL, Pearce WH et al. Enhanced production of the chernotactic cytokines interleukin-8 and monocyte chemoattractant protein- I in human abdominal aortic aneurysms. Am J Path 1993; 142:1423-1431.

50. Szekanecz Z, Shah MR, Pearce WH et al. Human atherosclerotic abdominal aortic aneurysms produce interleukin-6 and interferon-gamma but not IL-2 and IL-4: the possible role for IL-6 and interferon-gamma in vascular inflammation. Agents Actions 1994; 42:159-162.

51. Davis CD III, Pearce WH, Shah MR et al. Increased ICAM-1 expression in aortic disease. J Vasc Surg 1992; 16:474-475.

52. Leung DY, Collins T, Lapierre LA et al. Immunoglobin M antibodies present in the acute phase of Kawasaki syndrome lyse cultured vascular endothelial cells stimulated by gamma interferon. J Clin Invest 1986; 77:1428-1435.

53. Lin CY, Lin CC, Hwang B et al. Cytokines predict coronary aneurysm formation in Kawasaki disease patients. Eur J Pediatr 1993; 152:309-312.

54. Kobayashi R, Mizutani A, Hidaka H. Isolation and characterization of a 36-kDa microfibril-associated glycoprotein by the newly synthesized isoquinolinesulfonamide affinity chromatography. Biochem Biophys Res Comm 1994; 198:1262-1266.

55. Xia S, Ozsvath K, Hirose H et al. Partial amino acid sequence of a novel 40-kDa human aortic protein, with vitronectin-like, fibrinogen-like, and calcium binding domains: aortic aneurysm-associated protein-40 (AAAP-40) [Human MAGP-3, proposed]. Biochem Biophys Res Comm 1996; 219:36-39.

56. Hirose H, Ozsvath KJ, Xia S et al. Molecular cloning and expression of the cDNA for a putative aortic aneurysm-antigenic protein (AAAP) In: Proceedings of the SVS-ISCVS, 50th Annual Meeting, June 7, 1996.

57. Clifton MA. Familial abdominal aortic aneurysms. Br J Surg 1977; 64:765-766.

58. Tilson MD, Seashore MR. Fifty families with abdominal aortic aneurysms in two or more first-order relatives. Am J Surg 1984; 147:551-553.

59. Tilson MD, Seashore MR. Human genetics of the abdominal aortic aneurysm. Surg Gynecol Obstet 1984; 158:129-132.

60. Darling RC III, Brewster DC, Darling RC et al. Are familial abdominal aortic aneurysms different? J Vasc Surg 1989; 1:39-43.

61. Johnston KW, Scobie TK. Multicenter prospective study of nonruptured abdominal aortic aneurysms. Population and operative management. J Vasc Surg 1988; 7:69-81.

62. Norrgard 0, Rais 0, Angquist K-A. Familial occurrence of abdominal aortic aneurysms. Surgery 1984; 95:650-656.

63. Verloes A, Sakalihasan N, Koulischer L et al. Aneurysms of the abdominal aorta: familial and genetic aspects in three hundred thirteen pedigrees. J Vasc Surg 1988; 7:69-71.

64. Kontusaari S, Tromp G, Kuivaniemi H et al. A mutation in the gene for Type III Procollagen (COL3Al) in a family with aortic aneurysms. J Clin Invest 1990:86:1465-1473.

65. Tromp G, Prockop DJ, Madhatheri SL et al. Sequencing of cDNA from 50 unrelated patients reveals that mutations in the triple-helical domain of type III procollagen are an infrequent cause of aortic aneurysms. J Clin Invest 1993; 91:2539-2545.

66. Cohen JR, Sarfati I, Ratner L et al. Alpha-1 anti-trypsin phenotypes in patients with abdominal aortic aneurysms. J Surg Res 1991; 49:319-321.

67. Tilson MD, Reilly JM, Brophy CM et al. Expression and sequence of the gene for tissue inhibitor of metalloproteinases in patients with abdominal aortic aneurysms. J Vasc Surg 1993; 18:266-270.

68. Tienari PJ. Multiple sclerosis: Multiple etiologies, multiple genes? Ann Med 1994; 26:259-269.

69. Wucherpfenning KW, Strominger JL. Molecular mimicry in T-cell mediated autoimmunity: Viral peptides activate human T-cell clones specific for myelin basic protein. Cell 1995; 80:695-705.

70. Stastny P. Association of the B-cell alloantigen Drw4 with rheumatoid arthritis. N Eng J Med 1978; 298:869-871.

71. Gregersen PK, Silver J, Winchester RJ. The shared epitope hypothesis: An approach to understanding the molecular genetics of susceptibility to rheumatoid arthritis. Arthritis Rheum 1987; 30:1205-1213.
72. Tanaka S, Komori K, Okadome K et al. Detection of active cytomegalovirus infection in inflammatory aortic aneurysms with RNA polymerase chain reaction. J Vasc Surg 1994; 20:235-243.
73. Ozvath KJ, Hirose H, Xia S et al. Molecular mimicry and the etiology of the nonspecific abdominal aortic aneurysm. 1999 (In Press).
74. Weynand CM, Hicok KC, Conn DL et al. The influence of HLA-DRB 1 genes on disease severity in rheumatoid arthritis. Ann Int Med 1992; 117:801-806.
75. Tsuji K, Aizawa M, Sasazuki T. HLA 1991: Proceedings of the 11th International Histocompatibility Workshop and Conference. Yokohama, Japan, November 6-13, 1991. 1991 Oxford; Oxford Univ Press; 1:Table 12:W1 5.1.
76. Weyand CM, Hicok KC, Hunder GG, et al. The HLA-DRB I locus as a genetic component in giant cell arteritis. Mapping of a disease-linked sequence motif to the antigen binding site of the HLA-DR molecule. J Clin Invest 1992; 90:2355-2361.
77. Woessner JF, Jr. Matrix metalloproteinases and their inhibitors in connective tissue remodelling. FASEB J 1991; 5:2145-2154.
78. GenBank accession number >pirIA44111: residues 144-154.
79. GenBank accession number >pirIA39808: residues 59-66.

CHAPTER 13

Search for the Aneurysm Susceptibility Gene(s)

Helena Kuivaniemi, Gerard Tromp

A ortic aneurysms are the thirteenth leading cause of death in the United States with about 15,000 deaths every year from the rupture of AAA.[1] Diagnosing an AAA is problematic in the sense that most AAAs do not produce symptoms and are discovered only after death. Rupture of AAAs is a significant cause of mortality and morbidity, and it has been estimated, based on autopsy studies, that 1–6% of the population in the U.S.A. and other industrialized countries harbor aneurysms.[1] Moreover, the incidence and the deaths attributed to aneurysm rupture are increasing.[1,2] Despite the major advances in vascular surgery, the survival rate after a ruptured aneurysm is still low.[1] The best therapy for aneurysms is early detection and presymptomatic, elective surgery. The emphasis of care is, therefore, best directed towards early diagnosis of aneurysms. If it were possible to predict who is at risk for developing an aneurysm, for example with the help of genetic testing, diagnostic efforts including ultrasonography could be directed towards the individuals at risk.

There is no consensus as to the cause of aortic aneurysms.[3,4] Hypertension exists in about half of patients and is obviously an aggravating condition. Tertiary syphilis was once an important cause of aneurysms, particularly of the ascending thoracic aorta, but is a less common cause now. Another cause frequently cited is arteriosclerosis. Still another cited cause, particularly when cystic medial necrosis is prominent, is activation of enzymes such as collagenases and elastases that degrade the extracellular matrix.

Recently, a number of careful studies have demonstrated that aortic aneurysms are frequently familial, even when they are not associated with heritable disorders such as the Marfan syndrome or the type IV variant of the Ehlers-Danlos syndrome. It has been demonstrated that approximately 15% of AAA patients have another first-degree relative with the disease[2,5-8] (Tables 13.1 and 13.2). No controlled twin studies have been carried out for the occurrence of AAA, but there are case reports in the literature describing four sets of twins with AAA.[9,10] A discovery of a family in which three brothers had AAAs that were diagnosed between the ages of 60 and 70 years led to the suggestion by Clifton[11] that there may be a heritable trait that caused aortic aneurysms. Tilson and Seashore[9] reported on 16 families with AAAs in first-degree relatives. From the pedigree data, the authors concluded that the mode of inheritance was either X-linked or autosomal dominant, with the X-linked mode being more common. Subsequently, Tilson and Seashore[12] reported on a collection of 50 families and favored an autosomal dominant mode of inheritance. To establish what fraction of AAAs were familial, patients and their first degree relatives were interviewed in ten different studies (Table 13.1). The highest familial incidence of AAAs was reported by Powell and Greenhalgh,[13] who interviewed 60 patients and were able to collect family data

Table 13.1. Familial prevalence of AAA based on interviews

Authors	No. of Patients Surveyed	No. of Patients with Positive History	Familial Prevalence (%)
Norrgärd et al 1984[72]	87	16	18.4
Johansen and Koepsell 1986[73]	250	48	19.2
Powell and Greenhalgh 1987[13]	56	20	35.7
Johnston and Scobie 1988[74]	666	41	6.1
Cole et al 1989[75]	305	34	11.1
Darling et al 1989[33]	542	82	15.1
Majumder et al 1991[53]	91	13	14.3
Verloes et al 1995[54]	313	39	12.5
Lawrence et al 1998[34]	86	19	22.1
Kuivaniemi et al 1998*	72	19	26.4
Total	2468	331	
Combined Prevalence	13.4		

*Our unpublished results from the Detroit Medical Center

Table 13.2. **Prevalence of AAA among first degree family members based on ultrasonography**

Authors	Country	Brothers[1]	Sisters[1]	Other[1]
Bengtsson et al 1989[17]	Sweden	10/35	3/52	
Collin & Walton 1989[19]	UK	4/16	0/15	
Webster et al 1991b[25]	USA	5/24	2/30	7/103
Adamson et al 1992[15]	UK	5/25	3/28	
Bengtsson et al 1992[18]	Sweden			9/62
van der Lugt et al 1992[24]	Netherlands	16/56	3/52	
Adams et al 1993[14]	UK	4/23	1/28	6/23
Moher et al 1994[23]	Canada	9/48		
Fitzgerald et al 1995[20]	Ireland	13/60	2/65	
Larcos et al 1995[22]	Australia			0/52
Baird et al 1995[16]	Canada	7/26	3/28	
Jaakkola et al 1996[21]	Finland	4/45	1/78	
Total		77/358	18/376	22/240
Prevalence		21.5%	4.8%	9.2%

[1]Number of individuals identified with AAA/number of individuals examined by ultrasonography. Other refers to relatives other than sisters and brothers of the AAA patient.

on 56, and found a positive family history in 20 patients (35.7%). Summing the numbers from all of the studies reported to date in which the family history of altogether 2,468 AAA-patients was obtained by questionnaires or by interviews, the combined familial prevalence of AAAs is about 13.4%. In addition, ultrasonography has been used in 12 studies[14-25] to determine the prevalence of undiagnosed aneurysms in families. In the family studies that are summarized in Table 13.2, altogether 358 brothers and 376 sisters were examined by ultrasonography. Seventy-seven brothers (21.5%) and 18 sisters (4.8%) had AAA.

Ultrasound screening to detect AAAs has been used in several studies of high-risk patient groups such as patients with hypertension, peripheral vascular disease, or coronary artery disease.[2,6] Due to the relatively high incidence of AAAs in the general population, the dramatic nature of the disease in that rupture often leads to sudden death of the patient, and a survival rate of greater than 90% after elective surgery for AAA, efforts have been made to carry out even wider population screening studies to detect AAAs before they rupture.[2] About 5.4% of the more than 22,000 men and 1.1% of the about 5,800 women examined by ultrasonography were found to have AAAs. When comparing the family and general population ultrasonography studies, the prevalence for AAAs was found to be 4.0- and 4.4-fold greater when brothers and sisters of patients with AAA were screened, respectively, than when men and women of the general population were screened. These numbers are likely to be underestimates of the risk of siblings of AAA patients to develop AAAs, since the age of the screened individuals in most of the population screening studies was over 65 years, whereas in the family studies family members between ages 40 and 80 years were included, and it is known that the prevalence of AAA increases with age.[2]

In this review we present different approaches to try to identify the susceptibility gene(s) for AAA. These approaches are summarized in Figure 13.1. We would like to note that, although there are no success stories to be presented at this moment for AAA using these approaches, the feasibility of genetic and molecular biological approaches to unravel the pathogenesis of AAA has increased drastically in the past five years due to the development of both statistical and laboratory techniques.

Candidate Gene Approach

The efforts to identify the gene(s) involved in the development of aortic aneurysms have been directed towards analyzing candidate genes of biological significance to the vessel tissue, since aortic aneurysms are characterized by expansion and thinning of all the layers of the arterial wall. The list of possible genetic factors involved in the development of aneurysms includes: (a) structural components of arteries like collagen and elastin,[5,26] (b) the enzymes responsible for the degradation of the structural molecules,[8,27-29] (c) the inhibitors of the proteolytic process,[30,31] and (d) cell-signaling or cell regulation pathways for invading monocytes. Many of the known extracellular matrix genes have served as candidate genes for the disease, but a number have been excluded one at a time. One of the most promising candidate genes that for type III procollagen, but detailed analysis of the cDNA sequences from more than 50 individuals with aortic aneurysms clearly demonstrated that mutations in the gene for type III procollagen are an infrequent cause of aortic aneurysms.[32]

The developments in molecular biology and especially that of PCR[35] have provided excellent tools for screening for mutations in candidate genes for diseases.[36] We have used the candidate gene approach extensively and below we provide examples:

Fibulin-2 Excluded as a Candidate Gene for AAA

We used conformation sensitive gel electrophoresis and direct sequencing of PCR products to screen for mutations in the cDNA for fibulin-2, an extracellular protein, from 11 patients with AAAs and two controls.[37] For the direct sequencing, overlapping RT-PCR products were generated. The sequence analysis yielded information on 3,502 nt (3,411 nt of coding sequences) from each individual. Since 9 out of the 13 individuals sequenced were

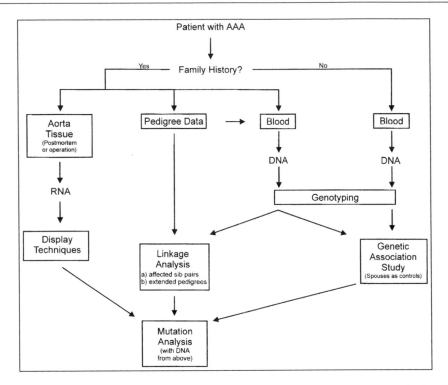

Fig. 13.1. Different approaches to identify the susceptibility gene(s) for AAA. See text for details on each of the approaches. As described in the text the mutation analyses will be carried out on positional candidate genes identified by DNA linkage or association studies or on candidate genes identified by display techniques.

heterozygous for at least one nucleotide, and in these nine individuals both alleles were therefore studied, a minimum of 77,044 nt of allelic sequences were analyzed. The number of allelic sequences analyzed could be as high as 91,052 nt (7,004 nt per individual), if it were assumed that the analysis included both alleles for all individuals studied. When compared to the published reference sequence,[38] a total of 14 single-base sequence variations were detected. Seven of the changes were neutral in that they did not result in an amino acid substitution. There were five missense changes at sites that were not conserved between human and mouse suggesting that they were not functionally important to the protein, and two missense changes at sites that were conserved between human and mouse. All but two of the sequence variants studied were also present in an additional set of 102 control alleles analyzed. One of these two changes was a missense mutation, but it did not segregate with AAAs in the family where it was identified, and the other change was neutral. In conclusion, fibulin-2 has a large number of sequence variations in comparison with our previous analyses of cDNA sequences for type III collagen. There were no obvious differences in the frequencies of the sequence variants of fibulin-2 among the aneurysm patient group and the control group consisting of 51 U.S. American blood donors.[37] These findings excluded fibulin-2 as the major candidate gene for aortic aneurysms, but suggested that fibulin-2 exhibits high degree of sequence variability. A possibility that the different variants contribute to the aneurysmal disease still exists, which could be addressed in association studies using the sequence variants identified.

TIMP1 and TIMP2 as Candidate Genes for AAA

The development of aneurysms is associated with remodeling of extracellular matrix (ECM), including breakdown of structural components of the vascular wall.[39,40] The activity of collagenase, a proteolytic enzyme responsible for the degradation of collagen, has been reported to be increased in ruptured aneurysmal aorta.[28] Another study found evidence for increased production of 92-kD gelatinase (MMP-9), an enzyme that can degrade numerous ECM components, in AAAs.[29] This increase in the proteolytic activity could be due to overexpression of the MMP-9 enzyme or downregulation of naturally occurring inhibitors of the collagenase. In fact, a preliminary study on decreased levels of tissue inhibitor of metalloproteinases (TIMPs) in AAAs has been reported.[30] Furthermore, the ratio of matrix metalloprotcinase (MMP) mRNA amount to TIMP mRNA was higher in AAA than in normal aortas when assayed using competitive RT-PCR with gene-specific external standards.[31]

TIMPs are major inhibitors of MMPs[39,40] and the relative deficiency observed in AAA tissues could be due to local tissue conditions inhibiting the expression or mutations in the primary structure of the TIMP genes. There are at least four members in the TIMP-family, namely, TIMP1,[41] TIMP2,[42] TIMP3,[43] and TIMP4.[44,45] All four have been cloned and sequenced, and the chromosomal localizations have been determined. TIMP1 resides on the X-chromosome,[46] TIMP2 on chromosome 17,[47] TIMP3 on chromosome 22,[43] and TIMP4 on chromosome 3.[45]

We studied the coding sequences of TIMP1 and TIMP2 in 12 unrelated patients with AAA to determine whether mutations in the TIMP genes are associated with AAAs.[48] All patients had a family history for the disease. Also, two clinically unaffected individuals were analyzed. The cDNA sequences of type III procollagen were previously determined for all of these individuals and were found to be normal.[32] Type III collagen was, therefore, excluded as a candidate gene for aneurysms in these individuals. In order to investigate the possibility that aneurysms were caused in these individuals by defects in the genes for TIMP1 or TIMP2, the sequences of the coding regions of TIMP1 and TIMP2 were determined in detail.

The sequence analysis provided 671 nt of TIMP1 cDNA sequences from each of the 14 unrelated individuals. The region analyzed included all the 621 nt of coding sequences. Only two sequence variations were found. One AAA patient was heterozygous for C/T change at nt 323. The sequence change converted the codon at amino acid position 87 from CCC for proline to CCT, also a proline codon. All the other 13 individuals had C at nt 323. Another sequence change was found at nt 434 (Fig. 13.2). Five individuals had C at this site, eight had T and one individual was heterozygous C/T. The sequence change converted the codon TTC at amino acid position 124 to TTT, but both of them are codons for phenylalanine, and this variant has also been reported previously.[49]

For TIMP2, we analyzed 750 nt which included all the 660 nt of coding sequences in the same 14 individuals as for TIMP1 analysis. Altogether we found two sequence variations. One AAA patient was heterozygous for C to T change at nt 306. Another difference, G to A transition occurred at nt 573 in one AAA patient who was heterozygous for this change (Fig. 13.3). Both variations occurred at the third positions of codons and they did not change the amino acids. All the sequence variants were confirmed by PCR-based restriction endonuclease assays, both on the cDNA and genomic level (Fig. 13.4).

In order to study the frequency of these variations, 102 control alleles and 168 alleles from AAA-patients (11 of the patients used for DNA sequencing and 73 additional unrelated AAA-patients) were analyzed by PCR-based restriction endonuclease assays using genomic DNA as a template. The TIMP1 gene includes four exons and three small introns.[50] The two variants identified in our study are in the second and third exon. Since TIMP1 is on X chromosome, the allele frequencies for males and females were analyzed separately.

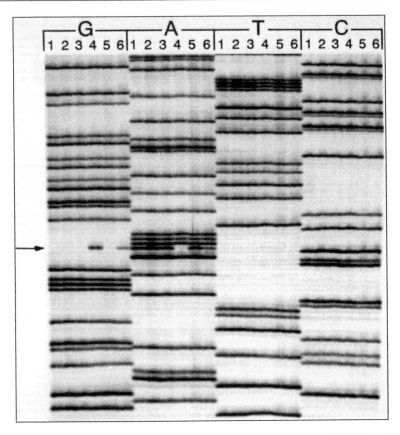

Fig. 13.2. Autoradiogram of a DNA sequencing gel with TIMP1 sequences. To facilitate the interpretation of results, G reactions from six (lanes 1-6) different individuals were loaded adjacent to
each other, followed by A, T, and C reactions, in this order, from the same individuals. Lanes G4
and G6 have a band (arrow) that is not present in the other G-lanes. Lane A4 is missing a band
that is present in other A-lanes. These changes all occur at nucleotide 434 in the TIMP1 cDNA
(Docherty et al 1985). The sequence is in antisense orientation.

Comparison of the data from controls and AAA patients found no significant differences
with the frequencies of the nt 323 polymorphism. The frequencies between the two female
groups were, however, significantly different (P=0.0019) for the nt 434 polymorphism.

The TIMP2 gene contains five exons and four introns.[51] The two sequence variants
were in exons one and three. The TIMP2 sequence variants were polymorphisms with no
significant differences between the control and AAA-group with nt 306 polymorphism. The
frequencies of nt 573 polymorphism were different between the control and AAA-groups in
males.[48]

In summary, patients with AAAs do not have any unique sequence variants in the TIMP1
and TIMP2 genes, and the transitions identified in these two genes do not change the amino
acids encoded. The differences in allele frequencies of nt 573 polymorphism of TIMP2 gene
between the control and AAA-groups are interesting preliminary findings that need further
follow-up with larger groups and carefully selected control groups for each ethnic group. It
is possible that a mutation in the gene for TIMP1 or TIMP2 contributes to the disease

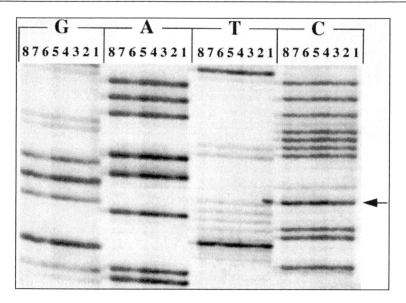

Fig. 13.3. Autoradiogram of a DNA sequencing gel with TIMP2 sequences. Reactions from eight (lanes 1-8) individuals were loaded as in Figure 2. Lane T1 has an extra band (arrow) indicating the presence of G to A change at nt 573 in one allele. The sequence is in antisense orientation.

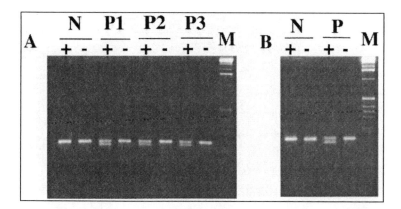

Fig. 13.4. Confirmation of TIMP2 sequence variants using agarose gel electrophoresis after restriction enzyme digestions. (A). Nt 573 sequence variant is present in three patients (B). Nt 306 sequence variant is present in one patient. Molecular marker is 1 Kb ladder (Gibco-BRL). N, control; P, patient; -, undigested; + digested PCR products.

process in patients with aneurysms. Such mutations could be mutations in the promoter sequence or large scale rearrangements in the genome, which could not have been easily detected by direct sequencing of RT-PCR products carried out here.

Search for New Candidate Genes Using Differential Display

Identification of Aorta-Specific cDNAs

In an attempt to identify aorta-specific cDNAs, we compared total RNA extracts from aorta with those from seven other tissue samples using differential display PCR. mRNA was first converted to cDNA by reverse transcriptase, using an oligo(dT) primer. Subsequent PCR amplifications with 32 combinations of 12 random primers were performed on all eight tissue samples. Each primer pair generated a distinct pattern of bands, and a total of 106 differentially expressed fragments, bands present in the aorta sample and not in the others, were identified among 18 of the 32 fingerprints. The excised bands were reamplified. Altogether 77 reamplified PCR products were cloned, and 178 individual clones were size-selected (to verify that the size of each insert corresponded to that of the original PCR product excised from the gel) and the sequences of their inserts determined. Twenty-two percent of the clones had identical or near identical sequences to at least one of the other cDNAs in our collection and were not studied further. Of the remaining 139 unique sequences, 63% were new, previously unidentified sequences, and 36% were known sequences and included e.g., elastin, heparin cofactor II, plasminogen activator inhibitor 1 and ATP synthase β-subunit.[52]

Further Characterization of EAo41, One of the Aorta-Specific Clones

One of the aorta-specific unique clones, named EAo41, was about 1 kb long. It was localized onto human chromosome 4q31.1-31.2 by human-rodent somatic cell hybrids and FISH (unpublished results). EAo41 was further studied using RT-PCR and EAo41-specific primers, and RNA isolated from various tissue samples. Control PCRs with β-actin primers were carried out to verify that the RNA samples and cDNA synthesis worked. The results demonstrated that the corresponding message is expressed in aorta and heart, to much lesser amount in skeletal muscle and not at all in the other samples. The expression level was, however, too low even in the aorta to be detectable by standard Northern blots. The gene for EAo41 was shown to be a single-copy gene in the human genome based on Southern blot experiments.

Amount of EAo41-mRNA is Lower in Aneurysmal Sac than in Non-Diseased Aortic Wall

To investigate the distribution of EAo41 expression in the different segments of human aorta, samples from ascending thoracic, descending thoracic and abdominal aorta were collected from autopsies, and the quantification was carried out as a ratio of signal produced in EAo41 RT-PCR to β-actin, or type III collagen (COL3A1) RT-PCR to β-actin signal. No differences in the EAo41/β-actin nor COL3A1/β-actin were detected in samples taken from different parts of the aorta or from donors of various ages. The EAo41/β-actin and the COL3A1/β-actin -ratios among eleven patients with aortic aneurysms were significantly lower than in 16 nonaneurysmal autopsy cases. This observation might be a secondary event due to the tissue destruction taking place in aortic dilatation and rupture. However, it does suggest that EAo41 is an integral part of the aortic tissue, and that EAo41 might play a role in disease processes involving aorta, but further studies are needed to elucidate its role in these diseases.

DNA Linkage Analyses to Identify Susceptibility Loci

The premise underlying the use of DNA linkage analyses is that AAAs are heritable. In addition to the case reports and family screening studies mentioned above, formal genetic analyses support the idea that AAAs have a genetic component in the pathogenesis. A statistical evaluation of the family data using an approach called segregation analysis by two independent research groups and from two different nationalities of AAA-patients, namely Belgian and US American, suggests that AAAs are caused by defects in a single gene in each family, but a possibility exists that there is heterogeneity between different families.[53,54]

Linkage is a term in genetics that indicates that a particular marker is co-inherited with a phenotype.[55-61] The marker may be a biochemical one such as blood type or histocompatibility antigens, a polymorphic DNA marker, or a phenotypic marker not part of the disease characteristics and coded for by a single locus. It is important to note that linkage is determined with respect to loci. A locus is a concept in genetics used to represent a region of DNA in the genome. The specific region may, but need not, be physically identifiable (e.g., a locus may be mapped to a band on a chromosome). Also, the locus concept, in terms of linkage, imparts no information about the DNA at the locus, nor does it specify that there is only one gene at the locus. Even if a particular gene that harbors mutations at a locus is identified by cloning and sequencing, there are likely to be many mutations that cause the disease, as has been the case in a large number of diseases.

Linkage of a disease phenotype and a marker means that the two are located closely on the same physical piece of DNA. The closer the two loci (the disease and the marker) occur on a piece of DNA, the less frequent crossovers are between them, and the more tightly the two are linked. The linkage between the disease and the marker can be expressed as a distance in genetic map units. Linkage distance (and map units) are measured as a function of the number of crossovers, which in turn is a function of the number of meioses (opportunities for crossover) and the distance between the loci. If two loci (disease causing gene and marker) are tightly linked, they are so close that the frequency of crossingover during meiosis is low or undetectable. Therefore, establishing linkage with the conventional approaches requires families with a large enough number of meioses. The most useful families for linkage studies have at least three generations, with affected individuals in all three generations and available for the study (alive and willing to give a blood sample for DNA isolation). The larger the number of offspring in each generation, the more meioses and, therefore, the more useful the family is for linkage studies, not only because it can provide good evidence for linkage but also because of the possibility that different families do not share the same genetic component (heterogeneity). Another requirement is that the disease is diagnosed accurately, because an undiagnosed individual will result, at best, in data that suggest a larger distance between the marker and the disease locus, and at worst, in data that suggest lack of linkage or exclusion of the gene in that family. It is desirable to obtain samples from as many as possible affected and unaffected individuals. The above constraints pose a particular problem for linkage studies of late-onset diseases such as AAAs. It is rare to find large families that have more than two generations of living individuals diagnosed with AAA, so that blood samples can be obtained for isolating genetic material. By the time an individual has been diagnosed with an aneurysm, his or her parents have died and the children of the AAA-patient are still too young to have aneurysms. Diagnosis is also a problem, since few individuals develop AAAs before the age of 50. Therefore, an individual who is currently between the ages of 50 and 55 and unaffected, may develop an AAA by the age 65 and linkage data obtained from such an individual's family would be skewed toward nonlinkage. Figure 13.5 shows a pedigree demonstrating the problems described above in traditional linkage studies.

To circumvent the problem of not being able to identify families with extended pedigrees and individuals with AAA in several generations, sib pair approach of genetic linkage can be used. Sib pairs are extremely useful in genetics of complex traits because of a feature that is unique to sib relationships (brother-brother; brother-sister; sister-sister): the degree of relationship between sibs for any given locus (position) on the genome can vary from 0 to 1,

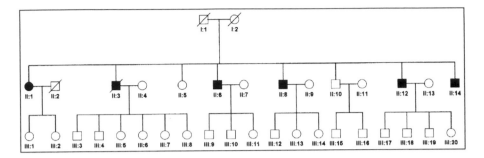

Fig. 13.5. A typical family with AAA. The black squares and circles indicate male and female members, respectively, diagnosed with AAA. Due to their relatively young age (30-40y) none of the children of the AAA-patients are known to have the disease. Also, the grandparents have died and no information is available on the exact cause of the death. As discussed in the text, such families are best suited to affected sib pair linkage analyses, since diagnostic uncertainty of the third generation and unavailability of the first generation make it difficult to analyze extended pedigrees using traditional linkage analyses.

where 0 means that they share no alleles and 1 means they share both alleles, i.e., they can share 0, 1 or 2 alleles at the locus. The distribution of the relationships expected by chance is simple and leads to predictions that can be tested without specifying the mode of inheritance (called model free methods). Sib pairs are easier to collect than extended families and usually there are more of these relationships available than many or most other extended family relationships. It is, however, less powerful than extended pedigree approach and large number of sib pairs might be needed to identify susceptibility loci. It is also worth noting that collecting samples from parents improves the power and certainty, since with the genotyping information from the parents it will be possible to say whether the sibs share alleles identical-by-descent or identical-by-state. There are two major types of sib pair analyses: affected sib pair analysis (both sibs have AAA) and discordant sib pair analysis (one sib has AAA and the other does not). Affected sib pair approach is the most appropriate for AAA since it is not possible to say with certainty that a person is unaffected (has no AAA and will not get AAA later).[62-64] The genetic term for this phenomenon is called age-dependent penetrance. It is also possible that the disease shows incomplete penetrance meaning that an individual has the gene but is not showing the disease and has an affected offspring and an affected parent, which is demonstrated by the fact that the disease skips a generation. Thus when the sample used for linkage analyses consists of affected sib pairs, the penetrance issue becomes irrelevant since there is 100% penetrance in the pool of individuals used for the analyses. The analyses are then simplified because the degree of uncertainty is low. Uncertainty reduces power in statistical analyses because it is equivalent to making the sample noisy, i.e., the signal-to-noise ratio is decreased. The only uncertainties left to deal with are the uncertainties introduced by phenocopies (which means that AAA can also be caused by other than genetic, for example by environmental factors) and heterogeneity (AAA caused by defects in multiple genes in the population level and each of the genes contributing to varying degrees in individual AAA patients).

Genome-Wide Search of Susceptibility Loci for Intracranial Aneurysms

As an example of affected sib-pair analysis, we would like to present our recent studies on intracranial aneurysms (IA). We have initiated similar kind of studies on AAA, but they are not yet complete. For the IA study we used patient material from Finland. Studying the

Finnish population as compared to outbred US population offers several advantages in a complex disease like aneurysms:[65] 1) the Finnish populations is genetically more homogenous; 2) the medical records are easily available; and 3) The Finnish Evangelical Lutheran Church has excellent family records that can be extended with Regional Archival records to about year 1600 making it possible to look for common ancestors between two or more separate families. Still another advantage in the IA project was the funds provided by the Finnish government to screen with magnetic resonance imaging about 600 asymptomatic first-degree relatives of IA patients.[66] The screening study increased the number of those family members known to have IA and available for the genetic study. In an analogous way, screening of asymptomatic family members of AAA patients with ultrasonography should detect those individuals who have dilatations or AAA but have not suffered a rupture nor had any symptoms caused by the AAA.

Previous epidemiological studies suggested that IAs cluster in families,[67,68] and we estimated from the Finnish IA-families that the relative risk to sibs is between 9 and 16.[66] In the sib-pair linkage analysis to identify susceptibility loci for IA, we used 48 affected sib pairs from 22 extended families and typed their DNA samples as well as those of selected relatives for 386 highly polymorphic simple repeat markers corresponding to a 10 cM average marker-marker distance. Six chromosomal regions gave positive results that exceeded the threshold of 0.8 as a multipoint lod score.[69] Two of the regions gave scores that exceeded 2.0, one on chromosome 19 and the other on chromosome X. The highest scores were detected with markers mapped onto 19q13.2, a gene-rich area containing such excellent candidate genes related to cerebrovascular, cardiovascular or membrane physiology and pathobiology as apolipoprotein E, cardiac troponin I, and human brain specific sodium-dependent inorganic phosphate cotransporter. Future studies will include confirming the results by collecting another set of affected sib pairs and analyzing them for the identified candidate loci, and using markers located more closely to each other (2uM) than the ones used in the first screening. Then candidate genes located in the support interval, will be analyzed for the presence of mutations (genetic alterations). If such mutations are detected, simple DNA tests can be designed to test individuals at risk in each family to identify those harboring mutations and being, therefore, at higher risk to develop an IA.

Genetic Association Studies as Tools to Determine Genetic Components of AAA

In association studies the purpose is to test candidate genes for their role in the disease pathogenesis by simply taking a sequence variant in the gene of interest, present in high enough frequency in the population so that the study is feasible and comparing the frequencies of the sequence variant (allele frequencies) and the various genotypes among the affected and unaffected groups.[70,71] Traditional association studies in which independent samples of affecteds and unaffecteds are collected (case-control studies) are plagued by the problem of mathematical associations that are due to such factors as stratification of the population and have no relationship to the underlying genetics. The problem is to collect truly representative samples. One approach that has recently been proposed and has gained wide acceptance among statistical geneticists is the use of spouses as controls, since spouses tend to be from the same ethnic and socioeconomic groups. In addition, for late-onset diseases, spouses will have shared many of the environmental factors. These samples are also probably the easiest to collect and do not require extended relationships or more than one affected per family. The power obtained from traditional association studies is the smallest of the approaches presented here. If the parents of the individual with AAA were available, the so called trios approach could be taken. In trios studies, the nontransmitted alleles from the parents can serve as independent alleles for estimation of allele frequencies, without the concern of samples stratification. Additionally, tests determining preferential transmission or nontransmission of alleles to the affected offspring using the transmission disequilibrium test provide increased power to detect association. The trios approach has recently become a popular approach for complex diseases, but it is not very suitable to AAA, since parents are often no longer alive when the patient is diagnosed with an AAA.

Concluding Remarks

Two important questions that we have not answered yet are: (1) What has prevented us (i.e., the AAA research community) from finding the susceptibility gene(s), and (2) can we expect to find them in the future? Firstly, a number of the techniques that we have described above are still relatively new, and many of them have not been fully implemented in the AAA research community. Secondly, approaches like the DNA linkage analyses and genetic association studies require a collection of DNA samples from a large number of AAA patients and their family members. For example for the affected sib-pair approach, samples from at least 200 sib pairs are needed. In order to succeed in collecting samples from such a large number of families, the initial AAA patient pool from which to gather these samples has to be in the order of 2,000–3,000 AAA patients when assuming that about 15% (300–450) of the patients have positive family history for AAA and about 50% of the affected relatives are alive and that only some of the affected relatives are sibs. Therefore, it would be important to examine with ultrasonography all the sibs that are over 55 years of age, which is expected to increase the number of sibs known to have an AAA and available for genetic studies. Using such a systematic approach should yield the samples necessary to perform genome scan with affected sib pairs and we would not have to worry about incomplete penetrance, inheritance modes and diagnostic uncertainty. Some of these parameters can be estimated for identified loci from the sib-pair data. After the genome scan has revealed, possible suggestive genetic loci, biologically and physiologically relevant candidate genes located in the candidate intervals could then rapidly be screened for mutations. This is called the positional candidate gene approach, and it has become a feasible approach thanks to the excellent progress of the human genome project in identifying and mapping a large number of the genes to date and all the human genes in the near future.

Acknowledgments

The original work carried out in our laboratories was supported in part by grants from NIH (NS34395 to GT) and American Heart Association—Michigan Affiliate (to HK and GT), and funds from Wayne State University School of Medicine. We would also likely thank our graduate student Xiaoju Wang for kindly providing Figures 13.3 and 13.4.

References

1. Ernst CB. Abdominal aortic aneurysm. N Engl J Med 1993; 328:1167-1172.
2. Bengtsson H, Sonesson B, Bergqvist D. Incidence and prevalence of abdominal aortic aneurysms, estimated by necropsy studies and population screening by ultrasound. Ann NY Acad Sci 1996; 800:1-24.
3. Roberts WC. Pathology of arterial aneurysms. In: Bergan JJ, Yao JST, eds. Aneurysms: Diagnosis and Treatment. New York: Grune & Stratton, 1982; 17-42.
4. Dzau VJ, Creager MA. Diseases of the aorta. In: Fauci AS, Braunwald E, Isselbacher KJ, Wilson JD, Martin JB, Kasper DL, Hauseer SL, Longo DL, eds. Harrison's Principles of Internal Medicine. 14th ed. New York: McGraw-Hill Book Co., 1998; 1394-1398.
5. Kuivaniemi H, Tromp G, Prockop DJ. Genetic causes of aortic aneurysms. Unlearning at least part of what the textbooks say. J Clin Invest 1991; 88:1441-1444.

6. Kuivaniemi H, Tromp G, Prockop DJ. Genetic causes of aortic aneurysms. In: Molecular Genetics and Gene Therapy of Cardiovascular Diseases. Stephen C. Mockrin, editor, New York: Marcel Dekker Inc., 1996;209-218.

7. Norrgärd O. Looking for the familial connection in aortic aneurysm. In: Greenhalgh RM, Mannick JA, Powell JT, eds. The cause and management of aneurysms. London: W.B. Saunders Company, 1990; 29-36.

8. Reilly JM, Tilson MD. Incidence and etiology of abdominal aortic aneurysms. Surg Clin North Amer 1989; 69:705-711.

9. Tilson MD, Seashore MR. Human genetics of the abdominal aortic aneurysm. Surg Gynecol Obstet 1984; 158:129-132.

10. Borkett-Jones HJ, Stewart G, Chilvers AS. Abdominal aortic aneurysms in identical twins. J Royal Soc Med 1988; 81:471-472.

11. Clifton MA. Familial abdominal aortic aneurysms. Br J Surg 1977; 64:765-766.

12. Tilson MD, Seashore MR. Fifty families with abdominal aortic aneurysms in two or more first-order relatives. Am J Surg 1984; 147:551-553.

13. Powell JT, Greenhalgh RM. Multifactorial inheritance of abdominal aortic aneurysm. Eur J Vasc Surg 1987; 1:29-31.

14. Adams DC, Tulloh BR, Galloway SW et al. Familial abdominal aortic aneurysm: prevalence and implications for screening. Eur J Vasc Surg 1993; 7:709-712.

15. Adamson J, Powell JT, Greenhalgh RM. Selection for screening for familial aortic aneurysms. Br J Surg 1992; 79:897-898.

16. Baird PA, Sadovnick AD, Yee IML et al. Sibling risks of abdominal aortic aneurysm. Lancet 1995; 346:601-604.

17. Bengtsson H, Sonesson B, Bergqvist D. Incidence and prevalence of abdominal aortic aneurysms, estimated by necropsy studies and population screening by ultrasound. Ann NY Acad Sci 1996; 800:1-24.

18. Bengtsson H, Sonesson B, Länne T et al. Prevalence of abdominal aortic aneurysm in the offspring of patients dying from aneurysm rupture. Br J Surg 1992; 79:1142-1143.

19. Collin J, Walton J. Is abdominal aortic aneurysm familial? Br Med J 1989; 299:493.

20. Fitzgerald P, Ramsbottom D, Burke P et al. Abdominal aortic aneurysm in the Irish population: A familial screening study. Br J Surg 1995; 82:483-486.

21. Jaakkola P, Kuivaniemi H, Partanen K et al. Familial abdominal aortic aneurysms: Screening of 71 families of patients with abdominal aortic aneurysms. Eur J Surg 1996; 162:611-617.

22. Larcos G, Gruenewald SM, Flether JP. Ultrasound screening of families with abdominal aortic aneurysm. Austr Radiol 1995; 39:254-256.

23. Moher D, Cole W, Hill GB. Definition and management of abdominal aortic aneurysm: Results from a Canadian survey. Can J Surg 1994; 37:29-32.

24. van der Lugt A, Kranendonk SE, Baars AM. Screening for familial occurrence of abdominal aortic aneurysm (in Dutch). Ned Tijdschr Geneeskd 1992; 136:1910-1913.

25. Webster MW, Ferrell RE, St. Jean PL et al. Ultrasound screening of first-degree relatives of patients with an abdominal aortic aneurysm. J Vasc Surg 1991; 13:9-14.

26. Menashi S, Campa JS, Greenhalgh RM et al. Collagen in abdominal aortic aneurysm: Typing, content, and degradation. J Vasc Surg 1987; 6:578-582.

27. Carmeliet P, Moons L, Lijnen R et al. Urokinase-generated plasmin activates matrix metalloproteinases during aneurysm formation. Nat Genet 1997; 17:439-444.

28. Webster MW, McAuley CE, Steed DL et al. Collagen stability and collagenolytic activity in the normal and aneurysmal human abdominal aorta. Am J Surg 1991; 161:635-638.

29. Thompson RW, Holmes DR, Mertens RA et al. Production and localization of 92-kilodalton gelatinase in abdominal aortic aneurysms. An elastolytic metalloproteinase expressed by aneurysm-infiltrating macrophages. J Clin Invest 1995; 96:318-326.

30. Brophy CM, Sumpio B, Reilly JM et al. Decreased tissue inhibitor of metalloproteinases (TIMP) in abdominal aortic aneurysmal tissue: A preliminary report. J Surg Res 1991; 50:653-657.

31. Tamarina NA, McMillan WD, Shively VP et al. Expression of matrix metalloproteinases and their inhibitors in aneurysms and normal aorta. Surgery 1997; 122:264-271.

32. Tromp G, W, Y, Prockop DJ, Madhatheri SL et al. Sequencing of cDNA from 50 unrelated patients reveals that mutations in the triple-helical domain of type III procollagen are an infrequent cause of aortic aneurysms. J Clin Invest 1993; 91:2539-2545.

33. Darling RC, III, Brewster DC, Darling RC et al. Are familial abdominal aortic aneurysms different? J Vasc Surg 1989; 10:39-43.

34. Lawrence PF, Wallis C, Dobrin PB et al. Peripheral aneurysms and arteriomegaly: Is there a familial pattern? J Vasc Surg 1998; 28:599-605.

35. Saiki RK, Scharf S, Faloona FV. Enzymatic amplification of b-globin genomic sequences and restriction site analysis for diagnosis of sickle cell anemia. Science 1985; 230:1350-1354.

36. Shikata H, Utsumi N, Kuivaniemi Het al. DNA-based diagnostics in the study of heritable and acquired disorders. J Lab Clin Med 1995; 125:421-432.

37. Kuivaniemi H, Marshall A, Ganguly A, Chu M-L, Tromp, G. Fibulin-2 exhibits high degree of variability but has no structural changes concordant with abdominal aortic aneurysms. Eur J Hum Genet (in press).

38. Zhang R-Z, Pan T-C, Zhang Z-Y et al. Fibulin-2 (FBLN2): Human cDNA sequence, mRNA expression, and mapping of the gene on human and mouse chromosomes. Genomics 1994; 22:425-430.

39. Dollery CM, McEwan JR, Henney AM. Matrix metalloproteinases and cardiovascular disease. Circ Res 1995; 77:863-868.

40. Werb Z. ECM and cell surface proteolysis: Regulating cellular ecology. Cell 1997; 91:439-442.

41. Docherty AJP, Lyons A, Smith Bj et al. Sequence of human tissue inhibitor of metalloproteinases and its identity to erythroid-potentiating activity. Nature (London) 1985; 7:66-69.

42. Stetler-Stevenson WG, Brown PD, Onisto M et al. Tissue inhibitor of metalloproteinases-2 (TIMP-2) mRNA expression in tumor cell lines and human tumor tissues. J Biol Chem 1990; 265:13933-13938.

43. Apte SS, Mattei M-G, Olsen BR. Cloning of the cDNA encoding human tissue inhibitor of metalloproteinases-3 (TIMP-3) and mapping of the TIMP3 gene to chromosome 22. Genomics 1994; 19:86-90.

44. Greene J, Wang M, Liu YE et al. Molecular cloning and characterization of human tissue inhibitor of metalloproteinase 4. J Biol Chem 1996; 271:30375-30380.

45. Olson TM, Hirohata S, Ye J et al. Cloning of the tissue inhibitor of metalloproteinase-4 gene (TIMP4) and localization of the TIMP4 and Timp4 genes to human chromosome 3p25 and mouse chromosome 6, respectively. Genomics 1998; 51:148-151.

46. Spurr NK, Goodfellow PN, Docherty A J. Chromosomal assignment of the gene encoding the human tissue inhibitor of metalloproteinases to Xp11.1-p11.4. Ann Hum Genet 1987; 51:189-194.

47. DeClerck Y, Szpirer C, Aly M et al. The gene for tissue inhibitor of metalloproteinases-2 is located on human chromosome arm 17q25. Genomics 1992; 14:782-784.

48. Wang X, Tromp G, Cole CW et al. Analysis of coding sequences for tissue inhibitor of metalloproteinases 1 (TIMP1) and 2 (TIMP2) in patients with aneurysms. Matrix Biology (in press).

49. Tilson MD, Reilly JM, Brophy CM et al. Expression and sequence of the gene for tissue inhibitor of metalloproteinases in patients with abdominal aortic aneurysms. J Vasc Surg 1993; 18:266-270.

50. Matsuda T, Kohno K, Kuwano M. (1992). Genbank accession D11139.

51. Hammani K, Blakis A, Morsette D et al. Structure and characterization of the human tissue inhibitor of metalloproteinases-2 gene. J Biol Chem 1996; 271:25498-25505.

52. Kuivaniemi H, Watton SJ, Price SJ et al. Candidate genes for abdominal aortic aneurysms. Annals NY Acad Sci 1996; 800:186-197.

53. Majumder PP, St Jean PL, Ferrell RE et al. On the inheritance of abdominal aortic aneurysm. Am J Hum Genet 1991; 48:164-170.

54. Verloes A, Sakalihasan N, Koulischer L et al. Aneurysms of the abdominal aorta: Familial and genetic aspects in three hundred thirteen pedigrees. J Vasc Surg 1995; 21:646-655.

55. Elston RC, Stewart J. A general model for the genetic analysis of pedigree data. Hum Hered 1971; 21:523-542.
56. Lathrop GM, Lalouel JM, Julier C et al. Strategies for multilocus linkage analysis in humans. Proc Natl Acad Sci 1984; 81:3443-3446.
57. Lathrop GM, Lalouel JM. Easy calculations of LOD scores and genetic risks on small computers. Am J Hum Genet 1984; 36:460-465.
58. Ott J. Computer simulation methods in human linkage analysis. Proc Natl Acad Sci 1989; 86:4175-4178.
59. Ott J. Analysis of human genetic linkage. The John Hopkins University Press, Baltimore, MD, 1991
60. Risch N. Genetic linkage: Interpreting lod scores. Science 1992; 255:803-804.
61. Terwilliger JD, Ott J. Handbook of genetic linkage. The Johns Hopkins University Press, Baltimore, MD, 1994.
62. Weeks DE, Lange K. The affected-pedigree-member method of linkage analysis. Am J Hum Genet 1988; 42:315-326.
63. Weeks DE, Lange K. A multilocus extension of the affected-pedigree-member method of linkage analysis. Am J Hum Genet 1992; 50:859-868.
64. Brown DL, Gorin MB, Weeks DE. Efficient strategies for genomic searching using the affected-pedigree-member method of linkage analysis. Am J Hum Genet 1994; 54:544-552.
65. de la Chapelle A. Disease gene mapping in isolated human populations: The example of Finland. J Med Genet 1993; 30:857-865.
66. Ronkainen A, Hernesniemi J, Puranen M et al. Familial intracranial aneurysms. Lancer 349; 380-384.
67. Schievink WI. Intracranial aneurysms. N Engl J Med 1997; 336:28-40.
68. Schievink WI, Schaid DJ, Rogers HM et al. On the inheritance of intracranial aneurysms. Stroke 1994; 25:2028-2037.
69. Olson JM, Vongpunsawad S, Kuivaniemi H et al. Genome scan for intracranial aneurysm susceptibility loci using Finnish families. Am J Hum Genet 1998; 63S:A17.
70. Cox NJ, Bell GI. Perspectives in Diabetes. Disease associations. Chance, artifact, or susceptibility genes? Diabetes 1989; 38:947-950.
71. Elston RC. Linkage and association. Genet Epidemiol 1998; 15:565-576.
72. Norrgård Ö, Rais O, Ängquist KA. Familial occurrence of abdominal aortic aneurysms. Surgery 1984; 95:650-656.
73. Johansen K, Koepsell T. Familial tendency for abdominal aortic aneurysms. JAMA 1986; 256:1934-1936.
74. Johnston KW, Scobie TK. Multicenter prospective study of nonruptured abdominal aortic aneurysms. I. Population and operative management. J Vasc Surg 1988; 7:69-81.
75. Cole CW, Barber GG, Bouchard AG et al. Abdominal aortic aneurysm: Consequences of a positive family history. Can J Surg 1989; 32:117-120.

Index